Augustine Ikelegbe (Ed.)

Oil, Environment and Resource Conflicts in Nigeria

Politics and Economics in Africa

Series Editors:

Robert Kappel
(Universität Leipzig/German Institute
of Global and Area Studies, Hamburg)

Ulf Engel
(Universität Leipzig)

volume 7

LIT

Oil, Environment and Resource Conflicts in Nigeria

edited by

Augustine Ikelegbe

LIT

Bibliographic information published by the Deutsche Nationalbibliothek
The Deutsche Nationalbibliothek lists this publication in the Deutsche Nationalbibliografie; detailed bibliographic data are available in the Internet at http://dnb.d-nb.de.

ISBN 978-3-643-90315-0

A catalogue record for this book is available from the British Library

Klosbachstr. 107
CH-8032 Zürich
Tel. +41 (0) 44-251 75 05
Fax +41 (0) 44-251 75 06
E-Mail: zuerich@lit-verlag.ch
http://www.lit-verlag.ch

LIT VERLAG Dr. W. Hopf
Berlin 2013
Fresnostr. 2
D-48159 Münster
Tel. +49 (0) 2 51-62 03 20
Fax +49 (0) 2 51-23 19 72
E-Mail: lit@lit-verlag.de
http://www.lit-verlag.de

Distribution:

In Germany: LIT Verlag Fresnostr. 2, D-48159 Münster
Tel. +49 (0) 2 51-620 32 22, Fax +49 (0) 2 51-922 60 99, E-mail: vertrieb@lit-verlag.de

In Austria: Medienlogistik Pichler-ÖBZ, e-mail: mlo@medien-logistik.at
In Switzerland: B + M Buch- und Medienvertrieb, e-mail: order@buch-medien.ch
In the UK: Global Book Marketing, e-mail: mo@centralbooks.com
In North America: International Specialized Book Services, e-mail: orders@isbs.com
e-books are available at www.litwebshop.de

Inhalt

Dedication

Dedicated to individual and group activists; youth, women, communal and ethnic groupings, local and international civil society and non governmental organizations, whose voices, labours, tears, wounds and blood have and keep watering the struggles of a people for equity and justice in the exploitation, management and distribution of their resources in a multi-ethnic federal State whose operation and use disadvantage them,

and

Omena and Oghenemaro Ikelegbe.

Acknowledgements

My profound gratitude goes to Emeritus Professor Andrew Onokerhoraye and the Centre for Population and Development, Benin City. The centre hosted the mid term International Round Table Conference on *Public Policy, Oil and Conflict.* I am grateful to the colleagues from various universities and research centres in Nigeria who responded to our conference flyers and made presentations. The selected and reviewed papers from the presentations constitute the chapters here. The research on the Niger Delta of which the mid term conference was part, was facilitated by a research grant from the United States Institute of Peace, Washington DC.

I should thank the Editors of the Politics and Economics in Africa Series of Lit Verlag under which this book is published for facilitating the publication. The concerns and inquiries of some contributors kept the work continually on the agenda of my uncompleted intellectual works. I have to thank my research assistants, Messrs Jude Momoh, Osumah Oarhe, Sonny Nwachukwu, Emeka Osemekha, and my colleagues at the Department for their contributions to the work. My wife, an academic herself, did not only support the project morally but keyed in to researching on the Niger Delta and contributed to this work. Over-all, the grace and tenacity to complete this work can only be attributed to the Lord God Almighty.

Augustine Ikelegbe
Professor of Comparative Politics and Public Policy, Dept. of Political Science & Public Administration,
University of Benin, Benin City, Nigeria.
and Adjunct Research Professor of Conflict and Peace Studies, Centre for Population and Environmental Development, Benin City, Nigeria.

Foreword

The Niger Delta crisis has lingered on for over four decades. The huge oil and gas resources of the region rather than being a blessing, has by its operations, practices, effects and dynamics become perilous. The complaints began as mere petitions by the host communities to Nigeria's oil industry in the 1970s, but have been transformed into militant agitation and violent confrontations since the late 1990s. The effects of the conflict on Nigeria's oil production, revenues, economy and security have been vastly adverse. The region is not only a challenge to Nigeria but to international energy supplies and regional and international security.

This book is welcome as part of efforts to provide deep insights, understanding and analytical explanations about this complex and lingering conflict. It comprises contributions by Nigerian scholars drawn from various universities and research institutes. The work presents research findings and scholarly works on diverse issues and dimensions germane to understanding and analyzing the conflict. These range from the environment, public policy, resource distribution and benefits, relations between the local people and the transnational oil companies to national security. The authors are concerned besides research and scholarship with how the conflicts can be resolved.

In many ways, the environmental and resource conflicts in the Niger Delta require indepth research analyses and presentations as this. The events, situations, incidences and even state policy and responses have been quite dynamic and rapidly changing. The interests of the international community, particularly those of foreign governments, international organizations and development agencies, donor agencies, international civil society and non governmental organizations have rapidly grown. Numerous policy and practical frameworks, efforts and actions are being packaged at different levels to address various dimensions and manifestations of the conflict. This is the time for clear, informed, indepth and current studies and analyses to provide broad understanding, explanations and prescriptions that can guide efforts and activities.

There remains a huge challenge to re-examine and critically interrogate the issues of resources and sustainable development. In Nigeria, the entire oil economy needs critical examination in terms of the actual impacts and consequences for the local peoples of the region; their livelihoods, fortunes and future. Is development being fostered and is the development on offer sustainable? Since oil and gas are exhaustible resources, what happens when they are exhausted and when no revenue inflows would be expected to the region? What would happen in the next thirty to forty years as the land and waters of the region are devastated, livelihood sources have been destroyed, social lifestyles have been perverted and productive attitudes and orientations have been disoriented? What challenges of rehabilitation and reconstruction would exist and how could we expect the Nigerian State that is reluctant with resource flows and developmental transformation of the region, to drive the processes of rehabilitation and reconstruction without oil and gas revenues. In other words, what are the consequences and future ramifications of the current state of un-sustainable development practices and conditions and how can a stable future be secured given that the region is currently conflict ridden and lies prostrate with poverty and under-development?

At the Centre for Population and Environmental Development (CPED), we are committed to supporting local organizations on peace building in the Niger Delta region.

I warmly commend this book to all those who seek to understand the complexities and dynamics of the conflict situations, actors and incidences of the Niger Delta. The work is an essential reading to all those involved in the Nigerian oil economy and the broad and concerted efforts to manage the conflicts in the Niger Delta and build the peace.

Andrew Onokerhoraye, Oon
Professor Emeritus, University of Benin and
Executive Director
Centre for Population and Environmental Development (CPED) Benin City, Nigeria.

1. The resource and environmental conflicts in the niger delta region of nigeria: an overview

Augustine IKELEGBE

1.1. The niger delta region, oil and conflict

Oil has been variously described as Nigeria's lifeline, lifeblood, soul and gum-stay as the country is sustained and held together by it. It is the centrepiece of the drive for state revenues, economic growth, foreign direct investments and debt repayment. It is also at the centre of the struggles for hegemony and inter and intra class power struggles just as it underpins conflicts, leadership crisis, instability and violence. The Niger Delta region is an important constituent unit of Nigeria. As Nigeria's oil belt, it is the main revenue and foreign exchange generator. Oil has since 1975 contributed over 90% of Nigeria's export earnings and over 70% of the federal budget. The region is strategically important to Nigeria, multinational oil and oil servicing companies, world energy supplies and the security of the emerging Gulf of Guinea oil potentials.

But the Niger Delta is today an epic case of a resource curse; resource exploitation without environmental, social and sustainable developmental responsibilities and how these spawn poverty, social-economic disruptions, internecine conflicts, violence, insecurity and instability. Since 1997, the region has become a virtual battlefield in which communal and ethnic groupings and militias have periodically engaged the Nigerian security agencies and military in very bloody encounters. In consequence, the region has been a large garrisoned command, with large and persisting naval and army deployments engaged in active service of containing the implosion.

The region is itself a paradox of several sorts. While it generates the bulk of state revenues, there is so little resource inflow that it is one of the most neglected, poor and least developed regions in Nigeria. Though it holds the bulk of the nations biodiversity, the Nigerian state has accommodated or allowed so much abuse and degradation that its fora and fauna are dying. Though it is a region of utmost opportunities for petro-business based investments and profits, it is also that of utmost threat to security of lives, livelihoods, property, business and local economies because of pervasive negative externalities and violent contestations against the Nigerian state and oil companies and between communities and groups. The region's oil and gas has brought so much revenue, profits, wealth and development for the multinational oil companies (MNOC), the Nigerian state and the ruling class but it has been perilous to the region, having brought so much poverty, immisseration, socio-economic disruptions, erosion of traditional and primary livelihood sources and degradation of the environment that sustains their livelihoods. Though the Niger Delta people live besides the oil wells, flow stations, oil platforms and particularly the well developed and modern oil company staff camps and quarters, they are shut out of basic social infrastructure, amenities and social services and live in misery, poverty and want.

The conflict is a crisis of complex and diverse dimensions. It is a crisis of environmental degradation and despoliation and a struggle for environmental responsibility, sustainability, remediation and restitution as the nature of the oil-environment nexus as managed by the state and oil companies has generated conflicts and violence between the region's communities and groupings and the multinational oil companies. It is a crisis of the character of governance and the state particularly as it pertains to political representation, minority rights and the national question that has fostered minority nationalism and self-determination struggles. It is a crisis of resource distribution, allocation and application as perceived insignificant benefits from oil resources has fueled agitation for resource control, regional autonomy, self determination, minority rights, state reforms, federal re-structuring and constitutional reform. It is a crisis of human insecurity as state responses and management of the conflict has subjected the region to human and civil rights abuses, extra-judicial killings, destruction of properties and communities and huge internal displacements. These dimensions have fuelled mass social discontent, social restiveness, communal, ethnic and sectional mobilization, sustained opposition, popular violence, separatist tendencies, intermittent rebellion and militia movements.

1.2. Issues and dimensions of the conflict

The conflict began with mere petitions and peaceful protests of host communities in the 1960s and 1970s. Increased awareness and mobilization of communities, civil and political society, youth and women since the 1980s has turned it into a huge melting pot of diverse protests and engagements. Particularly, the youth, with a militia arrowhead has been the vanguards and foot soldiers since the late 1990s. The region has been immersed in diverse resource based conflicts between and within communities, ethnic groups and groupings and between them, the MNOCs and the Nigerian state. The region is today a huge militarized and garrisoned command, with the most extensive and sustained military occupation, operations and actions since the Nigerian civil war. In fact, the conflict is the greatest and most persisting threat to Nigeria's fledgling democracy, nationhood, national unity and integration and stability and security.

The consequences of the conflict have been dire. The entire on shore and off shore oil and gas installations and operations has been subjected to sustained attacks, disruptions and destruction. The state has suffered huge losses of revenues, deflected investments, economic disruptions, huge costs of military hardware and operations and serious threats to energy supply and security. The region has been mired in repression, militarization, violence, socio-economic disruptions, damaged social services, carnage and insecurity of lives and property. This is apart from the degradation and devastation of its land and water and consequently livelihood sources.

At the core of the conflicts are the issues of the environment and resources. Oil exploitation has taken much land and water resources and made them scarce. It has polluted and degraded land and water and made them infertile for farming and fishing. It has raised the value of land and violent contestations for ownership. These have generated struggles for land rights, compensation for acquisitions, compensation for degraded land and water, remediation of degraded environment and empowerment for alternate non primary occupations. Petroleum resource endowment was expected to generate benefits particularly in terms of participation in the oil economy, infrastructure, industrial development and resource inflows. But the region has remained at the periphery of the oil economy and scarcely benefits from it. Besides, the region has suffered persistent developmental neglect. The region was disinherited by the 1969 petroleum decree, the land use decree and several other legislations and overall, very little resources have been ploughed back in terms of revenues. These realities have raised a generalized agitation and mobilization for resource control and state reforms, and fuelled diverse forms of coerced appropriations from the oil economy, MNOCs and the state. Beyond the regional struggles for resource control are diverse layers of resource control struggles as individuals, communities, communal leaderships and groupings, youth, militias, armed bands, oil theft syndicates and other criminal bands fight themselves and others in the competition for space, access and privileges to legal and compelled appropriations of benefits from the oil economy.

The Nigerian state has had a securitized conception and policy since the 1960s. But instead of suppression, it has rather raised counter violent challenges to a level where the conflict has slipped into a low intensity war. The federal and state governments have however established several interventionist development agencies and technical and policy committees, but these have been short in commitment, poorly funded and managed and have had feeble effects. Though the MNOCs initially hid behind the shield of state security agencies, they have since marginally improved on their corporate governance systems and social responsibility performances particularly through community development projects and memoranda of understanding. However, in spite of these, the MNOCs have not made substantial and adequate contributions and remedies and have not been able to buy back their social licence to operate or build a clement environment of business. The region is today a cynosure of attention and activities as governments, donor and development agencies, international organizations, local and international civil society and non governmental organizations are engaged in efforts at skills development, economic empowerment, governance systems building and peace building as part of comprehensive efforts to manage the conflicts.

1.3. Purpose and significance

The work seeks to provide a broad investigation and analysis of the critical issues, dimensions and ramifications of the conflict. In doing this, it investigates the critical issues of the environment and how oil and gas exploitation impacts on it; the conflicts that involve the environment, resources and the nature of governance; the actors- the Nigerian state, multinational oil companies and the host communities and their roles; and the nature of management of security and the conflict. The work also seeks to analyze the dynamics of the conflict by presenting the conflict situations, the nature of interactions and more specifically encounters between the state and oil companies on the one hand and the ethnic groups, communities and their popular groupings on the other, the nature of state and multinational oil companies (MNC) responses to the conflict, the implications of the nature of management of relations and demands and the implications of the conflict for national security. These issues that the work engages provide a broad understanding of the conflict and a broad elaboration of its significant dimensions and ramifications. Further, the work seeks ways out of the current conflict situations examined, by identifying existing responses and rapprochements and making recommendations and specifically policy, research and management strategies by which the conflict could be resolved.

A critical interrogation of the Niger Delta conflict, undertaken as it is here, from the perspectives of environment, resources and conflict is significant. It places a strategic region to world energy supplies and Nigeria's survival and a conflict situation with huge consequences for national and regional security under the spotlight. The study interrogates and demonstrates how the nature of management of the environment and resources spawns conflict. In placing the conflict under scholarly analysis, the book enables information, insights and understanding of the conflict configuration, trajectories, dynamics and outcomes. Beyond the issues of causality and dynamics, the work contributes to the generation of strategies, methods and policies for resolving the conflict.

1.4. Conceptual and methodological issues

Most studies of the paradoxes and predicaments of resource rich regions and resource conflicts have been undertaken at national and therefore macro levels. Analyses and explanations have also been macro, utilizing such conceptual and theoretical perspectives as Dutch disease, state rentierism, distributive politics, poor policy and regulation and greed in resource struggles and conflicts. Certainly, macro issues in the analysis of the complex conflicts in resource rich regions are important. The considerations of state and corporate policy, environments and national security for example, beyond explanations of resource predicaments are critical to our understanding of the local issues of the economies and struggles of local people in resource rich regions. But most of these studies have neglected the local peoples, host communities, local environments and economies. The dynamics and realities of their social existence, livelihoods; their environmental challenges; how they engage the state and oil companies and what nature of struggles and conflicts have emerged have also been neglected. This work charts a middle course in that it is a mix of macro and micro studies. While interrogating issues of policy, governance, security and effects which tend towards the macro, it brings in micro issues such as the environment, local economies and the dynamics of state and corporate relations with host communities.

A complex mix of theoretical and analytical perspectives is deployed in the work. Most of the contributors situate their study on some framework of concepts and theories. Critical concepts and theoretical perspectives as hegemony, internal colonialism, the rentier state, political economy, youth, corporate responsibility, state failure, conflict, governance and security are clarified and elaborated. Theoretical considerations of linkages or interfaces between critical variables such as environment, economy, health, conflict and security are interrogated. The work is also a mix of empirical and descriptive studies. While some of the chapters are based on empirical studies, others utilize mainly secondary data. Section 1 is particularly based on fieldwork and some quantitative analyses. Sections 2, 3 and 4 with some exceptions rely heavily on secondary data and descriptive analyses. Section 5 brings back survey research and empirical findings into the work.

1.5. Overview of organization and content

The work is divided into five sections. In Section 1, *Oil, Environment and Sustainable Development,* four chapters address the consequences of oil exploitation activities on the ecosystem, primary occupations, health care and conflict. The section begins with the investigation of the impact of gas flares, which consumes about 76% of the Nigeria's gas output in Chapter 2. The country has the unenviable record of the highest flare in the world. Peter Odjugo's study is based on the collection and analysis of rainwater and secondary data of health cases in selected flare site communities and a control site. In Chapter 3, Onovughe Ikelegbe investigates the effects of the negative externalities of oil and gas exploration on the primary local occupations and livelihood sources of the region, particularly agriculture, fishing, traditional craft and forest resources. Based on survey research in some communities, the study identifies adverse impacts and the direction of effects on the local economies. Chapter 4 uses a GIS based risk assessment approach to investigate the interface between the ecological crisis and health care status in the region. Amos Abu identifies the ways in which oil and gas exploitation has devastated and compromised the physical, biological, human, social and health environment and the probable interconnections between the ensuing ecological crisis and the disease profile. The interrogation of the ecology-security nexus engages the attention of the authors of Chapter 5. Here, Iyabo Olojede, Banji Fajonyomi and Ighodalo Akakpe situate the conflict on the environmental pollution and management in the region and the attendant communal dissatisfaction and reactions.

In Section 2, *State, Oil and Conflict,* the work shifts the analysis from the environment to the actors in the conflicts; the Nigerian state, the host communities and the youth. The nature of state policy and management of the conflict; the effects of the oil economy on the host communities and the youth and their responses are investigated. In Chapter 6, Orji Nwachukwu investigates the nature of state policy towards the Niger Delta. State repression and pacification and their forms are particularly highlighted. In Chapter 7, A. Agagu takes the policy investigation further by focusing on interventionist development agencies and efforts. Chapter 8 shifts the analysis from the state and policy dimensions of the conflict to the host communities and the resource curse that has been their misfortune. Here Augustine Ikelegbe examines the oil economy and the nature of host community responses that has shifted from initial welcome and accommodation to hostility and militant agitation. The issues and methods of agitation, the ensuing state repression and its consequences particularly as it relates to the emergence of counter violence are interrogated. Anthony Ogbemi-Ifediora in Chapter 9 examines the youth problematic in the conflict as it relates to the urban economy and governance.

Section 3 *The Crisis of Relations and Social Responsibility* brings in the multinational oil companies (MNOC) and the nature of their relations and responsibilities to host communities into the analysis. In chapter 10, Augustine Ikelegbe examines the crisis in the relations between the MNOCs and the host communities. The issues, causes and methods of the conflict as well as the responses and strategies of the MNOC and host communities are examined. In chapter 11, a theoretical basis for examining corporate social responsibility is presented by Remi Medupin. It examines different perspectives of state roles in the economy, the principle of social responsibility and locates corporate social irresponsibility in the region in state failure. Chapter 12 is more specific. In it Dauda Garuba examines the corporate responsibility of the MNOCs in Nigeria.

Section 4 *Oil, Conflict and National Security* introduces security into the work stream by examining conceptions, linkages and implications. Chapter 13 identifies the conceptual clash of security by different stakeholders and situates the Nigerian state conception of insecurity and in fact the crisis in traditional state centrism. Christopher Isike argues for a paradigm shift in conception and policy towards non-traditionalism and human security. In Chapter 14, Audu Gambo interrogates the political, economic and social implications of the conflict as it relates to national security.

In the last section, the *Challenges: Charting the Way Forward,* four chapters address and suggest ways of resolving the environmental and resource crises. Chapter 15 brings back the environment and presents some detailed environmental diseconomies of the petroleum industry as well as the legal and administrative framework of environmental management. Christopher Orubu, Ayodele Odusola and William Ehwarieme recommend a co-operative framework that brings in the communities in sustainable partnership for sustainable environmental management. In Chapter 16, E. Okojie and M. Ailemen shift the search for solutions to the crisis to a non-farm occupation and employment given the context of the huge ecological devastation and current youth attitudes and preferences. Chapter 17 focuses on conflicts and conflict management. A.S. Akpotor and Julius Azelama argue in favour of preventive measures of conflict management and contend that it is more effective against the backdrop of the current nature of communal and ethnic conflicts in the region. The final chapter brings in the utility of research in understanding and resolving the environmental conflicts in the region. B.A. Chokor uses the Niger Delta Environmental Survey findings and action plan to demonstrate the opportunities that research holds for conflict resolution.

Section I
oil, environment and sustainable development

2. The impact of gas flaring on rainwater quality in erhorike, delta state

Odjugo, P. A. O.

2.1. Introduction

Almost everything the human being does on a large scale influences the chemistry of the atmosphere. Some of the airborne substances introduced by human activity are beneficial to life in general and human life in particular (McDonald, 1985). They provide oxygen and other nutrients that are essential for humans and for the growth of crops and forests (Woodan and Cowling, 1987). Other substances are inert biologically. Still others are detrimental because they cause stress in plants, animals and microorganisms (Cowling, 1991). They also alter surface and groundwater quality, aggravate nutrient deficiencies in soils, or accelerate the soiling, weathering or corrosion of engineering and cultural materials (Olaitan & Lombin, 1984; Costa & Forley, 2000; Odjugo, 2002).

In Nigeria, anthropogenic sources of gas emission into the atmosphere are more from the transport sector, bush burning and gas flaring among others (Okoh, 2000). Nigeria has an estimated 100-120 trillion cubic feet of proven natural gas, making it the 9th largest concentrations in the world (Olagoke 2001). Associated gas with crude oil is routinely flared in the course of producing and processing Petroleum oil. This is a common practice in the oil production process that i not restricted to Nigeria alone.

However, the Nigeria case attracts more attention given the volume of gas flared since the beginning of commercial oil production about 45 years ago.

2.2. the impact of gas flaring on rainwater quality in erhorike, delta state

Statistics indicate that in Nigeria, an average of 4.9 million standard cubic metres of gas is being flared daily out of the 5.6 million standard cubic metres produced daily with crude oil. (Otoikiti, 2001). This flared gas volume when translated into crude oil is the equivalent of some 259,000 barrels. This means, that Nigeria flares more than 76% of its natural gas output, which is the highest anywhere in the world. Libya is a distant second, flaring 21%, Saudi Arabia 20%, Iran 19%, Canada 8% and Britain, Algeria and Mexico flare about 5% each of their natural gas output (Gbafade and Gowon 2002). Nigeria accounts for 28% of total gas flared in the world.

The World Bank in 2000, estimated that gas flaring in the Niger-Delta region of Nigeria where the crude oil is drilled amounted to some 35 million tonnes of carbon dioxide (CO_2) annually into the air. Other gases associated with gas flaring include oxides of nitrogen (NOx), Oxide of sulphur (SOx), oxides of carbon (COx), lead (Pb), Nickel (Ni) and Zinc (Zn) among others. The reaction of some of these gases with other airborne chemicals such as oxygen, ammonia and moisture give rise to acid rain – a condition of polluted rainwater, which adversely affects plants, animals and humans.

Studies on the impact of oil industries on environmental quality of the Niger – Delta show contrasting results. NDES (1997) revealed that almost all the organic, inorganic, metals and the physical properties of the flare in the air and water were below the Federal Environmental Protection Agency (FEPA) and World Health Organisation (WHO) stipulated standard. Okoh (2000), on the other hand, noted that pH, TSPM, all the physical parameters, and most of the heavy metals were above WHO maximum limit for residential areas. On health, while Tizhe (2001), did not notice any significant effects of gas flaring on health in Ubogo community of Delta State, Efekodo (2001), observed a significant effect on the health of people and the environment in Igbide and Olomoro in Delta State. With these conflicting results, one cannot really say much with a high degree of certainty about the exact magnitude of the effects of gas flaring in the Niger – Delta except more independent and quantitative studies from different parts of the oil producing area are carried out. It is on this premise that this paper is designed to assess the impact of gas flaring on rainwater quality and its implications on human health and the environment in Erhorike community of Delta State.

2.3. Area of study

The area of study is the gas-flaring site in Erhorike, Delta State. It is located on Latitude 5.30^{O}N and Longitude 5.57^{O}E (Fig. 1). It is about 6km east of Kokori and 25km south west of Ughelli. It has a population of about 5,000 inhabitants and the major occupation is farming. The prevailing climatic condition is that of equatorial type with annual mean rainfall of about 3000mm and annual temperature of 27°C. The vegetation is evergreen tropical rainforest.

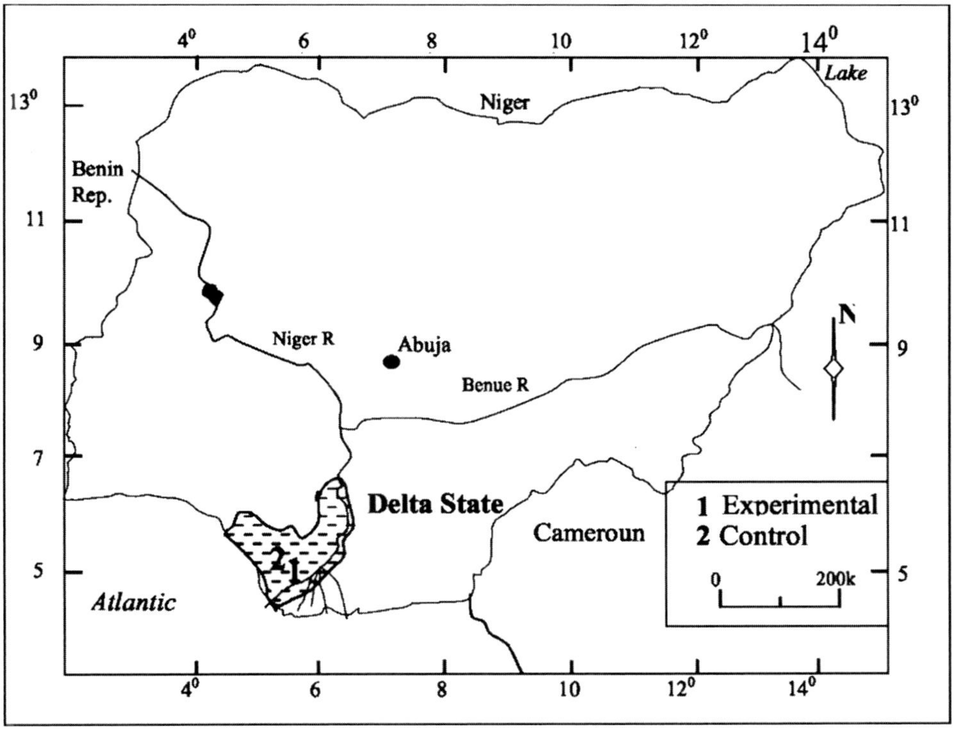

Fig. 1: Nigeria – Showing the Study Area

2.4. Materials and methods

Primary and secondary data were collected for this study. The primary data is the rainwater. The first rain in each month was collected for a period between January and December 2001, from seven sites with the raingauge. Out of the seven sites, three were situated on the southwest (Windward, March-October, and Leeward, November-February), three were located on the northeast (Windward, November-February, and Leeward, March-October), while the remaining site was the control located at Okpara Inland about 18km from the flare site. The control distance of 18km is selected because the results of Onuorah (2000) show that the impacts of gas flaring on the atmosphere are insignificant at a distance of 15km from the flare site. The experimental sites were located at 200m, 500m and 1000m from the bundwalls of the flare. Inhabitants of Erhorike reside within these distances even though the minimum limit allowed by the Environmental Impact Assessment Decree No. 86 of 1992, is 3km.

The rainwater collected was sent to a laboratory for analysis. Parameters analysed for include the; pH, Physical parameters ((Total dissolved solids, turbidity, colour, taste and odour); inorganic / organic parameters (CL^-, NO_3, SO_4, PO_4, Na, K, Mg and Ca); and heavy metals (Pb, V, N and Zn). The laboratory result was compared with the FEPA and WHO stipulated standards for domestic and commercial centres in order to determine the level of pollution and the possible effects. The compliance level was computed using Clerk (1984) equation.

where	PC	=	Percentage compliance
	N1	=	Number of times parameter complied with stated standard.
	P	=	100% (Assumed maximum compliance limit).
	N	=	Total number of Measurements.

He concluded that for healthy living compliance level must attain 70% or above.

The secondary data used were the health cases collected from the Health centre at Erhorike and the St Francis Hospital at Okpara Inland. The data are compared with known health problems associated with gas flaring. Five hundred questionnaires were distributed to elicit the opinion of the inhabitants on the impact of gas flaring on their health. These were analysed using simple percentages.

The results indicate that the mean total dissolved solid (TDS) of 1113.2 mg/L is 11.3% higher than the WHO 1000 mg/L maximum permissible limit (Table I). While the rainy season value of 949.75 ug/m^3 is marginally within the recommended standard, the Dry season (1440 mg/L) is above the limit. For 6 months (October to April), TDS, level exceeded the stipulated minimum limit. The stipulated limit of turbidity is 5 NTU.

The mean turbidity for the year (6.54 NTU) and that of dry season (9.58 NTU) exceeded the acceptable limit, but the rainy season (4.91 NTU) is within the limit. For 8 months (October – May) turbidity is higher than the limit putting the compliance level at 33.3%. In seven months, (October – March) the colour of the water is higher than the (15 TCU) stipulated limit thus having a compliance level of 4.6%. While the mean (16 TCU) and dry season (22.5 TCU) exceeded this limit, the rainy season (13 TCU) is found to be lower than the stipulated limits. As for taste and odour, both the mean (2.15 TON), dry season (2.9 TON) or rainy season (1.76 TON) were lower than the 3 TON recommended standard. All the physical properties and pH in the control sites were within the permissible limit. This implies that outside the flare sites vicinity, atmospheric pollution is very mild.

2.5. Rain water analysis

Table 1: Acidity and Physical Parameters of Rain Water

	Total Dissolved Solids (mg/2)	*Turbidity (NTU)*	*PH*	*Colour (TCU)*	*Taste and Odour (TON)*
Jan.	1582	10.8	3.0	25	4.61
Feb.	1391	9.2	3.1	22	3.02
Mar.	1176	6.3	3.6	20	2.40
April	1052	6.1	4.0	16	2.12
May	963	5.2	4.8	12	1.84
June	860	4.3	5.1	9	1.55
July	746	4.0	5.6	7	1.14
Aug	859	3.1	5.2	9	1.30
Sept	863	4.6	5.0	14	1.81
Oct.	1079	5.7	4.5	17	1.88
Nov.	1299	9.3	4.1	20	2.09
Dec.	1488	9.0	3.4	23	2.67
Total	13358	78.5	51.4	199	37.42
Dry Season	1140	9.58	3.4	22.5	2.9
Rainy Season	9493.75	4.91	4.7	13.00	1.76
Who 1996 Standard	1000	5NTU	6.58.5	15	3
Mean	1113.2	6.54	4.28	16.00	2.15
Control mean	766	2.4	6.6	44	0.8

It could be deduced from the physical parameters that the water losses its aesthetic value since it is dirty and coloured. This was confirmed by 92% o the respondents who revealed that the rainwater is not good for drinking because it is blackish and dirty in nature almost throughout the year. Though the test shows that the taste is within the acceptable limit, 86% of the respondents revealed that the water is rusty if left for two ormore days. This claim was confirmed personally by the researcher.

The permissible pH range for domestic drinking water is 6.5-8.5. The mean pH for this study was 4.28, while the dry and rainy seasons were 3.4 and 4.7 respectively (Table 1). This shows a high level of acidity especially during the dry season months of November to February. The pH range for the study period is 3.0-5.6. The 6.1 pH value recorded in the control site shows that it slightly falls below the recommended standard, thus making atmosphere to be mildly acidic.

Ogefere (1999) also noticed coloured and highly acidic waters a Sapele Gas Plant station.

With this level of high acidity, it could be reasoned that building or sculptures containing calcium or magnesium carbonates could be readily attacked by acid rain with the formation of a soluble sulphate. $CaCO_3$ + H2 SO4 Æ Ca^{2+} + SO_4^{2-} + H20 + $C0_2$. If the building stone is limestone or marble (both $CaCO_3$) or sandstone in which the quartz grains are held together by a coating of Calcium carbonate or iron oxide, the iron oxide will also dissolve in acid water. Fe2O3 + $3H_2SO_4$ Æ $2Fe^{3+}$ + 3 SO_4^{2-} + $3H_2O$.

Jan.	118.6	62.6	326.1	0.82	8.6	0.7	1.38	2.5
Feb.	116.3	64.2	317.4	0.91	7.4	0.52	1.51	2.27
Mar.	110.9	55.6	266.7	0.83	6.9	0.49	1.27	1.82
April	111.3	58.2	227.3	0.42	6.8	0.43	1.18	1.30
May	109.1	53.0	231.2	0.48	6.5	0.40	0.79	1.80
June	106.4	51.6	203.9	0.24	6.0	0.39	0.73	1.12
July	107.5	48.3	217.4	0.36	4.8	0.41	0.53	0.81
Aug.	101.7	49.7	226.6	0.53	4.9	0.35	0.50	0.99
Sept.	110.9	53.2	233.7	0.59	5.9	0.46	0.57	0.91
Oct.	114.8	50.6	259.5	0.69	5.6	0.59	0.63	1.02
Nov.	112.3	58.9	281.2	0.73	6.2	0.63	0.82	1.48
Dec.	121.7	62.3	220.0	0.83	7.3	0.66	1.42	2.0
Total	1341.5	668.2	3085	7.42	76.9	6.08	11.32	18.0
Mean	111.8	55.68	257.08	0.62	6.41	0.51	0.94	1.5
Dry Season	117.23	62	311.18	0.84	7.3	0.63	1.3	2.06
Rainy Season	109.8	52.53	232.19	0.52	5.9	0.44	0.78	1.22
Control	115.4	42	204	0.28	2.4	0.31	0.46	0.35
WHO 1996 standard	250	50	250		200			
FEPA 1991				5		50	50	75

Table 2 reveals that the mean value of all the organic and inorganic parameters analysed fall with the acceptable limit of either WHO or (FEPA) except NO3 and *SO4.* Table 2: Inorganic Parameters (mg/L).

On seasonal basis, NO3 was above the 50 mg/L maximum limit in both seasons. While SO4 was above the 250mg/L permissible limit during the dry season

(257.08 mg/L), it is lower during the rainy season (232.19 mg/L). This puts the compliance level of CL^-, PO_4, Na, K, Mg and Ca2 to be 100% while that of NO3 and SO4 were 16.7% and 50% respectively. All the values of the control site were within FEPA and WHO permissible limit and thus have 100% compliance. Uche (1997) revealed that staying in a region where NO3 is above 0. 1PPM (26mg/L) or SO4 higher than 0.5 PPM (130 mg/L) for years could result to great risk of deteriorating health. Okoh (2000) specifically stated that living in an environment where NO3 is above 50mg/L (0.2PPM) for a long period leads to infantile methaemoglobinaemia occurring more for children below 6 months. O'neill (1993) suggested a further potential hazard of forming carcinogenic nitrosomines, in the human digestive system by the conversion of nitrate to nitrite and subsequent reaction with amino acids. NRC (1977), NAS (1986) and Cowling (1990) noted that SOx, NOx, and VOC have the highest range of detrimental effects on sanity and for this reason, these three primary pollutants are major keys to the proper management of air quality in most industrial regions of the world.

Table 3: Metals in Rain water (mg/L)

	Lead (Pb)	*Vanadium (V)*	*Nickel (Ni)*	*Zinc (Zn)*
Jan.	0.019	0.0019	0.0211	0.0674
Feb.	0.026	0.0018	0.0220	0.0692
March	0.018	0.0016	0.0203	0.0568
April	0.015	0.0016	0.0201	0.0329
May	0.012	0.0013	0.0197	0.0317
June	0.010	0.001	0.0186	0.0236
July	0.008	0.0008	0.0175	0.0219
Aug.	0.008	0.0008	0.0194	0.0346
Sept.	0.010	0.001	0.0205	0.0432
Oct.	0.010	0.0009	0.0209	0.0586
Nov.	0.012	0.0014	0.0211	0.0618
Dec.	0.015	0.0017	0.0216	0.0638
Total	0.163	0.158	0.2438	0.5655
Mean	0.0135	0.0013	0.02032	0.0471
Dry Season	0.018	0.0017	0.02 157	0.0656
Rainy season	0.013	0.0011	0.0196	0.0379
Control	0.006	0.0006	0.009	0.02 1
WHO 1996	0.01	0.001	0.02	3

The recorded mean values of Pb (0.0 135), VC (0.00 13) and Ni (0.02032) were higher than the recommended 0.01, 0.001 and 0.02 mg/L respectively, while Zn (0.04713), is within the recommended standard of 3 mg/L (Table 3). With the exception of Zn, other elements like Pb, V and Ni were also found to be higher than the approved limit at all seasons (rainy and dry), but the values were generally higher during the dry season. This work is in contrast with the findings of Asiweh (1999) who showed that all the heavy metals were within the acceptable WHO limit. The variation in his findings with this work could be attributed to distance differential from the flare sites where the studies were carried out. While this work takes care of the immediate vicinity of the flare site (1km radius), that of Asiweh was a bit remote (10km from the flare site).

With the high concentration of heavy metals, inhabitants of these areas are exposed to a very high risk of health hazards associated with high dose of heavy metals. This includes impaired blood synthesis, hypertension, hyperactivity and brain damage. In this vein, O'neill (1993) noted that the intelligence of children is being lowered in urban areas in which there are high inputs of lead from motor vehicles.

2.6. Health implications

Table 4: Reported Cases of Diseases

Diseases	*Erhorike*	*Okpara Inland*
Malaria fever	32	27
Respiratory diseases	11	2
Eye diseases	8	4
Hypertension	7	5
Heat related diseases	7	1
Intestinal diseases	5	10
Thyphoid fever	5	20
Cardiac diseases	4	3
Measles	4	12
Skin diseases	4	9
Cancer	3	1
Stroke	2	1
Brain disorder	1	1
Others	7	4

2.6. Health implications

Malaria fever topped the list of diseases that affected the inhabitants of the flare site. This is the commonest disease in the tropical forest area of Nigeria and is not peculiar to areas of gas flaring. But gas flaring might have increased the incidence of malaria fever directly and/or indirectly. Odjugo (2000) noted that temperature guides the sporogony of the protozoa causing malaria and that sporogony accelerates from 25 days at 10^oC to 8 days at 32^oC and that the activity and rate of mosquito bites increases with increasing temperature until 35^oC and thereafter decreasing bites set in. These favourable temperatures for reproduction and bites are what the flare provides. Moreover, the high temperatures created by the flare make indoors uncomfortable especially during the early hours of the night. This forces the villagers who do not have temperature-modifying gadgets, either due to poverty or lack of electricity to operate them, to stay outdoors till late at night. Such outdoor stay noted Odjugo (2002) exposes them to more mosquito bites and increased incidence of malaria fever.

The high atmospheric concentration of smoke density, total suspended particulate matter, $_{NOx,}$SOx and heavy metals among others, must have accounted for the high incidence of respiratory diseases such as aggravated asthma and cough; eye diseases such as eye irritation, cataracts, glaucoma; and high blood pressure and associated diseases. The relatively high incidence of intestinal diseases may not be unconnected with the polluted water that they drink since the village under study has no potable water. The main sources of water are hand-dug wells, river and rainwater. Heat and cardiac related diseases are also common.

Comparing the reported health cases in the flare site (Erhorike) and the control site (Okpara Inland), a reasonable degree of variation is noticed. While the highest five diseases in the flare site are malaria fever, respiratory diseases, eye diseases, intestinal diseases and hypertension, that of the control site include malaria fever, thyphoid fever, measles, intestinal and skin diseases. Respiratory and eye diseases which are common to areas of heavy air pollution (O'neill, 1993, Oke, 1995 and Uche 1997), are evident in a high degree in Erhorike while the percentage is very negligible in Okpara Inland. The variation in the disease pattern between the flare site and the control is further buttressed by Kendall's correlation value of 0.321, which is not statistically significant. The implication is that diseases that ranked highest in the flare site are relatively low in the control site.

From the foregoing, one may not be able to say categorically that gas flaring is the sole cause of the above diseases. However, the pattern displayed could lead to the conclusion with the high level of confidence that gas flaring has directly or indirectly aggravated the magnitude of the occurrence of the diseases.

2.7. Conclusion and recommendations

From the physico-chemical analysis of the rainwater, it was observed that acidity level is high with mean pH value of 4.2. Apart from the acidic nature of the rainwater, it also lost its aesthetic value since the physical parameters such as total dissolved solids, (11 13.2mg/L), turbidity (6.54mg/L) and colour (1 6.6TCU) were higher than the maximum permissible limit. It is only the taste and Odour 2.15 (TON) components of the physical parameters that are within the recommended limit. The inorganic parameters have higher compliance level. CL^-(111.8 mg/L), PO4 (0.62mg/L), Na(6.4mg/L), K (0.5 1mg/L), Mg (0.94 mg/L), (Ca^2(1 .5mg/L) were all within the acceptable limit while only NO3 (55.68 mg/L) and SO4 (257.08 mg/L) exceeded the recommended level. Apart from Zn (0.047) among the heavy metals, which is below the recommended standard, Pb (0.0135 mg/L), V (0.013 mg/L) and Ni (O.0203 2 mg/L) were greater than the WHO recommended standard. With this result, it is clear that within 1000 metres from the bundwalls of the flare site, the atmosphere is heavily polluted and not conducive for habitation either for domestic or residential purposes.

For operations of the oil industry to be sustainable and environment friendly, better environmental and waste management practices are needed. To start with, the atmospheric pollution through gas flaring must be stopped or reduced to the barest minimum and the gas re-injected into the underground gas reservoir. More gas and petrochemical industries should be established to produce cooking gas. Adequate arrangement and contacts should be made with West African countries and beyond to supply their gas needs. Available evidence shows that Nigeria can currently supply the gas need of the entire West African countries if the gas industry is well developed. This will not only cleanse the atmosphere but also provide much revenue to the government and employment to the numerous unemployed and underemployed Nigerians. Moreover, more states in the Niger Delta in particular and the entire nation in general should convert from the current total dependence on hydroelectric power supply to thermal electricity by using gas as is currently done by the Rivers State government.

References

Asiweh, C.P.A. (1999) Environmental and Socio-Economic Implications of the Oil Industry in Rivers State. Paper presented at the *Annual Conference of the EHOs*, held at Asaba, 18-2 1 August, 1999.

Clerk, R. T. (1984) *Mining and Environment:* Ibadan: ONAKS Press

Cowling, B. E. (1991) Acid Rain and other Airborne Pollutants: Their Human Causes and Consequences. In Davis K. and M.S. Bernstam (eds) *Resources, Environment and Population:* A Supplement to Vol. 16, 1990.

References

Efekodo, G. (2001). "The impact of Oil Exploration and Exploitation in Isoko Land", An. M.Sc. thesis presented to the Faculty of Social and Management Sciences, Ondo State University, Akungba-Akoko.

Gbafade, U. and P. Gowon (2002) Gas Flaring: Why burning the Resources? *Atmospheric Review,* Vol. 14 (2): 82-103

McDonald G. J. (1985) *Climate Change and Acid Rain.* McLean, Virginia: Mitre Corp. National Academic of Sciences (NAS) 1986. *Acid Deposition: Long-Term Trends.* Washington D.C. National Academic Press

National Research Council (NRC) (1977) "Effects of Nitrogen Oxides on Vegetation"
In *Nitrogen Oxides.* Washington D. C.: National Academy of Science. 197-2 14. Niger Delta Environmental Survey (NDES) (1997) Environment and Socio-Economic
Characteristics: Vol. 1, Phase 1.

Odjugo P.A.O. (2000) Global Warming and Human Health: Current And Projected Effects. *Environmental Analar No. 5: 428-449.* (University of Uyo, Uyo).

Odjugo, P.A.O. (2 002) "Health, Safety and Environment: Challenges of Environmental
Health Officers in the Present Democratic Dispensation". Paper presented at the
3rd *Delta State Triennial Delegates Conference/NEC Meeting,* Asaba 13-15, June 2002

Ogefere O. C. (1999) "Environmental Impact Evaluation of the Sapele Flow Station and Gas Plant". Paper presented at the *Annual Conference of the EHOs,* held at Asaba, 18-21 August, 1999.

Okoh, N. V. (2000) Environmental Impacts of Oil Exploration, *Atmospheric Review,* Vol. 12 (1): 68-80

Olagoke, C. (2001) Nigeria Economy: The Oil Sector, *OPC: Bulletin No. 12.*

Olaitan, E. O. and Lombin, G, (1984) Introduction *to Tropical Soil Science,* London: Macmillan Pubs

O'neill, P. (1993) *Environmental Chemistry,* London: Chapman and Hall.

Onuorah, A. O. (2000) The Oil Industry and Environmental Degradation *Journal of Atmospheric Science.* Vol. 10(2): 42-63.

Otoikiti, S. A. (2001) Economic Cost of Gas Flaring in Nigeria, *Journal of Economic Development,* 25(1): 18-35.

Tizhe, N. H. (2001) "Socio-Economic Implications of Oil Exploitation in Ubogo Community", An. M.Sc. thesis presented to the Faculty of Social and Management Sciences. Ondo State University, Akungba-Akoko

Uche, J. I. (1997) Air Pollution. In Odjugo, P.A.O. and J.E. Evwoh (Eds). *Physical Components of the Environment.* Agbor: Pon Pubs.

Woodman J. N. and Cowling, E. B. (1987) Airborne Chemicals and Forest Health. *Environmental Science and Technology* 21: 120-126.

3. The effects of oil exploration on the primary economy of the niger delta

Onovughe O. IKELEGBE

3.1. Introduction

Oil exploration and extraction activities in the Niger Delta have been invaluable to the growth and development of the country. This is reflected in both spatial and aspatial terms. The oil boom of the 1970's led to a period of economic and state expansion and till date, oil accounts for about 75-80% of state revenues. Consequently the oil sector of the economy has formed the bedrock of overall national development. The result of such a mono-cultural economic profile has been massive exploration and exploitation of oil and gas resources by oil companies to generate more resources for development.

Actually, oil exploitation has had positive and negative impacts on both the ecology and general economic life of host communities. Whilst the positive impacts of the oil exploitation are manifested in the growth and development of the entire nation, the negative effects are often borne by the host communities. Oil wealth has been coupled with intractable environmental problems in the host communities. Largely, the effect is more devastating on the ecology and in the primary sector of the economy. Consequently, the entire existence of such oil communities has been threatened.

The environmental crisis has led to incessant agitations and repressions as well as the growth of pressure groups in the area. There has been as a result, a regional crisis, which has its roots in political alienation, economic deprivation and environmental devastation. Since the primary economy of the region largely hinges on agricultural and fishing activities, there is a dire concern for a proper conservation of its natural resources in order to ensure a sustainable development.

The paper critically analyzes the effects of oil exploitation on the primary economy of host communities and the resulting crisis. Further, it examines the perception of host communities to the palliative measures of oil companies to redress the effects of oil exploitation and address the grievances of the host communities. Finally an attempt is made to proffer suggestions that would enhance sustainable development, manage the environmental crisis as well as guide public policy.

3.2. Oil, environment and conflict in the niger delta

The environmental problems that arise from oil exploitation activities are related to the physiography, vegetation/ flora and fauna. That these have been impacted negatively and have in turn devastated the region's primary economy has been a major concern in recent years. In sum, oil has distorted and disrupted the socio-economic lives of the Niger Delta communities as well as constitutes a major threat to their sustenance and existence.

Exploration, production, transportation and marketing activities of oil companies result in environmental degradation in the terrestrial, marine and atmospheric life of the Niger Delta. These largely manifest in form of oil spills and gas flaring. The latter has led to the incidence of acid rain as well as the acceleration of the global warming phenomenon. The emission of gases of varying composition and toxicity are detrimental to plants and animals. While crank case oil disposal from workshops, industries, power stations and commercial houses are not only aesthetically unpleasant but also prevents growth of soil flora and fauna as well as contaminates surface and ground water. These tend to affect agriculture, which is the economic base of the Niger Delta.

The economic and social life of communities may be dislocated as a result of these devastations. Aina (1996:2) identifies the general ecological impact of oil spills as including decrease of fishery resources and damage to wildlife such as sea birds and marine mammals, human hazards through eating of contaminated sea foods, decrease of aesthetic values due to unsightly slicks on "oiled beaches" and modification of habitat leading to delay in re-colonization and succession. The negative effects are too numerous to mention. Land, air, rivers and creeks, vegetation and cultural sites have all been devastated.

This is more so because of the scale and extent of environmental degradatory activities of oil. Egborge (1999:23) notes that about 2, 369 471, barrels of crude oil was spilled in the Niger Delta between 1976 and 1996 out of which 1, 820,411 barrels were not recovered and absorbed. About 479, 392 mm3 of gas has also been flared between 1958 and 1994 and as at 1993, Nigeria was the world leader of gas flaring. Physical, political, socio-economic and cultural impacts also abound. The results have been described by activists as economic strangulation or economic disempowerment of the people. Disempowerment here is viewed not only in terms of destruction and deprivation but also the absence of commensurate observable gains in development.

In addition to these physical devastations, the Niger Delta has been abandoned, marginalized and relegated in policy and developmental focus. There is absence of social amenities and poverty and unemployment have become so rampant that the people have become frustrated and violent (Nzemeke 1999). The result has been aggression towards oil companies, which is indicated in disruptions of oil production and vandalization of oil facilities and pipelines. There is also popular agitation, increased activism of pressure groups as well as violence and state repressions in the region. The manifestations of anger against corporate and public policies and practices in the region continue with huge implications for the safety and stability of the oil industry and national revenues.

The dilemma is how to continue to harness the regions valuable oil resources, which destroy its rich natural resources and pollute its environment and at the same time ensure sustainability in the environment. This point in the direction of how to ensure that the environment despite oil exploitation activities is still ecologically balanced to ensure livability for flora, fauna and man. The Niger Delta Environmental Study (NDES) (1997: 261-267) has noted salient issues that presently constitute a constraint to development in this region. These are; inadequate diffusion of appropriate technologies, poverty, institutional weakness, disruptive economic forces, policy distortions and inequity, inadequate attention to community participation, inadequate education and information, lack of appropriate methods for resource valuation and the limited empowerment of women.

The absence of an acceptable method of measuring the value of damaged natural resources is central to the continuous environmental degradation in the Niger Delta. When destruction of a region's natural resources is equated with compensation, particularly in monetary terms, the consequence is that of more environmental degradation. Oil firms see the insignificant compensation as a license for a continued devastation of the region's natural resources. Added to this, is the fact that these compensations have further encouraged intra/inter communal clashes within the region. Arguably, a given natural resource (such as biodiversity) given its bequest value may be of diverse economic value to the community for several years. This may not be commensurate to such compensations. Compensations in themselves are not a sin qua non to the sustainability of an industrial environment. It only acts as a short term or mirage solution to a life long impending issue that in real terms need to be dealt with.

The Nigerian government, oil companies and various researchers have acknowledged the need for sustainable development in the Niger Delta. The government has set up several policy initiatives as well as agencies to address the issue. There was the establishment of the Federal Environmental and Protection Agency (FEPA) in 1988. The Niger Delta River Basin Authority (NDRBA), the Oil Mineral Producing Areas Development Commission (OMPADEC) 1992, and the Niger Delta Development Commission (NDDC) 2000, were also established. These agencies have had or have mandates for resource management, environmental protection and the development of the Niger Delta. While their achievement require an empirical survey overtime, the paper acknowledges that their achievement at present, in terms of the adverse effects of oil exploration of the host communities is still minimal. Perhaps this is an area for further research.

The inclusion of environmental considerations into national economic development has been of utmost concern to policy in recent times. This is clearly manifested in the Fourth National Development plan, which called for"steps to ensure environmental planning as an integral part of project planning; to enhance environmental conservation and prevent degradation". In this regard the plan called for "the development of an Environmental Impact Statement (EIS) for all projects that are geared towards mitigating adverse environment problems" (Nigeria: 1985)

The need for assessments of the environmental impacts of large-scale industrial projects like those of the oil companies cannot be overemphasized. Such is under-scored by the interdependence between economic systems and the ecosystems, the compatibility between resource use, conservation and economic growth and the role of the natural environment in sustainable development. Oil companies and other researchers have in the present carried out several studies in this regard.

Most environmental impact assessments have largely focused on the bio-physical and chemical aspects of environmental pollution rather than their socio-economic impact. Chokor (1996: 152) argues that both the environmental and socio-economic considerations are important for a more balanced and integrate view of the impacts of the oil industries. This is the main thrust of this paper. The paper attempts to analyze critical environmental and sociocultural concerns of the selected host communities. Emphasis is on the effect of oil exploitation on environmental degradation, and the primary economy as well as the issue of compensation as perceived by host communities.

3.3. Methodology

The study notes the difficulty of identifying relevant variables, criteria and methods of assessment as well as the subjectivity in value judgment associated with socio economic descriptions. In relation to the latter, the study adopted the two basic ways of redress, suggested by Chokor (1996: 157); the supplementing of the use of questionnaires with personal observation. Two hundred (200) Questionnaires comprising structured and unstructured items were administered in six (6) host communities to oil companies in the Niger Delta. The communities were selected by purposive sampling. The questions relate to the effects of oil exploitation on the economic base of the communities, their environment, political and social issues as well as their perception of the compensations presently being paid by these companies. Respondents were however selected randomly as all adult community members constituted the sample for study. There were high percentages (99%) of questionnaire returns although the interviewers indicated a negative attitude and apathy to the survey by host communities. This was depicted in the reluctance of a large number of the respondents in host Communities especially in Otu -Jeremi to give audience and answer questions posed by researchers.

3.4. Area of study

The host communities selected for this study are Otu-Jeremi, Oghior, Iwrekan, Otor-Udu, Ubeji and Ekpan. These are located in Ughelli South, Udu , Warri South and Uwvie local government areas of Delta State. Otu-Jeremi and Otor Udu are the headquarters of Ughelli South and Udu local government areas respectively. The area is situated in the lowland rain forest ecological zone. The total population of the towns are; Otu-Jeremi (2,593), Otor Udu (1,057), Oghior (1,428) and Iwhrekan (600) Ubeji (235,750) and Ekpan (250,000) persons.

3.5. The primary economy of the niger delta

The primary economy of the area is largely resource based. Over 80% of the population participates in the primary sector of the economy. The main activities are agriculture, fishing, forestry and traditional craft industries. These are briefly examined hereunder.

Agriculture: The farming system of the region is mainly the traditional peasant subsistence/crop farming method, with land and labour being the principal inputs of production. The major crops include cassava, yam, cocoyam, maize, plantain, groundnut, fruits and vegetables. Farmers in the area engage in double mixed and inter-cropping practices. Agricultural practices in the area depict a low level of adoption of fertilizers, agro – chemicals and improved seedlings. In Delta State, the NDES (1997) noted a lower fertilizer usage in the Southern zone (11.48%) where this study area is located as compared to the Central and Northern zones where there was (40-50%) utilisation. Fields may be cultivated for 1-2 years and left to fallow for 2-5 years. Plantation agriculture (largely raphia palm, oil palm, timber and rubber) is also practiced here, although at a very low scale due to the poor capital resources of the region's citizens.

Fisheries Production: The main ecosystems where fishes thrive in the region include inshore waters, brackish water, mangroves, fresh water swamps, flood plains as well as rivers. In terms of species, three broad fauna groupings are evident. They are coastal marine, brackish and fresh water stocks. These are harvested all year round (especially in the dry season) by locales with crude fishing nets, baskets and traps. In the coastal areas, industrial trawl fisheries are common.

Traditional Industries – These include canoe- carving, pottery, thatch making, mat making, palm oil processing, local gin (ogogoro) distillation and food processing (garri, fufu and starch from cassava). These industries are characteristically based on local skills and inputs. Sand and clay are also mined in the rivers and creeks of the Niger Delta. Sand is the primary raw material on which the existing glass industry is based. Clay is commonly used for local earthen pot makers.

Forestry/Forest Reserves – The economically significant timber resources in the Niger Delta are; sawn logs, transmission poles, building poles, fuel wood, bamboos and chewing stick. Very few areas presently remain as large hectares have been converted to farmlands and for urbanisation purposes, especially with the creation of new local government headquarters in these areas. The latter has enhanced status and facilitated rapid development. Some forests are still being preserved as natural groves or abode of divinities or 'evil forests'. Non – timber forest resources also abound in the Niger Delta. The main forest resources include snails, oil palm, rafia palm, and there are also various fruits such as bush mango (ogbono), spices, roots, tree barks and climbers among others.

3.6. Location, growth and development of the oil industry in the niger delta

Crude oil exploration started with Shell D'Arcy, now Shell Petroleum Development Company in 1938. The development of the oil industry in the Niger Delta has grown remarkably with the emergence of petroleum as a major resource in the country. There has been a remarkable increase in the rate of production for export from the late 1950s. In 1958, it accounted for 15% of total export value. It has accounted for more than 97% of total exports since the 1990s. It has also contributed immensely to government revenue. In 1980, it contributed about 81%. It presently fluctuates between 66% and 83% (NDES Report 4 (1): 52, Aina 1996:1).

The oil and gas industry has over 10,000km of pipelines, 100 flow stations, 10 gas plants and 3 refineries in the Niger Delta (Egborge 1999:23). The development of the oil industry has led to the growth of modern manufacturing industries, which are mainly petroleum based. Oil refining and petrochemical companies account for 70-75% of total industrial activity in the Niger Delta. There are three refineries in Tai/Eleme, (Port Harcourt) and Warri, two petrochemical plants in Warri and one at Eleme.

Despite its large hydrocarbon reserves, the region remains poor and its infrastructure is largely undeveloped. Unsustainable exploitation and environmental degradation are increasingly impairing the natural resource base on which the rural communities depend (World Bank Report 1995:1). But oil exploitation is still viewed as the life wire of the host communities who because of neglect by the Nigerian state expect development from the oil prospecting companies. In this regard oil companies have made several efforts to provide infrastructural facilities in host communities. These include the provision of pipe borne water, access roads, health facilities, electricity, jetties, markets, school buildings, employment and manpower development.

Apart from infrastructure provision for communities, there have been other benefits from the oil industry such as the development of human resources, the growth of a new breed of businessmen (oil field contractors) with 'contract jobs' in these companies and the development of skilled and unskilled labor (particularly porters, security and linesmen) and employment with high remuneration. This has also led to improved living standards and sharp increases in the cost of living particularly in the oil cities of Warri and Port Harcourt. Several companies have been involved in erosion control and land reclamation assistance in the Niger Delta.

3.7. Oil exploitation, ecology and primary economy

The effects of oil exploration on the Niger Delta can be viewed in two respects, the negative and the positive. While host communities have benefited from oil exploitation activities, the adverse effects on the ecology and consequently the primary economies of the communities seem to be more overt and notable. Oil pollution has damaged the physical or natural environment, which has in turn interfered with the social, economic and political environment.

In a study of the effects of oil exploration on the socio-economic life of Nigerians living in host communities, Ikeagwuani (1984:11) notes that at the exploration and exploitation stages, tracts of cultivated agricultural lands, forested areas and fish ponds are tampered with when networks of seismic lines are cut during geophysical exploration. Explosives buried and detonated along the paths kill fishes, oysters, crabs and other marine organisms. Where the seismic operation is near villages, the explosives tend to damage buildings. Destruction of the natural environment may result from the introduction of strange substances like;

i) Drilling muds and well cuttings

ii) Radio active materials used for well testing (neutron logs use radioactive sources and if lost in the well bore), could seriously pollute the ground water

iii) Gas flared during production

iv) Oil spillages caused by corrosion of pipes, over pressures, overflow, hose failures, sabotage and vibration.

v) Produced formation water and other effluents.

The air may be polluted through gas flaring, blowouts or direct evaporation of hydrocarbons on exposed surfaces or see page from under ground oil/gas reservoirs. The prevalence of artificial light in the ecosystem has had adverse effects on plants and animal habitat. The generation of constant heat around flare pit causes evaporation of produced water, which increases the salinity of pool water. Previous empirical studies in Port Harcourt have revealed that within 40m radius, the natural vegetation may be bare or dried up due to abnormal salinity while some natural vegetation and fruits further away from these sites may grow well but fail to fruit (Imevbore 1979). In Brass, Rivers State, the extinction of a variety of shrimp caught at night has also being noted. The inference is that the brightness from gas flares may have disturbed their mating habits, and hindered further multiplication.

The pollution of terrestrial and underground water results in high water salinity, while oily wastes and chemicals dumped into flowing and stagnant water may interfer with the natural aeration (due to oily film that coats water surfaces). This kills fishes and marine organisms because of lack adequate intake of oxygen. The toxic effects on marine life also affects the host communities whose fish and water supplies is largely from streams, creeks and rivers. The toxicity may also adversely affect agricultural land as soils and underground water are contaminated. The result is low agricultural crop yield. Economic crops such as rubber and timber as well as food crops are depleted. The destruction of sacred places, animals and things has also been noted in some host communities such as Ozoro, Aboh and Ndoni in Delta State and Eke-Idemili in Anambra state. These problems have further heightened poverty levels in the Niger Delta whose economy is largely resource based. Oil exploitation has thus made land degradation, water and air pollution, lack of access to portable water deforestation and lack of biodiversity to become major issues of concern in the Niger Delta.

But environmental management by the oil companies and the federal government has been poor. There is lack of appropriate institutional arrangements geared to environmental management, absence of effectively implemented policies and poor private sector involvement. Where states and local-government agencies are involved in environmental management, they are highly constrained by lack of funding, trained staff, technical expertise, adequate information, analytical capacity and other pre-requisites for implementing comprehensive policies and programmes (NDES 1997:81)

3.8. Data presentation and analysis

The study presents issues from the respondents' perception as elicited from the survey. Communal perception is important as it indicates the nature attitude to and relations with oil companies. Added to this is the fact that perceptions of host communities must be taken into account in order to achieve sustainable development.

3.9. Oil exploitation and the primary economy of selected communities in the niger delta.

The survey reveals that respondents are largely in the following occupations in area; Farming (22%), fishing (25%), trading (27%), mat and basket weaving (3%), palmwine tapping (2%), hunting (2%), rubber tapping (5%), civil servant (7%), student (6%) and others (1%). These activities largely constitute the primary sector of the economy in the study area. The paper attempts to critically examine the effects of oil exploration on these activities. The information elicited from the questionnaires is discussed here under.

It has been mentioned earlier that subsistence agriculture is the mainstay of economy of the region. Farmers' excesses are sold at scheduled periodic markets. These local communities are mainly inhabited by indigenes that have little or no preference for imported food items. Consequently their sustenance largely depend on land availability and the continuous production of foodstuffs from farms to meet the food demand of the region. All the selected communities with the exclusion of Ekpan, experience diverse problems such as induced community clashes, decline in fish catch, poor agricultural productivity, deforestation and hunger. Other consequences in Ekpan community are, the congestion of housing development. This has been accentuated by an attempt to maximize and fully utilize the available land resulting in poor ventilation and an increase in slum sites.

Oil exploration and exploitation activities have had adverse impacts on the agricultural sector. Among these are;

i) *Stunted growth of crops and low crop yield in localities:* Evidently, total crop yield has declined in these localities in recent years. This is clearly observable in the sizes of growing plants as well as harvested crops. For instance, cassava stems, which prior to oil exploration grew to a height of 6–7 feet with stems of about 4–6 centimeters in diameter, now grow to a height of only about 3-4 feet with stems of about 1–2 centimeters in diameter. Their leaves have also become yellowish – green and the final yields have also become much smaller in their sizes. The result is hikes in cost of these foodstuffs due to low outputs. A typical case is that of garri, which is a by-product of cassava and is the main staple food in the locality. It has increased from N3 (a bucket measure), to N150 o ver the years. The decline in agricultural productivity is largely attributable to poor and infertile soils, which is largely a consequence of land and air pollution resulting from oil activities. This has hindered the enabling environment for plant growth and high yields.

ii) *Shortage of farmland and destruction of raw materials/crops:* The shortage of farmlands, destruction of natural raw materials, deforestation and destruction of crops in this locality largely result from gas flaring and exploration activities. Sand and clay, which are basic raw materials for local industries are overheated, baked or polluted from underground seismic activities. Oil exploration activities create large burrow pits that are in most cases, abandoned afterwards. The pits often contain water covered by oily film. Apart from the fact that the land is rendered unavailable and unsuitable for farming activities because of its newly acquired physiographic characteristics, the oil permeates the surrounding lands. Consequently, flora, vegetation and crop production do not thrive well in such areas. The large expanse of land that bear oil installations and pipelines are also unavailable for farming purposes. Such land areas are taken over with little compensations that are not commensurate to the deprived returns from agricultural yields were such land to still in used for farming activities. In Ekpan community, farming is fast becoming extinct as the mainstay of the economy due to rapid land development. Former farmlands have now been acquired for development of other operational structures. This has also facilitated a consequent shift of a large population of the displaced farmers to the trading sector of the economy in order to ensure sustenance.

iii) *Aquatic Environment and Fish Depletion:* Explosives detonated in rivers and creeks due to seismic surveys kills fishes and fast moving seismic boats often destroy fish traps as they sail along. Seismic disturbance and water pollution, through the see page of oil wastes and toxic chemicals into underground waters also kill fishes. Fishes tend to be more susceptible to these attacks when they are in small waters (which are predominant in the study area) than larger rivers where they can swim into unpolluted areas for survival. Results from the field survey indicate that there is the migration of fishes from smaller rivers and creeks to much larger rivers.

The incidence of oil spillage in the study area is very high (99%) For instance, in Ubeji, which is largely a fishing community, it has given rise to the growth of weeds, the death of mangrove trees and aquatic life. It has led to a rapid decline in fish catch. The Ubeji creek has become a mud of crude oil due to incessant oil spillage. This has also deprived the fishermen of their source of livelihood. The result is out migrations of fishermen to the high seas. This also has led to an increase in the cost of their catch. The result is that there have been hikes in the price of fish, especially of some species that have become very rare. Fish scarcity is compounded by the fact that the available manpower largely consists of elderly persons, who are illiterates and also lack modern expertise in their primary forms of production.]

iv) *Forestry and Plantation Agriculture:* Communal forests preserved for socio cultural cum spiritual purposes and privately owned forests and plantations have been depleted due to the activities of the oil industry. The construction of industrial sites such as the Utorogun gas plant at Otu -Jeremi and other oil installations in the area has resulted in the loss of prior existing natural preserved forest areas. Other losses result from seismic operations of oil installations and gas flaring sites. Respondents argued that the problems are two fold; that of deforestation and the extinction of animals such as the bush dog that thrives in natural forests, mangrove forests swamps and fresh water forests. Large expanse of these, have been the degraded in the area as a result of oil activities.

v) *Traditional industries:* This section of the primary economy has become heavily impoverished. Due to 'oil wealth,' youths tend to develop tastes for high-income employment in the skilled and unskilled sectors of the oil industry. The result is a high incidence of out migration from the study area to the oil city of Warri, which is located 6-10 kilometers away from these communities.

In general terms, the overall impact of environmental degradation on the primary economy in host communities is diverse and varied (see Table 1). For instance, in Ubeji, Otu-jeremi, Oghior, Ubeji and Udu communities, high levels of environmental degradation such as oil spills, gas flares, flooding, deforestation, soil infertility, ill health and noise pollution as well as minor earth tremors are currently being experienced. Ekpan community on the other hand, is associated with gas flares and flooding problems. The latter largely results from the rapid land developments, which involves the building of structures on water channels and the compartmentalization of housing structures. Also common here is the development of slum sites due to congestion.

3.10. Effects of the oil economy on the economy and environment of the niger delta region.

In the study area, activities of oil companies have hinged critically on the Natural Resource base – soils, biodiversity, water resources, fisheries and minerals. These have been taken for granted and harnessed as if they were finite. In real terms, limited stock exists and is subject to irreparable degradation. The attendant effects of these activities on the study area, in broad terms directly or indirectly affect human health, economic activities and the entire ecological system. In more specific terms, the major environmental concerns are land degradation, fisheries stock depletion, deforestation, forest degradation and bio-diversity loss.

The study categorised the observable effects of the oil economy on the host communities into two; positive and negative. The positive effects include road construction, maintenance, building of schools, the provision of agricultural and fishing equipment's, financial aids, light, water, employment and scholarships among others. The negative effects include deforestation, gas flaring, oil spillage, biochemical degradation, stunted plant growth, low crop yield, earth tremors, noise pollution and negative impacts on human health. The results from the field survey are shown in Table 2 below.

The study notes very low scores for the positive effects of the oil industries in the host communities. These scores are even lower in the areas related to the growth of the primary economy although much higher scores are indicated in the areas of the provision of portable water as well as the development of human resources. The negative effects on host communities have very high rating from the respondents. This further emphasizes the issue of economic strangulation and unsustainable development of the host communities.

The respondents were asked to identify the sources of threats to the sustainability of their communities. The results are presented in Table 3 below.

The result indicates that unbalanced growth, destruction of properties and probable extinction are identified as threatening the sustainability of the communities.

TABLE 1: SUMMARY OF EFFECTS OF OIL EXPLORATION ON THE PRIMARY ECONOMY

Economic Sector	*Effects on the primary economy*
Environment	Pollution of Environment Shortage of farmland
Agriculture	Low crop yield in localities Stunted growth of crops Destruction of raw material/crops Soil infertility
Forestry	Deforestation Extraction of animal species
Fishery	Lower yield Migration of fish to larger water Fish depletion due to pollution activities
Traditional/Craft Industry	Shift of youths from primary sectors to casual jobs in oil companies
Social economic	Urban migration Impoverishment of communities High tastes for imported urban goods Unemployment

Source: Fieldwork 2002

TABLE 2: EFFECTS OF OIL EXPLORATION ON THE ENVIRONMENT

Positive effects			*Negative effects*		
Provision of Farm tools	32	16	Deforestation	184	92
Provision of fishing equipment	34	17	Gas flaring	198	99
Financial aids	62	31	Oil spillage	198	99
Provision of light	94	47	Bio-chemical land degradation	190	95
Provision of water	160	80	Extinction of animal species	170	85
Construction of roads	120	60	Stunted plant growth	184	92
Maintenance of roads	90	45	Low plant yield	192	96
Provision of Scholarship	182	91	Earth tremors	164	82
Provision of employment	22	11	Noise pollution	196	98
Building of Schools	100	50	Effects on human health	182	91
Others	54	27	Flooding	66	33
			Ill health	32	

Source: Fieldwork 2002.

3.11. Effects of the oil economy on the niger delta society

We investigated the effects of the oil economy on the politics, economy, society, culture and environment of the region. The results are presented in the following sub-categories.

TABLE 3: PERCEPTION OF THE SUSTAINABILITY OF THE ENVIRONMENT OF HOST COMMUNITIES

Perception of respondents	*Number of respondents*	*Percentage %*
Extinction of host communities due to fires and consequent loss of lives	22	11
Fear of earthquakes due to present tremors being experienced	20	10
Destruction of property due to burst pipers blow out	36	18
Low plant and animal yield	20	10
Unbalanced growth and development in the region.	40	20
Non responses	62	31
Total	200	100

Source: Fieldwork 2002

Political Effects

The oil economy has generated increased political awareness in the communities. This has led to the formation of pressure groups, such as youth and women groups among others. The various groups tend to focus on how the communitie s could benefit from the oil companies. This is mainly in terms of human resource, physical and economic development. The sole aim is to constitute viable pressure groups that would mount a struggle against the prevailing inequity in production relations within their individual communities.

Underneath this novel political awareness in the formation of pressure groups who struggle for their rights, is the issue of youth restiveness which has become a recent phenomena throughout the Niger Delta region

Socio – Cultural Effects

Social decadence, which has become a serious problem in the region, has been attributed to "oil wealth". There is increase in crime rate and social disorder, particularly among youths. The influence of foreign cultures on host communities has resulted in a decline of social norms and distortion of cultural values. The oil economy has promoted the growth of prostitution. Young females have been lured into hovering around areas of oil operations in search of income as commercial sex workers. The result has been negligence of their educational pursuit and upsurge in the transfer of veneral diseases. Other social problems evident in these areas are increase in the incidence of smoking and use of hard drugs, infidelity and marital instability.

The awareness of oil money has also accentuated intra and inter community clashes especially as it relates to compensations paid to host communities. Further, there is the dislocation of cultural shrines and practices. In Otu-Jeremi, Iwrekan and Oghior, some groves and sacred places / shrines have been demolished and relocated because of oil exploitation activities. The communities now believe that such interferences as evacuation from their abode and subsequent relocation of worshipper meeting sites into towns have resulted in the loss of the "gods" powers and strong influences on the people.

3.12. Effects on the environment

Oil exploitation has created an environment that is poor in aesthetic value and visibility and that is noisy and polluted. The effects of these on agriculture have earlier been discussed. In Out-Jeremi, Iwekan, Udu and Oghior communities, the large presence of burrows pits as well as oil pipe lines that criss – cross land areas reduce the aesthetic value of the environment. The burrow pits not only alter the topography of these communities but also tend to accentuate gully erosion as their sizes increase with heavy rainfalls on alluvial soils. The large depressions also constitute a threat to hunters who search for game at night. There is loss of land for other economic and non – economic purposes. Oil pipelines in the communities run through vacant land, family compounds and in some cases even separate toilets from their residence. The pipelines also endanger the lives of fauna, flora and members of the communities particularly when spills occur. There is also poor visibility because of smoke that is exuded from oil activities. This results in poor sight and other health hazards such as lung diseases.

The sustainability of the ecology of the Otu – Jeremi community is presently endangered by excessive heat in its environment. The town is presently entrapped by intense heat from both the flow stations located within the town and the Utorogun gas plant located at its periphery. The latter is also believed to be the cause of vibrations that lead to cracks in the walls of houses. The soil texture and structure are also adversely affected. There is a general bio – chemical degradation of land areas.

Respondents believe that the vibrations from seismic activities cause their walls and houses to crack. Secondly, they argue that there is a rampant incidence of acid rain in these communities. This they believe results in acute corrosion of their roofing sheets and pollution of rain and river water which some communities still depend on for domestic uses.

Economic Effects

Apart from the adverse effects on the primary sector of the economy as earlier discussed, oil activities have also resulted in a boost in the service sector of the economy. The increase in population due to the in – migration of oil workers have boosted the threshold population required to sustain this sector of the economy. Traders and other participants of the informal sector of the economy, in their responses, have acknowledged an increase in their activities the employment of indigenes as casual workers has also resulted in a general increase in household incomes. The consequence of higher cash flow in these hitherto subsistence economy coupled by the fact that the inhabitants, have developed higher tastes and preferences for exotic goods due to contact with workers in the oil companies have combined to generate a growth of commerce and modernism.

3.13. Perception of the palliative measures of oil companies

In order to pacify the aggressiveness and hostilities in host communities, various oil companies have made invaluable efforts towards community development in their areas of operation. These are mainly in the area of manpower development and the provision of various infrastructures. The former focuses on the award of scholarships to indigenes of host communities at the secondary and tertiary levels of education. The number of these beneficiaries is often proportional to the volume of oil produced by the respective communities.

Infrastructural facilities provided for some host communities include health facilities, roads and related infrastructures such as jetties, electricity, markets, portable water and school buildings. Others are the provision of erosion control and land reclamation assistance especially in the fresh water swamp where many villages are seasonally flooded giving rise to river erosion during the rainy season. Agricultural practices in the area have had some boost in terms of the provision of farm tools, fishing equipments and financial aid. Specifically, the selected communities have all benefited from the oil companies in this regard. There is also the provision of employment opportunities.

These efforts although are welcome, are seen as grossly inadequate. The increasing awareness of pitiable benefits relative to induced degradation and large oil profits by the companies have heightened the aggression and hostilities of community members in recent times. Additionally, these development efforts have also furthered regional disparities in the Niger Delta region as pockets of non-oil producing communities within the same regional landscape remain abandoned, neglected and undeveloped.

Responses from the host communities reveal that although the citizenry lauds these efforts, they are much aware and still very apprehensive of the consequences of the continued effects of oil exploration activities in their communities. The consequence is an insatiable desire for benefits from the oil economy and the growth of pressure groups that make endless demands for compensations. Nevertheless the indigenes also acknowledge the fact that the compensations cannot be equated with the consequences of unsustainable industrial development practices being employed nor to the ecological devastations already caused by the oil companies.

Rather the compensations are seen and regarded as mere palliative measures that the existing power class among the indigenes uses as a means of self-sustenance as well as personal enrichment. This is more so as the said compensations in most cases do not reach the entire members of the community. In some communities, this has further contributed to the general restiveness amongst the populace. A typical case is that of communal clashes between the ruling class/chiefs and the youths of the Ozoro community in Isoko North local government area of Delta State, which led to the loss of several lives and property in 2002. Such induced communal clashes have also resulted from other socio – economic imbalances in oil company's policies as they seek to appease the host communities.

The uncontrolled exploitation and lack of policies to ensure a balance between the ecology and industrial development continues to be an endless fear in the minds of the indigenes such that, most frequently, there are reactions, aggressions and violence towards the oil companies despite their attempts to pay compensations and provide facilities. In sum, the palliative measures of oil companies to host communities, although a welcome initiative, are not regarded as a solution to the impending threat to their environment. The paper therefore makes suggestions in this regard in order to ameliorate the present ecological problems.

3.14. Recommendations: the way forward

Efforts to ensure the prevention and control of pollution particularly by the oil companies should be intensified. There should be constant monitoring of the industrial activities to check if they conform to existing rules and regulations of already established and existing Nigerian environmental agencies. Environmental impact assessments should be enforced in the oil industry.

All organisations responsible for environmental management in the Niger Delta should adopt an integrated approach. In this regard, bodies such as the OMPADEC, NDDC and FEPA should co-ordinate, integrate and streamline their policy objectives since they are set up to achieve the same goals. There is also the need to improve the present industrial development strategies and technologies that will ensure sustainability.

Rather than oil companies giving compensations to host communities, efforts should be geared towards adopting a community based approach. In this regard, the paper proffers that although exploitation licenses are given by the Federal Government, individual communities should be involved in the decision making processes relating to the exploitation of natural resources, in a spatio– temporal dimension. This is important because of the growing awareness of environmental consequences. The oil companies and in particular the Federal Government should endeavor to provide alternate forms of 'development' in such zones such that as the primary economy is being devastated, the large number of people who loose their jobs, particularly in the primary sector of the economy, should have alternative forms of employment.

In line with the NDES (1997) recommendations, the author also proffers the promotion of a "grassroot approach" in environmental management in the Niger Delta, with a view to aligning the goals of conservation with the needs of the local people. NDES recommended an incorporation of indigenous resource management systems into modern principles and methods of resource conservation and environmental management in order to enhance environmental sustainability. The local people should be involved in the planning, project choice, location, funding and other aspects of environmental management and socio-economic development in order to generate local support for sustainable operations and development.

References

Aina, E.O.A (1996) Towards Environmentally Sustainable Development in the Petroleum Industry in Aina E.O.A. and Adedipe N.O. (eds) *The Petroleum Industry and the Environmental Impact in Nigeria.* (Ibadan) F.E.P.A Monograph pp. 1-6

Chokor, B. A. (1996) Assessment of the Environmental and Socio-Economic impact of the Petroleum Industry: A conceptual and methodological framework in Aina E.O.A. and Adedipe, N.O. (eds) *The Petroleum Industry and the Environmental Impact in Nigeria.* (Ibadan) F.E.P.A Monograph pp. 152 –157.

Egborge, A.B.M (1 999) "Forty Years of O il and Gas: Environment and Sustainable development. Proceedings of the First Urhobo Economic Summit. " *Vanguard,* January 6. p. 23

Ikeaguani, F.D. (1984) Petroleum Exploration and Exploitation: Impact on the Socio – Economic life of Nigerians living in oil producing areas. *A Publication of the Nigerian National Petroleum Development Company,* Inspectorate Division, Lagos pp. 11-16, 22.

Obibi, C. (1998) "Niger Delta Needs: Challenges of the new Oil Mineral Producing Areas Development Agency," Lagos: *The Guardian,* August 11, p 15.

Orhiunu, O.O. (2003) Impact of Oil Industry on Ubeji and Ekpan Communities in Delta State-An Unpublished B.Sc project submitted to Department of Geography and Regional Planning, University of Benin, Benin City, Nigeria.

Niger Delta Environmental Survey (1997) Abridged Version of Findings and Recommendations of Phase 1 Report, Volume 4, pp 15-100

Niger Delta Environmental Survey (1997) Op cit. Phase 1 Report, Vol. 1, pp. 261-267

Nzemeke, R. (1999) "Haronu Adamu Blames Niger Delta Crises on the Military." Lagos: *The Guardian,* September 22.

Uhuegbulam B. S. and H.N Data (1981) Handling of production effluents in fresh water environments: Seminar on Petroleum Industry and the Niger Delta Environment. p. 108 citied in Ikeaguani (1984) p 16.

4. Ecological crisis and health care delivery in the niger delta region: towards a gis-based risk assessment approach

Amos Abu

4.1. Introduction

In her 1988 Christmas message, Queen Beatrix of the Netherlands states, *"The Earth is slowly dying and the inconceivable... the end of life itself is actually becoming conceivable. We human beings ourselves have become a threat to our planet..."* The fact that this message comes from the Queen of the homeland of Shell makes the import doubly instructive.

Health and Environment are synergistically related. There is ample evidence that human-induced ecological crisis are major health hazards. However, links between environment and health are not always clear because of our incomplete knowledge of disease epidemiology and pathogenesis. This is worsened by the ever-increasing complexities and subtleties in environmental degradation and pollution. Such a situation may be perilous to the repository local population in a setting characterized by insensitive and regressive health care coupled with the impunity with which the environment is devastated. This grim picture is true for the marginalized people of the Niger Delta Region.

This paper attempts to highlight the hazards of a denuded environment and inappropriate and inequitable health care delivery system. It also advances superior algorithms and heuristics within Spatial Decision Support System (SDSS) based on Geographical Information System (GIS) to the problems of ecological crisis and health care delivery in the Niger Delta.

The remainder of this paper is divided into 8 sections. In order to properly situate the discussion that follows, the next section provides background information about the Niger Delta. Section 3 follows this, which is on Data Sources for this paper. Section 4 provides the Theoretical Framework. Sections 5 and 6 respectively, deal with the Structure of Ecological Crisis and Health Care/Disease profile of the study area. Section 7 provides a working solution to these problems. The policy and practical implications are also highlighted in this Section. In Section 8, we peep into the future and reach a conclusion.

4.2. The niger delta region

The Niger Delta is perhaps the most difficult area in Nigeria. Its problem is tied to the relief, which limits the development potential of the area. The region is traversed by an intricate network of creeks and rivers, which together with the heavy rainfall makes it a massive swamp. Generally speaking, soils here are strongly acidic and deficient in phosphorus, and in some areas saline. Farming is therefore limited. The most important activity is fishing (Ikporukpo 1983).

Perhaps the greatest implication of the nature of the terrain is its limitation on transport development. The many creeks and rivers that cut across the area together with the swamps make movement generally difficult. This is especially the case as the area is subject to flooding daily in areas by the coast and seasonally in more inland areas. This rugged terrain does not in anyway belittle the importance of the region to the very survival of the Nigerian economy.

Of the 36 states in Nigeria those in the Niger Delta and particularly Rivers, Bayelsa, Edo, Delta and Akwa Ibom States account for virtually all of the oil production. These 5 States are the homeland of minority groups who have very little or no access to national political power. Indeed it is this scenario where the country depends so much on resources found in an area whose indigenous people have little or no control or access to political power that gives meaning to the level of deprivation in the Niger Delta.

4.3. Data sources

In the face of data paucity and tardiness in reporting, this paper utilizes information from several sources including federal office of statistics, technical reports, publication in journals, memos; text books, unpublished theses and dissertations as well as other sundry sources.

4.4. Theoretical framework

In order to provide a context for appreciation of the issues at hand and the analyses and discussions that would follow, this section is meant to give a theoretical umbrella. Germane to this paper is the need for conceptual clarification on the nature of the interrelationship between environment and health.

Health and Environment are synergistically related with a considerable degree of isomorphism between them. However, in specific terms, the contribution of environmental factors to the causation of disease is difficult to isolate for several reasons. For instance, for any disease to occur several factors operating concurrently may be involved (Fox, Hall and Elveback 1970). Besides, these factors seldom impact health directly. In spite of these difficulties, the impact of environmental factors in disease causation is widely reported in the literature and is depicted in Figure 1.

Figure 1 portrays the environment as the all inclusive external condition that surrounds man. It is made up of physical, biological and socio-economic components. We shall albeit very briefly discuss them.

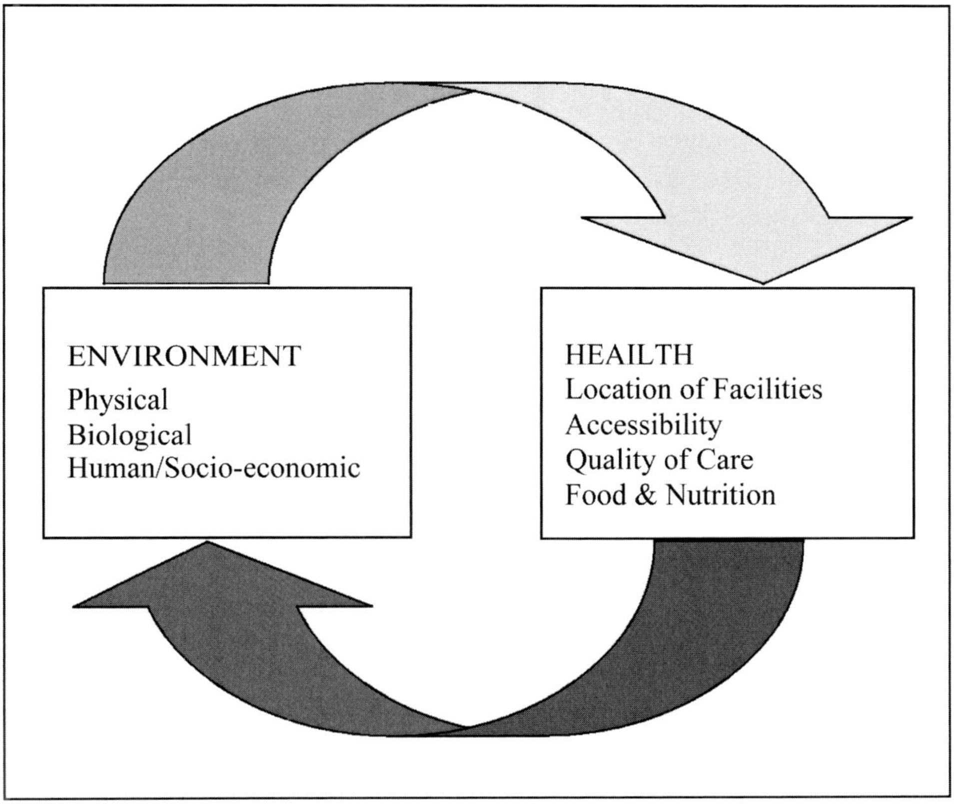

Fig. 1 Health and Environment Inter-Relation

Diseases are known to proliferate in places where both the physical and especially the biological environment have been seriously modified or destroyed by man. Such environmental modification or destruction may result from mans varied activities as farming, mining, pollution (water, land, air, plants) and deforestation. In any or all of these instances man's health is placed at jeopardy. Disease arises as a consequence of the interaction with disruptive environmental inputs as received by respiratory, digestive, dermal and other systems in humans at the interface of the environment with the living organism (Pedersen, 1996).

4.5. Physical environment

Man's physical environment refers to landscape air, water (rivers, lagoons, creeks, brooks, sea), land, climate etc. It is the surrounding inanimate milieu that extends considerable influence on the existence, spread and flourishing of diseases and pestilence. Man thus strives to live in equilibrium reached with his immediate physical environment. However, the unending desire for material acquisition brought man into open conflict with the physical environment. The result is a compromised physical environment that in turn affects the health of man for the worse. For instance, predicted changes in global temperature and depletion of stratospheric ozone are likely to precipitate major health hazards, an increased frequency of floods and storms; changes in the availability of food and good quality domestic water supplies and climate – related changes in the ecology of insect vectors for diseases (Betham, 1992).

4.6. Biological environment

The biological environment is made up of (flora) plants, (fauna) animals and unknown numbers of microorganisms. These may influence human health favourably or unfavourably, in many direct and indirect ways (Fox, Hall and Elveback 1970). Diseases are known to proliferate in places where both the physical and especially the biological environment have been seriously modified or destroyed by man. Such environmental modification or destruction may result from mans varied activities as farming, mining, pollution (water, land, air, plants) and deforestation. In any or all of these instances man's health is placed at jeopardy. Disease arises as a consequence of the interaction with disruptive environmental inputs as received by respiratory, digestive, dermal and other systems in humans at the interface of the environment with the living organism (Pedersen, 1996).

4.7. Human/socio-economic environment

The human, or socio-cultural milieu is essentially man-made and relates to the density, geographical distribution and mobility of populations, to socio-economic status, political systems, housing, diets, habits, customs and general life-styles (Haines, 1990). Given the multiple threats to well being posed by the characteristics of the host, disease agent and the environment, it is not surprising that all societies have developed some provision for health care (Foster, 1992). However, the critical issues are the location, accessibility, quality of services and utilization by the intended population. Accessibility could be conceptualized in terms of the ability to get to a place quickly and cheaply and the availability and affordability of a utility (Ikporukpo, 2002). In Nigeria, especially in the Niger Delta Region, there is a growing concern that utilities are usually less available to those who need them more. This is what informed Hart's (1971) inverse care law which states that "the availability of good medical care tends to vary inversely with the need of the population served". This problem is well articulated with the radical structural paradigm of deprivation. It sees the problem of disadvantaged areas and peoples as resulting from conflicts inherent in capitalist societies. The critical factor is access to power in a wider societal context. According to Smith (1977) "community or neighbourhood differences in wealth, rewards, knowledge and other bases of power are likely to be reflected in differences in well being, as the redistributive effects of private economic activity and public policy favour those with power to influence the process". Thus the encompassing political economy is seen as constituting the prime reference (Okafor, 1989).

4.8. Health and environment

Health as a state of 'complete physical, mental social well being' implies only a healthy population can make use of the environment in a sustainable way. Plant and animal, water and soils impacted by oil spillages are bound to be less able to support the human population that depends on these resources for survival. Such impoverished people are in turn bound to "over tax their environment" just to elk out a living in an endless cycle of environmental decay – poverty – ill health.

In sum, the environment plays a decisive role at influencing health. However, the impunity with which this life support system is been devastated by oil related activities in the region have meant swift deterioration in living and health situation of the population of Niger-Delta. Specifically, there exists copious evidence that environmentally induced stress and deterioration are rising concurrently with instances of mental, physical and social malaise among the people at rates that exceeds what obtains in other parts of the country.

4.9. The structure of ecological crisis in the niger delta

Oil spillages do not just occur. They chiefly result from operational errors and oversight in production, transportation and storage. They often uncontrolled movement of spilled crude oil results in the pollution of rivers, creeks, ponds and farmland (Ikporukpo, 1983). The problem is worsened by the nature of the terrain and the size and frequency of occurrence. For instance between 1970-1980 there were 18 spillage involving more than 1 million barrels of crude oil.

Air Pollution and Health: The hallmark of oil exploration and exploitation in Nigeria is the unprecedented flaring of gas. The resultant emission of carbon dioxide, lead, methane and other poisonous gases and residues can impact health. They could result in respiratory diseases (asthma, bronchitis etc.); eye infections (glaucoma, cataract); and cancer of various sites, especially lung cancer etc. Furthermore, the area is affected by acid rain.

Water Pollution and Health: The problem of pollution of water bodies from crude oil is captured by West Africa (1980:427) during spillage at Apoi. *"At the beginning fish floated dead on the surface of the river water and were washed up on the beaches. Now there are almost no fish"*. It is important to recognize that it is not only fish and other aquatic life forms that are affected by water pollution. It has direct health implication for the people who depend on such polluted water for drinking, cooking and other chores. It is believed to be responsible for the increase in diarrhoea and other gastro intestinal diseases in the region (Iyun and Oke, 2000).

Noise Pollution and Health: The level of noise (unwanted sound) pollution in the region is considerable. The sound intensity that human ear is exposed to is usually measured in decibels (dB). In spite of the paucity of data there is ample evidence on the ground to show clearly that noise emanating from oil related activities in Niger Delta exceed the tolerable human limit of 60 decibels by ten fold. The associated health problems included deafness, irritation and mental stress.

Solid Waste Pollutants and Health: According to the National Nuclear Regulatory Agency (NNRA) the whole Niger Delta is at high risk of nuclear explosion because oil firms were abandoning materials in the fields and dumping dangerous radioactive wastes in their areas of operation in the region. This warning came on the heels of an earlier warning by the presidential adviser on Petroleum matters. In July 2002, the nation was alerted of plans to dump 35,000 tonnes of oil drill cuttings in the Niger Delta, which already has a similar tonnage of the same waste stockpiled in abandoned locations (Asuelimen, 2002). Meanwhile, Shell and AGIP have been accused of dumping about 10,000 tonnes of oil waste cuttings at 23 well sites in Eleme, Etche, Mgbede and Ogba communities in River State. Of grave concern is the radioactive waste dump at Mgbede (25km from Port Harcourt) waste treatment plant in Etche. Needless to say that 'continuous unscientific disposal of radioactive oil wastes will, in future lead to large scale birth of babies with congenital deformities, morons, stunted growth and miscarriages'. Other effects are high rate of terminal diseases such as cancer, diabetes, hypertension and renal failures. In the interim, epidemic of sores, spasm tumor, hearing and other problems are on the increase in the region . Trees in the area are reportedly withering.

Degraded Soil and Vegetal Resources: Perhaps the soil and vegetal resources in the Niger Delta have borne most the hazards of oil spills. This is because during oil spills, soils become less fertile as nutrients essential to plant growth become scarce while those that are toxic to plant become more available (Odu, 1977). Petroleum pollution reduces plants, productivity (Kinako, 1981). Consequently, the vegetal communities in the region become relics and anthropogenic derivatives of former climax plant communities. Today the ecological problems precipitated by petroleum mining activities have reached a crisis level. Greenpeace (1994) succinctly articulated this problem thus;

> the most notorious action of both companies (Shell and Chevron) has been the flaring of gas Sometimes in the middle of villages, such as in Dere (Bomu oil field) or very close to human habitation as in the Yaola and Korokoro oil-fields... this action has destroyed wildlife and plant life, poisoned the atmosphere and therefore the inhabitants in the surrounding areas and made the residents half deaf and prone to respiratory diseases whenever it rains...., all we have is acid rain which further poisons water courses, streams, creeks and agricultural land.

Indirect Impacts: The foregoing discussion does not mean that the impact of a degraded environment is obvious in all occasions. For the most part, the consequences of environmental degradation on health are insidious and indirect. However, since changes in the environment often lead to change in ecological regime, they no doubt impact the delicate balance of nature and indirectly initiate, promote, exacerbate and foster health problems. The mere fact that under such circumstances, our well being, including certainty of shelter and privacy, "the sine qua non" of life is jeopardized, our health will no doubt be adversely affected (Iyun, 1990). In particular, Ikporukpo (1983) observed significant (15 and 18%) reduction in farming and fishing communities in Ogunlagba and Odimodi as a result of oil spillages. This may be indirectly responsible for the very high levels of malnutrition among children and women in the region. The nutritional problems that can be easily identified using FAO anthropometrics indicators include: stunting, wasting and a combination of both. Nutritional problems in turn predispose people to assault from other diseases such as malaria.

4.10. Health care delivery and disease profile

There is a growing hypothesis that health care delivery system is both a cause and effect of the disease profile in any region (Abu, 2002; Abu 1992). In an ideal situation, health care facilities are located to give the greatest access to those in greatest need for health care. Services are carefully packaged to reflect the health needs of the people. In the context of the Niger Delta, the location and services rendered are a negation of this ideal state. Three levels of spatial inequalities and inequities can be easily identified. First, there is the rural – urban dichotomy in favour of urban areas consequently, big cities such as Port Harcourt and Warri, have a disproportionate share of available health resources to the neglect of rural communities where most of the people live and where the need for health care is greatest. Second, there are intra-city or neighbourhood differentials in the location of health care facilities. For the most part, major and strategic health facilities are located in or on the fringe of rich neigbourhoods such as GRAs and opulent areas as against inner city and migrant suburbs and slums'. Third, are the rural – rural disparities. This finds expressions in two forms: (a) villages where oil companies have their oil wells installations tend to be better served by health facilities than others without the physical presence of oil companies. (a) Villages with oil platforms usually have health and other social amenities to serve oil workers to the neglect of the indigenes.

Disease Profile: Fever seems to be widespread and it may be because of the fact that the region is heavily infested by mosquitoes (Iyun, 1984). The increase in vector borne disease as a result of ecological changes is well documented in the literature (Verhasselt, 1987; Ebene 1983; Adekolu – John 1983; Hunter 1982; Kloos et. al. 1981 Roundy 1980.

The occurrence of stomach ache may not be unconnected with polluted water sources. Of grave concern is the incidence of unknown or undiagnosed diseases, which is as high as 10.3%. They may include malignant and nonmalignant tumours and neoplasm, chronic and degenerative diseases and sexually transmitted diseases including of course HIV and AIDS.

4.11. HIV/AIDS/STI in the niger delta: a situation report

Of all the existing theories on HIV/AIDS, perhaps the most plausible and relevant to the situation in the Niger Delta is diffusion (Shannon and Pyle,1989). According to the Sentinel report of HIV/AIDS and STIs in Nigeria (Table 2) there is unprecedented increase in prevalence rates. Table 2 depicts an interesting spatio-temporal pattern. However, the worrisome aspect of the disease and virus is the trend in the Niger Delta. For instance, in 1991 the prevalence rate was 0.0% for Cross Rivers and Edo. It increased to 4.1% in 1993, 6.1% in 1995 and 8.0% by 2001 for Cross River State. In Delta State the prevalence rate, which was 0.8% in 1991, was as high as 3.1% in 1995 and stood at 5.8% by 2001. For River States, the prevalence rate which was 3.3 by 1991 more than doubled (7.7%) by 2001.

Table 1: Disease Profile

Complaints	*Number of Cases*	*% of Total*
Fever	87	46.5
Stomachache	19	10.2
Rheumatism	10	5.3
Eye Trouble	9	4.8
Weakness	9	4.8
Headache	9	4.8
Chest pains	8	4.3
Pregnancy – diseases	7	3.7
Backache	4	3.2
Cold		2.1
Others		10.3

Source: Iyun (1984)

Table 2: Comparison of the HIV Prevalence in Selected States from 1991-2001

S/N	*State*	*1991*	*1993*	*1995*	*1999*	*2001*
1.	Adamawa	0.3	1.3	5.3	5.0	4.5
2.	Anambra	0.4	2.4	3.4	6.0	6.5
3.	Benue	1.6	4.7	6.7	16.8	13.5
4.	Borno	4.4	6.4	1.3	4.5	4.5
5.	Cross Rivers	0.0	4.1	6.1	5.8	8.0
6.	Delta	0.8	5.1	3.1	4.2	5.8
7.	Edo	0.0	1.8	0.0	5.9	5.7
8.	Enugu	1.3	3.7	9.2	4.7	5.2
9.	Kaduna	0.9	4.6	7.5 (estimate)	11.6	5.6
10.	Kano	0.0	0.4	2.5 (estimate)	4.3	3.8
11.	Kwara	0.4	2.4	1.9	3.2	4.3
12.	Lagos	1.9	6.8	8.8	6.7	3.5
13.	Osun	0.0	1.4	1.6	3.7	4.3
14.	Oyo	0.1	0.2	2.7	3.5	4.2
15.	Plateau	6.2	8.2	3.8	6.1	8.5
16.	Sokoto	1.8	3.8	5.8	2.7	2.8
17.	Abia	Not done	Not done	Not done	3.0	3.3
18.	Ebonyi	Not done	Not done	Not done	9.3	6.2
19.	Imo	Not done	Not done	Not done	7.8	4.3
20.	Ekiti	Not done	Not done	Not done	2.2	3.2
21.	Ogun	Not done	Not done	Not done	2.5	3.5
22.	Ondo	Not done	Not done	Not done	2.9	6.7
23.	Jigawa	Not done	Not done	Not done	1.7	1.8
24.	Kastina	Not done	Not done	Not done	2.3	3.5

25.	Kebbi	Not done	Not done	Not done	3.7	4.0
26.	Zamfara	Not done	Not done	Not done	2.7	3.5
27.	Bauchi	Not done	Not done	Not done	3.0	6.8
28.	Gombe	Not done	Not done	Not done	4.7	8.2
29.	Taraba	Not done	Not done	Not done	5.5	6.2
30.	Yobe	Not done	Not done	Not done	1.9	3.5
31.	FCT	Not done	Not done	Not done	7.2	10.2
32.	Kogi	Not done	Not done	Not done	5.2	5.7
33.	Nasarawa	Not done	Not done	Not done	10.8	8.1
34.	Niger	Not done	Not done	Not done	6.7	4.5
35.	Akwa Ibom	Not done	Not done	Not done	12.5	10.7
36.	Bayelsa	Not done	Not done	Not done	4.3	7.2
37.	Rivers	Not done	Not done	Not done	3.3	7.7

Source: Sentinel Report on HIV Prevalence

It is difficult to explain the decreasing trend of HIV in Kaduna, Kano, Lagos, Adamawa, Benue and also the static trend in Borno and Sokoto. It may be related to the effects of the government policies, practices and location of interventions that are regressive and do not favour the states in the Niger Delta region.

Field discussion with the local people reveals a booming sex market in the region. For the most part, Oil Company workers (foreigners and non-indigenous Nigerian from other parts of the country) entice and lure young girls to sex for financial reward. The poverty level of the region means such girls can hardly say no to such overtures. Other sexually transmitted infections are also on the increase in the Niger Delta. They include Gonorrhea, Syphilis, Herpes, Genital Ulcers, Fungi and Bacteria infections etc. Besides, vectored disease complexes such as *Schistosomiasis, Filariasis, Onchocerciasis and Dracunculiasis,* have become endemic in the area (FMOH Epidemiological Bulletin, 2000).

4.12. What to do and policy implications

It is not realistic to envisage an instantaneous stoppage of oil activities in the area. What is needed is a more ecologically friendly approach. Such an approach would critically depend on early warning and early action systems for early detection and mitigation of environmental hazards in the perspective of prevention. To this end, the role of Geographical Information Systems (GIS) would loom large. It is particularly suitable for the rugged terrain since it can utilize information from remotely sensed data and imageries. The ability of GIS to integrate spatial data with their attributes in current and easily retrievable forms is an added advantage.

In spite of the much talked about epidemiological transition (Abu, 1994), the major diseases of the region are parasitic and infectious diseases. These diseases are better prevented than cured. The need to redress our health policy to reflect this need cannot be overemphasized. This would in effect mean a movement from current over-emphasis on curative medicine to preventive medicine. In the setting of the region, the provision of portable water, VIP toilets, fumigated mosquito nets and health education would be more cost effective than attempts at cure.

More importantly, diseases are asymmetrically distributed in space. It makes intuitive sense that measures aimed at arresting their menace should not be randomly distributed. The involved processes of determining the most effective and efficient location –allocation of health facilities transcends what can be done manually. These issues can now be resolved through the use of computers. There now exist vast analytical tools such as algorithms and heuristics within a Spatial Decision Support System (SDSS). Before going into these in detail, it is important to reflect on the underpinning philosophy. The underlying belief upon which this paper's argument is based is the need for equity, social and territorial justices in the location and allocation of health services. The Risk Approach (RA) operationalizes this philosophy.

The goal of the risk approach is to identify individuals, households and areas that are more susceptible to, or are affected more frequently by particular morbidity than are others with the idea of paying greater attention to the former (WHO 1984; Hayes 1991;Abu, 2002). The foundation is the struggle for equity in health, and the strategy is reallocation of resources to meet unmet needs for health care so that those who need more get more (Hayes, 1991). To wit, the aim of RA is to provide services on the basis of something for everyone but more for those in need and in proportion to that need (WHO 1984, Abu, 2002, Araya et. al. 1991). This is of profound importance in areas where health resources are scarce and where people display considerable degree of ambivalence in health seeking behaviours (Lambo, 1962). Luckily, Abu (2002) has developed very useful indicators for identifying risk at households and communities for concentrated attention.

However, the RA does not preclude the importance of location of health services. For one thing, even when preventive measures are fully operational, the need to make use of health facilities will still be there. In what follows, we shall discuss the relevance of GIS/SDSS for determining the best location of health facilities subject to certain constraints. Although, the problems of the optimum location of public facilities has straddled over 200 years and engaged the minds of countless researchers, two basic approaches are in vogue. They are algorithm and heuristic approaches. In the algorithms, all locations and their spatial and non-spatial attributes are usually considered and analysed to arrive at the best locations. The major advantage of algorithms is that the location arrived at is usually optimal. The major limitations are that they are costly and time consuming. Prior to the era of jet speed computers, the analysis can endure for a decade and more. With heuristics, based on your knowledge of the area, certain locations that are not viable /feasible may be eliminated before the running of the analysis proper. Thus time is saved. However, locations arrived at may be less than optimal.

GIS and SDSS combine the fine points of both approaches in a very creative manner to arrive at best locations. In particular, SDSS incorporates the perspectives of all stakeholders; the population, the analyst, decision maker and politician in a creative way. GIS/SDSS are blessed with additional advantages. They can capture, store, retrieve, analyse, manipulate and modulate at unimaginable speeds and resolution scales. However, they must be seen as tools that aid decision and not ends in themselves.

4.13. Conclusion

The massive modification in the physical environment in combination with social, political, economic and medical neglect have resulted in disease explosion, which in turn, is having major environmental effects. Ironically, in Nigeria and in particular the Niger Delta, health delivery is based on coverage. Consequently, the available health resources are spread thinly with little success. Further, the health resources are not equitably nor efficiently distributed. When these are juxtaposed with a system that still places premium on curative rather than preventive medicine, the health situation in the region is better imagined than experienced. Clearly, it is the increase in disease complexes, occasioned by ecological crisis and manifest shortcomings in health care delivery that provides the context for the understanding of the health status of the people of the Niger Delta. Beyond this, GIS/SDSS provides us with a formidable tool for prompt detection and mitigation of ecological hazards and effective and creative location-allocation of scarce health resources.

We are persuaded to hope that the compelling global movement and ethical demand for the State to pay added attention to those peoples and areas in the greatest need would place environmental and health issues at the nexus of government policies and programmes in the Niger Delta. That way we may be able to save human kind and his environment from slow death.

References

Abu, A.(2002) Environmental and Behavioural Risk Assessment for Childhood Diarrhea Intervention. Unpublished Ph.D. Thesis, Department of Geography, University of Ibadan, Nigeria.

...... (1994) Urban Environment Quality and the Incidence of Cardiovascular Diseases in Third World Cities... Paper presented at IGU Regional Conference. Prague August 22-26.

Adekolu – John, E. E. (1983) The Impact of Lake Creation on Guinea-worm Transmission in Nigeria in the Eastern Side of Kainji Lake *International Journal for Parasitology* Vol. 13, No. 5,

Amstrong, R. W., (1983) Medical Geography and Risk Assessment. In R. Akhtar and Y. Verhasselt. (eds) *Disease Ecology and Health* Readings in Medical Geography Rawat Publications, Jaipur India.

Araya et. al. (1991) Persistent Diarrhea in the Community: Characteristics and Risk Factors. *Acta Pae diatrica scandinavian* Vol. 80

Asuelime, S. E. (2002) "Oil Poisons Weak DPR" in *The Punch,* November 21, 2002, Bentham, G. (1992) "Global Climate Change and Human Health" *Geojournal* Vol. 26, No. 1

Dever, G. E. A. (1980) Community Health Analysis German town, MD: Aspen Systems Corporation.

Ebene, R. T. (1983) "Health Risks to the Population of Gezira/Sudan Following Irrigation of Cotton Fields" *Geojournal* Vol. 7, No. 3,

FOH/NACA (2002) 2002 HIV/Syphilis Sentinel Sero-Prevalence Survey in Nigeria. Foster, H. D. (1992) Health, Disease and the Environment. CRC Press Inc. Boca Raton Florida.

Fox, J. P., Hall, C. E. and Elveback, L. (1970) Epidemiology; Man and Disease Ch. 3 Macmillan, New York

Greenpeace (1994) "Shell Shocked The Environmental and Social Costs of Living with Shell in Nigeria, Amsterdam. Greenpeace International.

Haines, A. (1990) The Implications for Health in Leggett, J. (ed) Global Warming Hart, J. T. (1971) The Inverse Care Law. *Lancet* 1

Hayes, M.V. (1992) On the Epistemology of Risk: Language, Logic and Social Science. *Social Science and Medicine* Vol. 35, No. 4.

Hunter, J. et. al (1982) "Man-made lakes and man-made disease: Towards a Policy Resolution". *Social Science and Medicine* Vol. 16.

Ikporukpo, C. O. (1983) Environmental Deterioration and Public Policy in Nigeria. *Applied Geography* Vol. 3.

Ikporukpo, C. O. (2002) Spatial Engineering and Accessibility. An Inaugural Lecture, University of Ibadan

Iyun, B. F. and Oke, E. A. (2002) Ecological and Cultural Barriers to Treatment of Childhood Diarrhea in Riverine Areas of Ondo State, Nigeria *Social Science and Medicine* Vol. 50.

Kinako, F. D. S. (1981) "Short-term Effects of Oil Pollution on Species Numbers and Productivity of Single Terrestrial Ecosystem" *Environmental Pollution,* 26
Kloos, H. et. al. (1981) Intestinal Parasitism Ethiopia, with special emphasis on ecological and cultural associations *Social Science and Medicine* Vol. 1 5B. Knox, P. L. (1979) Medical Deprivation, Area Deprivation and Public Policy, *Social Science and Medicine* Vol. 1 3D

Legget (1990) Christmas Message of Queen Beatrix of the Netherlands, Global Warming; the Green Report.

May, J. (1958) The Ecology of Human Disease, M.D. Publications Inc., New York.

Odu, C. T. I. (1977) Oil Pollution and the Environment. *Bulletin of the Science Association of Nigeria.* Vol. 3.

Okafor, S. I. (1989) The Limits of Spatial Engineering Staff/Postgraduate Seminar Department of Geography, University of Ibadan.

Pedersen, D. (1996) Disease Ecology at Crossroads Man-Made Environment, Human Rights and Perpetual Development Utopias. *Social Science and Medicine,* Vol. 21, No. 2.

Roundy, R. W. (1980) The influence of vegetational changes on disease pattern. In: M.S. Meade (Ed.) *Conceptual and Methodological Issues in Medical Geography* Univ. N. Carouna.

Shannon, G. W. and Pyle G. F. (1989) The Origin and Diffusion of AIDS: A View from Medical Geography. *Annals of the Association of American Geographers* Vol. 79, No. 1.

Smith, D. M. (1997) Human Geography: A Welfare Approach. London, Edward Arnold.

Stanley, W. R. (1990) Socio-economic Impact of Oil in Nigeria. *Geojournal,* Vol. 22 No. 1.

Technical Report. Support by WHO, DFID, UNICEF, UNAIDS.

Verhasselt, Y. (1987) Environmental Changes and Health in Tropical Africa *Revenue Belge Geographie 111e annee 1987, Fascicules 3-4.*

WHO (1984) Risk Approach in Maternal and Child Health/Family Planning Care. Geneva FHE/MCH/854.

5. The ecology – security nexus: oil, environmental pollution, communal unrests and national security in nigeria

Iyabo OLOJEDE, Banji FAJONYOMI & Ighodalo AKHAKPE

5.1. Introduction

Oil exploration and the environmental hazards attendant to it, has in recent times elicited conflictual relationships among stakeholders in oil producing communities in the Niger-Delta region of Nigeria. The conflict stems from the mutual antagonistic positions held by stakeholders on their respective roles in relation to the production and distribution of the cost and benefits attendant on oil exploration in the Niger-Delta region. The oil producing communities for example, argue that they suffer double tragedy in terms of oil pollution and resource depletion. The state and oil companies on the other feel that there should be a stable and peaceful condition to explore and produce oil in order for there to be a conducive atmosphere to negotiate sharing of benefits accruing from oil exploration and production.

However, lack of sincerity on the part of the state and the oil companies operating in oil producing communities have generated extensive conflicts which if not speedily resolved could threaten the security not only of Nigeria but the entire West Africa sub-region. Indeed, in recent years, the spate of tensions and protests by restive inhabitants of oil-bearing communities in the Niger-Delta region of Nigeria has become a subject of scholarly investigation.

It is against this backdrop that the paper seeks to explore and examine the role of oil based pollution as an exacerbating factor in the oil related crises and conflicts in the Niger Delta region and the need to arrive at strategies to manage such conflicts first through containment and ultimately by resolving their root causes. The specter of oil spills in recent years and its attendant pollution portend enormous risk to human existence. This has been the cause of recent agitation for environmental justice and security in oil-bearing areas. It is our view here that the ability of conflict managers to diagnose early signals of conflict and move frontierly to resolve them will go a long way in conflict prevention. It is on this basis that this paper, which is both empirical and expository, explains, predicts and proffers solutions to the oil related conflicts.

The paper is structured as follows. The first section considers conceptual and theoretical issues germane to the subject of the study. The second section examines a brief historical background to the Nigerian state, which also embraces background information on the Niger-Delta. The third section deals with oil exploration, pollution and resource scarcity in the Niger-Delta. The fourth, discusses the role that these developments play in the violence in the Niger Delta region. The fifth section explores the causal linkages between oil pollution, environmental resource scarcity and national security. The paper ends with a suggested framework for resolving the crises and conflicts contingent on oil pollution and other environmental injustices in the oil producing communities in Nigeria.

5.2. Method and scope of the study

In order to carry out the study, both primary and secondary data sources were used. Primary data were obtained through the administration of questionnaires and focus group discussions in three selected oil producing communities – Afiesere (Deltas State), Bomu (River State) and Otuasega (Bayelsa State). Questionnaires were administered to randomly selected 240 residents in the three selected oil producing communities. Out of 240 questionnaires administered, 200 or 85% were returned. The questionnaire was designed to give an insight into the level of satisfaction or dissatisfaction with oil exploration activities in the communities.

The empirical data gathered were complemented by useful information from secondary data. Information gathered from this source were drawn from various documents of environmental civil societies, Shell, and the Federal Environmental Protection Agency (FEPA) etc. These were complemented with relevant materials sourced from magazines and national newspapers such as Tell, Newswatch, Concord, Thisday and The Guardian.

The three communities that cut across three major oil-producing state were selected based on what we consider as their strategic location in the politics of oil exploration and distribution. Afiesere is located in Ughelli North Local Government Area of Delta State. It has a population of about 3,429. The predominant occupations here are fishing and farming. Bomu is in Gokana Local Government of Ogoni-Land in Rivers State. It has an estimated population of 50,000. The people are predominantly fishermen and farmers. Being one of the earliest sites for oil exploration in Nigeria and arena of recent civil unrest, it was found useful. Otuasega is in Ogbia Local Government Area of Bayelsa State. It has a population of 7, 487 (1991 census figure). Its people engage in farming and fishing. It has been a site of oil spillage and a potential area of oil related conflict.

5.3. Conceptual and theoretical issues

Certain crucial variables central to the subject matter of our discourse need to be properly defined in other to avoid some ambiguities. These are oil pollution, the conflict it generates and national security.

In the course of exploration and distribution of oil, it is not uncommon for waste materials to be discharged into the environment where oil production and distributed activities take place (Cormack 1983:6). In the advanced countries, conscious efforts are made by the government and oil companies to reduce the effect of oil pollution and other environmental hazards contingent on oil production and operation (Cable & Cable 1995). Oil pollution can by defined as the release of oil and oil related waste materials into the ecosystem by continuous material seepage, refinery effluent, oil-rig discharge, e.t.c (Wardley 1979). Put differently, oil pollution refers to "the introduction of foreign materials or substances to air, land and sea (river) which have detrimental effects on the ecosystem" (Iwepe 1998). Two issues immediately come out from this definition. First, oil pollution could take place in the air, sea and/or land. Second, such occurrence of oil pollution exhibits variations based on the peculiarities of the environment in which they occur and by implication the strategies for resolving or addressing them (Wardley 1979).

Several scholars have identified different types of oil pollution and their effects on finite and non-finite resources (Cormack, 1983; J. Ward Ley, 1979; Ikporukpo, 1980). However, what remain crucial to environmental conflict management is how to manage the conflicts that these activities generate so as to prevent them from assuming unmanageable proportions that could threaten national and sub-regional security. To be sure, since the medieval period, man has had to contend with the management of waste materials ranging from scattered disposal of waste in the immediate living environment to the more complex industrial discharges and waste in contemporary times.

It can be said that what has made oil pollution in oil producing areas, crisis and conflict prone is the lack of juristic and juridical mechanisms aimed at redressing such occurrences (Ikhariale 1997:1). In many parts of the world, legislations are made to regulate every aspect of oil production, storage, transportation and the use of both crude oil and of its derivatives. However, in Nigeria-one of the largest producers of crude oil in the world, such legislation, where in existence, are rarely executed. It has to be pointed out that legislation can only stop gross and operational discharge. Attention has to be paid to other vital aspects of oil production activities such as:

> The training of crew, hours worked and probably the design construction and layout of all plant and equipment such as to minimize the chance of a wrong or careless operation leading to an oil spill (Wardley 1979).

It is the glaring absence of these measures that has exacerbated the already tense crisis situation among stakeholders and the push of the conflict into crisis. Conflict can be conceived of as a "state of tension which exist where one party that perceives that its goals, needs, desires or expectations are being blocked by another party" (De-Reuck n.d.). "Following Reuck's classification, conflicts assume higher dimensions just as its objective increases in value, the more numerous its objectives the greater the areas it covers and the more the belligerent parties, the larger the arena (De-Reuck n.d.). A crisis situation is said to have arisen, to use Gramble and Walton (1976) medical parlance, at the point in the progress of a disease at which an important development or change takes place which is decisive for the recovery or death".

The situation in oil-bearing communities has degenerated into conflictual relationship among stakeholders in the production and distribution of oil due to lack of sincerity and crude profit considerations by oil-multinationals and their business partner, the Nigerian state. Apart from the structural imbalance in the distribution of oil wealth in the country, there is the fact that "renewable natural resources such as arable land and aquifers are destroyed due to the extraction of a non-renewable, finite resource", (Obi 1999:433). As Obi (1999:433) explains, mainly some of the sporadic and episodic outburst of violence in the oil producing areas that has created threat to national security can be traced to scarcities of land and water resources. The situation has been compounded by increased oil exploitation consequent upon trade liberation, "the maximization of export earnings as a strategy of crises and debt management" and the use of coercive force to secure a "safe heaven" for oil profit expropriation by the ruling elite in connivance with oil multi-national companies (Obi 1997:8). These situations have made environmental conflicts a major threat to national and sub regional peace and security.

Oil pollution as it affects security at either national or regional levels can be gleaned from two angles. The first is to conceive of security in relation to the labour process and it shows the link between members of a given society and the environment they inhabit. In this context, it is expected that this relationship between man and his/her environment should be able to meet his/her basic needs for there to be human security (Ibeanu 1999:12). The second relates directly to its political connotation. This refers to the ability of state officials to protect its interests or values either internally or externally from threats, and to maintain order internally with minimal use of violence (Ibeanu 1997). A state is perceived of as strong or weak depending on its ability to mediate over conflicts between individuals and groups in society with minimum use of force (Ibeanu 1997).

Some scholars perceived the state as relatively autonomous of class conflicts in society. According to this articulation of the role of the state in society, the state requires such autonomy to enable it mediate the conflicts emanating from the competition among groups for scarce resources. Thus for this groups of scholars the state is a "settled terrain" (Osaghae 1995). However, for some scholars the state realm is contested and under this condition the state cannot act as a neutral umpire in the event of conflict among classes. Indeed, as in the case of the Niger Delta, it colludes with the ruling elites and the multi-national oil companies engaged in the exploration and production of oil, to exploit and marginalize the minorities of the region (Osaghae 1995:342).

In order to secure an avenue for the accelerated exploitation of oil in the Niger-Delta, the state has used violence and terrorism to silence restive indigenes of oil producing communities. This character of the state was brought into focus under military rule when state security apparatuses were used "in the subjugation of the uprising as well as... harass, arrest and detain community and group leaders. State repression in the region, has underscored the character of the state as an instrument of violence and terrorism (Osaghae 1995). In recent times, this character and attendant activities of the state has led to uprising in the region.

Therefore, rather than build on the material security of the people through protection of finite and non-finite resources, the state spends oil revenue on physical security needs such as equipping the military and paramilitary forces with instruments of coercion. This no doubt has exacerbated the conflict in oil-bearing communities.

5.4. Background: state, oil and politics

Nigeria is the most populous country in Black Africa and its enormous potentials both human and material has put it in a vantage position to play a pivotal role in the socio-economic and political development in the West-Africa sub-region and Africa. However, its ethnic and religious pluralism have often served as instrument of division mainly to serve ethnic, religious, group and sectional interests (Nnoli 1978).

With a population of 88.5 million people (1991 Census), the country has about 371 ethnic groups with the Hausa-Fulani, Yoruba and Igbo being the majority and dominant groups. Either by accident or design the three dominant ethnic groups have continued to monopolise political power since political independence in 1960 to the detriment of the minorities. Besides, this structural problem, the British colonial enterprise in Nigeria halted the gradual evolution of a stable political system that was emerging before the scramble and petition of Africa by major European powers (Federal Republic of Nigeria 1987:28). The British imperial government, motivated more by economic expediency rather than political unity, forged a tenous union of the various peoples of Nigeria into what has been severally described as an error of design (Ayoade 1997:28). Indeed, a one time

British Colonial Governor in Nigeria, Sir Hugh Clifford referred to the emergent Nigerian State as:

> ... a mere collection of self-contained and mutually independent native states separated from one another... by great distances, by differences of history and tradition and by ethnological, racial, tribal, political, social and religious barriers (Saro-Wiwa 1992:9).

This transformation of an hitherto mere geographical expression, to a foisted nation state, has made the country perhaps the most artificial of the many administrative units created in the course of colonization of Africa (Saro-Wiwa 1992).

The nature and character of the structure of the Nigerian state is no doubt a major problem affecting its politics and society. However, there are other exacerbating factor, which has elicited conflicts in the country. One of such problems is the crisis of governance. To be sure, the post independence years failed to address the myriads of problems neglected by the colonialist. But the failure of the emergent indigenous rulers to frontally address and redress these problems led to apathy, cynicism and restiveness by the people on everything "public" in the country. The ruling elites have spent more time pursuing ethnic politics and primitive accumulation of wealth rather than socio-economic and political empowerment of the people. The kind of legitimacy crisis the political economy has spurned has made violence, subversion and coercion legitimate instruments of statecraft (Ekeh 1989:1). Ekeh (1989) captured the salient features of this emergent political culture as characterized.

> ...by vast array of negative elements such as corruption, violence and mistrust. These are derivations of an arbitrary state that emerged from the destruction and wastage of centuries of slave trade and colonialism.

It is the inability of leaders of state to manage the crisis caused by structural and exigential factors that have made these negative elements of state arbitrariness, violence, and lack of social goods and environmental justice, the dominant phenomena in the Nigerian state and politics. Some of the effects of this development are the spectre of neglect, degradation and marginalization in the Niger-Delta and other minority areas. All these developments are better appreciated by looking at the historical context of the Niger-Delta.

The Niger-Delta region of Nigeria has attracted world attention in recent years for two main reasons. In the first place, it is "the goose that lays the golden eggs." The Niger-Delta region accounts for 80% of the country's oil industry. This has made the region an important constituent unit in the country's federal structure. According to Khan (1994:436), the Niger-Delta region has become increasingly relevant to the Nigeria state because since oil was discovered at Oloibiri, the

Niger-Delta basin has proved to be the most prolific producing area in Nigeria, containing seventy eight oil fields, including the largest Forcados and Yorkri.

Apart from its oil wealth, the Niger-Delta region is described as fragile and complex. This characterization is not unconnected with the "vast interface between land and water" (World Bank Report 1995). It comprises of 26 language groups located in the East-Central Coast of Nigeria. The region comprises of six core states, which include: Delta, Edo, Bayelsa, Rivers, Akwa Ibom and Cross Rivers States.

However, with the rising profile of oil as chief revenue earner for the country and the political leverage that goes with it, a political definition of the region is emerging with a broadened coverage that includes Ondo, Abia and Imo states.

TABLE 1: CONTRIBUTION OF OIL TO GOVERNMENT RE VENUE (1957-1992)

YEAR	*OIL REVENUE*	*YEAR*	*OIL REVENUE*	*YEAR*	*OIL REVENUE*
1957/58	0.0	1969/70	17.3	1981	75.1
1958/59	0.1	1970/71	26.0	1982	71.8
1959/60	1.0	1971/72	52.5	1983	69.1
1960/61	1.1	1972/73	41.5	1984	77.2
1961/62	7.4	1973/74	71.4	1985	77.8
1962/63	7.3	1974/75	80.8	1986	65.8
1963/63	4.0	1975/76	87.0	1987	76.0
1964/65	5.3	1976/77	84.0	1988	78.2
1965/66	9.1	1977/78	84.1	1989	82.3
1966/67	13.3	1978/79	80.0	1990	79.8
1967/68	14.0	1979/80	83.6	1991	78.1
1968/69	9.9	1980	83.8	1992	82.0

Source: Ikporukpo, C.O. (1996) Federalism, Political Power and the Economic Power Game: Conflict over Access to Petroleum Resources in Nigeria in 'Environmental and Planning: Government and Policy'. Vol.14.

Table 1 shows the increasing profile over the years of oil as the chief revenue earner for the federal government and the mainstay of the country's economy. Form 0.0 percent in 1957/58, it rose to 82.0 percent as at 1992.

Its four ecological zones comprise of coastal barrier islands, mangroves fresh water swamp, forests and lowland rainfall (World Bank Report 1995). Rainfall is usually high and river flooding occurs intermittently particularly during the rainy seasons. And when this is combined with "low flat terrain and poor drained soil", the resultant flooding and erosion make land and land scarcity a vexed issue in the region (World Bank Report 1995). The fragility and complexity of the geo-political structure of the Niger-Delta region has over the years been compounded by environmental degradation particularly pollution emanating from oil exploitation. This has rendered the region into what many would regard as 'wastelands' due to the deleterious effects of oil pollution on human and material resources.

Despite the rich potentials and abundant wealth such as rich agricultural land, fertile forests, excellent fisheries and the reservoir of oil and other minerals, the region remains one of the most backward in the country, due primarily to environmental injustice, majority ethnic hegemony and insecurity (World Bank Report 1995).

5.5. Oil exploration and environmental degradation in nigeria: the oil pollution-community dissatisfaction nexus

The first company to enter the oil sector in the country was the Royal Dutch Oil Company – Shell D'Arcy presently known as Shell Petroleum Development Company. It was given the license to prospect for and exploit oil over the mainland of the country in 1938 in an area covering about 367,000 square miles (Obi 1999:47). However, the dominance of Shell oil company in Nigerian oil sector was reduced when in 1957, the area of its operation was reduced to 16,000 square miles in other to create room for other oil corporations and the Nigerian State to enter the industry (Obi 1999:46). Even at this stage, the oil industry was insulated from the rest of the economy and the state contented itself with collection of rents in a 50-50 profit sharing formula entered into with other oil companies by virtue of the 1959 Petroleum Profit Tax Ordinance (Obi 1999:47). And as Amu (1982) rightly observed:

> Under the concessionaire era, the oil industry was dominated by the oil multinationals, -this (was) meant to be so under the colonial minerals ordinance of 1914 (and) government,s role was limited to collecting rents and royalties from oil companies and making laws to regulate the activities of the ministry (cited in Obi 1999).

The entry of the Nigerian state into the oil industry did not sway the center of power in the industry from the oil multinationals. This is because the Nigerian state relied overtly on oil rents to sustain a regime of statism and primitive accumulation by the ruling political elite. The Nigerian National Petroleum Corporation (NNPC) has lacked the technology and managerial ability to regulate the activities of the oil multinationals. In the circumstance, the Nigeria oil industry has remained mainly "extractive, dependant and foreign dominated in structure" (Obi 1999).

The fact that the oil industry is dominated by the big five-Shell, Mobil, EIF, AGIP and Texaco has made the struggle for and exploitation of oil in the Niger-Delta "Hobbesian." The crisis and conflict this spawns have been accentuated in

TABLE 2: MAJOR OIL COMPANIES IN NIGERIA

S/N	*OIL COMPANY*	*SHAREHOLDING*	*OPERATOR*
1.	Shell Petroleum Development Corporation	NNPC – 55% SHELL – 30% ELF – 10% AGIP – 5%	SHELL
2.	Mobil Producing Nigeria	NNPC – 50% Mobil – 42%	MOBIL
3.	Chevron Nigeria	NNPC – 60% Chevron – 40%	CHEVRON
4.	Nigeria AGIP Oil Company	NNPC – 60% AGIP – 40%	AGIP
5.	Elf Petroleum Nigeria	NNPC – 60% ELF – 40%	ELF
6.	Texaco Oversea Nigeria Petroleum	NNPC – 60% TEXACO – 20% CHEVRON – 20%	TEXACO

SOURCE: Iyayi Festus (1999) 'Oil Companies and the Politics of Community Relations on the Niger-Del ta' Being paper presented at the CDHR Workshop on the oil producing communities of the Niger-Delta, organized at the Gateway Hotel, Otta. September 20th – 23rd , 1999.

recent years by the dwindling resources and the inability of state operators to have free access to state funds to dispense patronage and other group sectional interests. While the state remains jittery on account of its problem with raising revenue to sustain patron-client networks, it has had to contend with the restive and marginalized ethnic minorities clamouring for environmental justice and resource security. Thus, rather than being the "primary mechanism for securing specific social, economic and political orders in particular geographic domains... (the state is) best seen as a site of conflict between these conflicting social forces. The state is not or has not been the coherent and homogenous representative of the nation's collective entitites" (Osaghae 1999:84).

The struggle to enhance state revenues, the inability to address socio-economic needs of the people and the stiff competition to optimally exploit and corner substantial portion of global oil market among oil multinationals, has exposed the oil producing minorities dangerously to environmental hazards and socio-economic and political marginalization.

The oil bearing and producing communities argue and the facts appear to be on their side that because they feel the full brunt of:

(Oil) pollution and environmental degradation attendant to oil production, they have a right to adequate compensation,a clean and safe environment, and a fair share of oil rents, while the state and its partners, the oil multinationals, insist on the optimization of rents and profits on the basis of modalities defined exclusively by the partnership (Obi 1997).

The ensuing conflictual relations and the tension they elicit have been exacerbated over the years by two factors. In the first place, the introduction of SAP and its regime of free market and export maximization left the country's economy more battered than hitherto the case, leading in the process to socio-economic misery among the people. Second, the authoritarianism of military rule made civil society particularly in the oil producing areas more vociferous and combatant (Obi 1997). The use of brute force by state security agencies made the already tense situation more conflict ridden.

In a field survey carried out in three (3) oil-producing communities in the Niger-Delta region, there was overwhelming evidence from members of the communities interviewed that oil exploration and its resultant oil pollution has induced resources scarcities in the region. The three oil producing communities surveyed were Afiesere in Delta state, Otuasega in Rivers state and Bomu in Cross Rivers State. The responses of members of these communities and their leaders are presented in the table below.

TABLE 3: OIL POLLUTION AND RESOURCE SCARCITIES

Oil Pollution Brought Resource Scarcity	*Afiesere Frequency*	*%*	*Bomu Frequency*	*%*	*Otusaga Frequency*	*%*
Yes	42	60	60	100	39	55
No	28	40	0	0	31	45

Source: Field Survey October, 1999.

As shown in Table 3 above, in Afiesere Community about 60% of respondents believed oil pollution has contributed to resource scarcity in the region. Also 53% of respondents in Otuasega expressed the same view. In Bomu, 100% of respondents observed that oil pipelines crisis-crossed their land, while their community had nothing to show for their contributions to the national purse.

Moreover, a greater percentage of respondents interviewed stated that oil based pollution is the major factor responsible for conflict in oil producing communities. The table below showed the pattern of responses to this question.

From the table above it is glaring that oil pollution (75%) ranked highest among respondents as the major factor disturbing the peace in the Niger-Delta.

TABLE 4: FACTOR RESPONSIBLE FOR CONFLICTS AND COMMUNITY UNREST

FACTORS	*TOTAL No. OF RESPONDENTS*	*FRE-QUENCY*	*%*
Oil Pollution	200	150	75
Failure/Delay in Clean Up exercises	200	100	50
Inadequate Compensation for Oil Spills	200	140	70
Lack of access to policy makers in Government and oil companies	200	120	60

Source: Field survey October 1999.

Indeed a respondent in Bomu aptly summarized this piteous situation that the people of oil producing communities find themselves. As he puts:

> If oil companies do not pollute our environment, we would continue profitably and peaceably as farmers and fishermen. We will not ask for compensation for destroyed land, crops and rivers. They also will have no ground to settle us for destroying our yams, pumpkins and fishes. Our youths will not protest. They will not kill us. (sic)

From facts gleaned from Shell Oil Company's publication, the company is not unaware of the biting effects of oil production and its accompanying pollution that host communities has had to contend with. However, from its periodicals one could see efforts it has made to alleviate the adverse effects of oil exploration on the people of the Niger-Delta in particular and Nigerian at large. Shell has articulated certain policy objectives, which it aims to realize among which are to establish a community development programme, to work with host communities and voluntary organizations and groups to develop their area of operation and to encourage host community participation in project planning, implementation and monitoring, open and maintain channels of communication with all social segments in the host communities in order to address their problems and to focus community development on vital areas with broad benefits to host community.

In furtherance of these objectives, Shell has over the years carried out community development programmes that covered vocational training, health care, education, agriculture development, gender programmes, microcredit/business development, water/sanitation and infrastructures. Also, in the health sector, the company in collaboration with government and host communities supported twenty-two hospitals in the Niger-Delta.

TABLE 5 SHELL ASSISTANCE IN EDUCATION SECTOR 1994-1998

Projects	*1994*	*1995*	*1996*	*1997*	*1998*
University Scholarships	430	430	427	430	430
Post Primary Scholarships	1,852	2,589	2,600	2,600	2,600
Science Teachers	20	66	66	252	383
Schools with Science Teachers	4	14	14	57	63
Youth Training Programme	-	5	4	7	2
Youth Oil & Gas Training Scheme	-	-	-	-	1

Source: People and the Environment Shell Petroleum Development Company of Nigeria Ltd, 1998.

Table 5 shows the level of Shell assistance to oil producing communities in Niger-Delta. As shown in the table below, 430 University Scholarship were awarded in 1994, 430, in 1995, 427 in 1996, 430 in 1997 and 430 in 1998. A total of 1, 652 post primary scholarships were awarded in 1994, 2,589 in 1995, 2,600 in 1996, 2,600 in 1997 and 2,600 in 1998. To promote the study of science, about 20 science teachers were employed for a number of schools in 1994 and this was increased to 383 in 1998. Essentially, these were efforts geared towards strengthening human resources in few oil producing communities. When juxtaposed against the enormous wealth Shell in particularly and other oil multinationals have gained from the Niger-Delta region, these assistance peter into insignificance.

Going by the spate of violent clashes between state security operatives and angry youths of oil producing communities, it would appear that the level of assistance they have received does not impress the people of these areas. This situation in oil producing communities is also compounded by the insensitivity of the state and federal government to the plights of the people whose ecosystem have been damaged by the operation of oil companies. In the section that follows we look at the ecology – security nexus and how this has generated national and global concerns over the years.

5.6. Environmental pollution and community

Dissatisfaction: the ecology – security nexus

The point has been made that it would seem that the global concern over environmental safety is informed more by selfish profit consideration rather than altruistic desire for environmental justice. However, Homer-Dixon has acknowledged that;

Scarcities of environmental resources – in particular cropland, fresh water, and forest, are contributing to mass violence in several areas of the world. While these 'environmental scarcities' do not cause wars between countries, they do sometimes aggregate stresses within countries, helping stimulate ethnic clashes, urban unrest, and insurgencies (Obi 1997 b).

This perspective also argues that environmental stresses, tensions and conflicts pose serious threat to the national and global security interests of Western countries. As Obi (1997B) puts it:

> This violence affects Western national interests by destabilizing trade and economic relations, provoking migrations and generating complex humanitarian disasters that divert militaries and absorb huge amount of aid.

The logic of capitalism and the cut throat competition for control of oil resources by groups have made the movement towards conflict resolution almost difficult. The scale of environmental pollution could have been avoided if only the federal government had taken effective regulatory control of the processes of oil production. However, its rentier character, lack of patriotism among the ruling elite and inadequate managerial skills has combined to frustrate any genuine attempt at such regulation.

In the wake of increasing dependence of the capitalist world on cheaper sources of fuel for industrial use, attention have been focused on Nigeria which offers one of the cheapest sources of fuel. The oil multinationals are further encouraged by the loose regulatory method of state institutions. As the search for and use of oil intensify, the environments where oil operations take place have come increasingly under stress, which portends danger to the ecosystem.

Oil based pollution has become a worrisome outcome of oil exploration because it threatens to destroy the sources of livelihood of the people and their social and health condition. Assessing the impact of oil pollution on their community, in a report submitted to the World Conference of Indigenous Peoples on Environment and Development at Rio's Earth summit in June 1992, prominent Chiefs from River State posited;

> Thus apart from the oil pollution from the oil industry's emissions and flares day and night, producing poisonous gases that are silently and systematically wiping out vulnerable airborne biota and otherwise endangering the life of plants, game and man himself, we have widespread water pollution and soil-land pollution that effectively result in the death of most acquatic eggs and juvenile stages of life of fishes and shell fish and sensible animals (like oysters) on one hand while on the other hand agricultural land contaminated with oil spills become dangerous for farming, even where they continue to produce any significant yields (Obi 1997).

The "water of the Niger-Delta contains at least 6 percent of Petroleum and often up to 60ppm" (Obi 1997). What this boils down to is that oil companies apply low environmental safety standards in Nigeria. This feeds into a crisis of governance, as the state tends to abdicate most of her responsibilities to the people. To compound matters, in spite of degraded environments and the generation of the nations wealth, the oil minorities, since the days of military rule, have continually been marginalized from not only political power but the distribution of oil wealth, which have been designed to favour the dominant ethnic groups in the country (Pearson 1970).

All these serve as trigger factors of conflict, which has seen, angry youths take on state institutions and oil company personnel and installations in their communities. It is against this backdrop of intermittent and sporadic attacks from aggrieved youths in oil producing communities that oil spills have been attributed or blamed on sabotage of oil pipelines by agitated inhabitants of oil producing communities (Obi 1997). It would appear that such acts of sabotage are meant to attract compensation from oil companies or government. And as Ikporukpo (1980:225) has noted:

> ...on a world scale sabotage which accounts for three percent of total spillage is not important. Only three countries – Nigeria, Nicaragua and USA recorded spills due to sabotage. That four of the seven sabotage caused spills were in Nigeria underscores the importance of sabotage in the country...

But when this figure is placed against the overall causes of oil spill, sabotage takes the back burner. For instance, "a report of the Nigeria Ministry of Petroleum Resources puts the largest causes of spill as equipment malfunction (38 per cent) and corrosion of equipment" (21 per cent). I t is therefore clear that substantial parts of oil spill in recent times are not caused by sabotage (Obi 1997a).

While this security concern has tended to encourage acquisition and use of military power against sometimes, unarmed inhabitants of host communities, non-military threats are hardly given serious attention. For example, worsening poverty and pauperization in the region particularly among the youths have not been given any consideration. That the local economies of the people of the region have been generally decimated and the people made increasingly pauperized and immiserable because of the depletion of non-finite natural resources of the people and hazards on the ecosystem arising from oil exploration is not being addressed. This condition of poverty and misery in the region and the management of security has been intensified over the years due to the failure of the post-colonial state "in crucial areas of citizen welfare and protection and, in particular, the violence and terrorism which it unleashes against its own citizens (Osaghae 1999

85). The Nigerian state has remained insensitive to the need to address the poverty and welfare in the region.

5.7. Oil pollution, community dissatisfaction and national security: towards a conflict management and resolution approach.

Conflict is an endemic phenomenon in every society. But like a keg of gunpowder, it must not be ignited or allowed to explode before it is detonated. Crises and conflicts in oil producing communities in the Niger-Delta must be viewed with serious concern and efforts must be made to address them by authorities and stakeholders in the region because of the snow-balling effect that they could have on national, sub-regional and global security notably in the areas of refugee crisis, human carnage and disruption of economic activities (both national and international).

The first area that should attract attention in conflict prevention and resolution in the Nigerian oil producing communities is to look at the structural basis of the Nigerian Federation. Although the fact that Nigeria is ethically divided or pluralistic in nature can not be regarded as a factor of conflict, its use as a means of ethnic domination by the Hausa – Fulani, Yoruba and Igbo has led to the clamour by minority ethnic groups for rights to political power, equity and equitable distribution of resources. These are areas where ethnic minorities have often been marginalized. Certain innovations in power sharing arrangement are required. Rotational presidency has been proposed. It should be tried for some years to ascertain its usefulness.

If federalism is seen as a means for power sharing, then its objective must be to integrate different units, not by making them loose their identities and relative autonomy but by providing for their differences in control and organization and guaranteeing that they all have a say in the conduct of its affairs (Osaghae 1997:7). Adamolekun (1991) in his prognosis for Nigeria federalism admonishes government "to continue policies and practices that emphasize accommodation and co-existence and to diffuse conflict to some degree through constitutional devolution and economic deregulation." Perhaps, Nigerians need to sit down to appraise the past and chart a course for the future in a national conference, be it sovereign or not. This is apposite at this juncture of the country's life because "the present Nigerian society is founded on injustice, lack of fair play, and denies individuals and groups equal opportunity and fair hearing in all adjudicative matters" (Tell magazine 08:06:1998:34).

If state creation, as a means of dousing ethnic tension has failed to produce the much desired peace among ethnic nationalities and ethnic majority domination of political power, it is because the use of the state as source of capital accumulation and reproduction of social forces have remained high on the national agenda. And as Suberu has noted:

> Because the state have become the structural means through which a considerate proportion of national wealth is made available to groups and individuals, in the form of state – controlled opportunities and amenities, the feeling is strong among ethnic communities especially those in the more ethnically mixed state, that only by having states of their own can they achieve modernization or benefit equitably from the 'national cake' (cited in Gboyega 1999:40)

Efforts of government therefore must be geared towards de-emphasizing the central government as the source of wealth for groups and individuals in society. This is an area where civil society groups engaged in promoting economic growth can be of immense importance. All these can only take place if the state is reconstituted or restructured. The revenue allocation formula has to be reviewed in such a way that the goose that lays the golden eggs is compensated in such a manner that will make life meaningful for them. The present 13% derivation principle is grossly inadequate. A review is therefore needed and urgently too.

Government as a matter of urgency should activate its various legal frameworks for environmental safety and muster the political will to prosecute those found to have violated environmental safety standards in the oil industry and the oil producing areas (Cable & Cable 1995:1-4). The first step toward this would be to carry out environmental impact assessment before any oil operation is carried out in an area. The reason why environmental stresses and crises do not transit to conflict and violence in the developed world is because there are legal regulations and enforceable laws controlling oil producing activities (Cable & Cable 1995). Besides, environmental consciousness among civil society organizations has also helped to draw attention to and sensitize the people to environmental hazards. As a measure for forestalling future crisis and conflicts in mineral producing communities, the federal state and local governments should ensure that there is an enshrined provision in the constitution dealing with the minimum environmental safety standards that mineral explorers must conform with in order to continue in business as it obtains in Europe and America.

Government and other stakeholders in oil related matters should learn not to wait for pre-dispositional factors of conflict to develop into trigger roles. (Adekanye 1999:107). Government should learn to explore channels of dialogue and communication in a tripartite pattern involving government, oil companies and host communities with a view to resolving potential conflict issues (Agin 1997:131-137). Most of the conflicts in oil producing areas are exacerbated due to communication gaps between major stakeholders. The communities and groups in oil-producing communities must be involved in decisions that affect their social existence as a people.

5.8. Conclusion

The state – oil nexus and corporate environmental practices has transformed the Niger Delta region into one characterized with conflict and a gradual sip into full-scale violence and terrorism. At the bottom of this conflict and tension are the perceived environmental injustice and a skewed federal system that has tended to favour the dominant ethnic groups in the country. In the absence of effective conflict prevention and regulatory mechanism, the inhabitants of oil producing areas have resorted to what appears to be their only option to bring world attention to their plight and force other stakeholders to address their pressing needs. However, the crisis in the international capitalist system and the subordinate role of the state in the international capitalist order has intensified the tension in the Niger-Delta oil producing communities.

Conflict resolution mechanism must emphasize the need not to wait for trigger roles in conflict before such conflict variables are addressed. Dynamic factors of conflicts and their precipitating role must be prevented. Power sharing and restructuring of the country's federal system may not be the solution. Moral leadership may well prove the magic wand to the resolution of the country's incessant conflicts and tension soaked federation. However, such leadership within a democratic system should come under the control and sanction of the people.

References

Achebe, C. (1983) *The Trouble With Nigeria,* Enugu: Fourth Dimension Publishers. Adamolekun, L. (1991) "The Federal Solution: Assessment and Prognosis for Nigeria and Africa." Publish: The Journal of Federalism (Vol. 21) Fall.

Adekanye, J. B (1999), "Conflict Prevention and Early Warning systems." S. Klason, E. Rothschild et al (eds.) *Common Security and Civil Society in Africa.* Nordiska Afrika Institutet.

Agin, C. (1997) "Understanding Community Relations in Nigeria's Oil Industry". V.E Eromosele (ed) (1997) *Nigerian Petroleum Business: A Handbook. Lagos: Advert Communication Limited.* pp. 131-137.

Ayoade, J.A.A (1997) *Nigeria and the Squandering of Hope.* An Inaugural Lecture delivered at the University of Ibadan, 4th October…

Cable, S. and Cable .C. (1995) *Environmental Problems: Grassroots Solutions* The Politics of Grassroots Environmental Conflict. New York: St Martins.

Clifford, H. cited in K. Saro Wiwa (1992) *Genocide in Nigeria: The Ogoni Tragedy.* Port Harcourt: Saros International Publishers.

Cormack, D. (1983) *Response to Oil and Chemical Marine Pollution,* London: Applied Science Publication.
De-Reuck, A. (n.d) "The Logic of Conflict: Its Origin, Development and Resolution" In M. Bank (ed) *Conflict in World Society. A new perspective on international relation,* London: Wheatshed Books…
Ekeh, P.P. (1989) "Emergent Political Culture." I n P. P. Ekeh et al (eds.) *Nigeria Since Independence: The First 25 Years.* Ibadan: Heinmann Ltd. Pg. 1.
Federal Republic of Nigeria (1987) *Report of the Political Bureau,* Abuja: MAMSER. Gamble, A. and P. Walton 1976, *Capitalism in Crisis:* London: Macmillan Press.
Ibeanu, O.O. (1999) "Oil Resource Flow and Rural Nigeria" in T. Granfelt (ed.) *Managing the Globalised Environment Local Strategies to Secure Livelihood:* IT Publications.
Ikhariale, M. (1997) "A Constitutional Imperative on the Environment: A Programme of Action for Nigeria" Paper Presented at the International Conference on Environmental Law and Policy March 19th – 21st
Ikporukpo, C. O. (1980) "Managing Oil Pollution in Nigeria: Towards an Alternative Approach" in P. O. Sada and F. O Odemerho (eds). *Environmental Issues and Management in Nigerian Development.* Ibadan: Evans.5
Iwepe, T. O. (1998) "Environmental Pollution as a Result of Oil Spillage in Niger-Delta Region: A Case Study of Warri" B.Sc Thesis Department of Geography and Planning, Lagos State University, Lagos.
Khan, S. (1994) *Nigeria: Economy of Oil* Oxford: Oxford University Press.
Nnoli, O. (1978) *Ethnic Politics in Nigeria,* Enugu: Fourth Dimension Publication.
Obi, C. (1997) "Oil, Environmental Conflict and National Security in Nigeria: Ramifications of the Ecology – Security Nexus for Sub-Regional Peace" ACDIS Occasional Paper.
Obi, C. (1997b) "Resources, Population and Conflicts: Two African Case Studies." Being Paper Presented to Laureates of the Council for the Development of Social Science Research in Africa (CODESRIA) on the theme: 'The Political Economy of Conflict in African' Dakar, Senegal, August 18-22.
Obi, C. (1999 a) Globalization and Environmental Conflict in Africa." *African Journal of Political Science,* vol. 4 (1) June.
Obi, C. (1999) "Oil, Environmental and Conflict in the Niger-Delta." *Quarterly Journal of Administration* vol. 30, No. 3 &4 September – January.
Osaghae, E.E. (1995) "The Ogoni Uprising, Oil Politics, Minority Agitation and the Future of the Nigerian State." *African Affairs (94).*
Osaghae, E.E. (1997) "The Federal Solution in Comparative Perspectives." *Politics* Vol. 16 (1)
Osaghae, E.E. (1999) "Exiting from the state in Nigeria." *African Journal of Political Science.* vol. 4 (1) June.
Pearson, S.R. (1970), *Petroleum and the Nigerian Economy,* Standard/California: Standard University Press.
Suberu, R. quoted in A. Gboyega (1999) "Inter-governmental Relations" in O. Oyediran and A. Agbaje (eds.), *Nigeria: Politics of Transition and Governance 1986-1996.* Dakar: Council for the Development of Social Science Research in Africa (CODESRIA) Pg. 240.

Tell Magazine (Lagos) June 8, 1998.
Wardley – Smith, J. (1979) "Sources of Pollution" in Wardley – Smith (Ed) *The Prevention of Oil Pollution* London: Graham and Trotman Ltd.
World Bank Report (1995) *Defining an Environmental Development Strategy for the Niger-Delta.*

Section II
state, oil and conflict

6. Oil, state policy and conflict in the niger delta

Orji NKWACHUKWU

6.1. Introduction

The nature of state policy on oil and its externalities in Nigeria is an important dimension of the crisis in the oil-rich Niger Delta. State policy, as we shall attempt to use it, refers to a purposive course of action followed by the state in dealing with a problem or matter of concern (Pal, 1997; Sambo, 1999). In this context, the matter of concern is the oil-based conflicts in the Niger Delta, which have arisen out of the relationship between the multinational oil companies and the oil producing communities. The oil companies, seeking to maximize profit from oil production, have inflicted severe damages on the ecosystem and sources of livelihood of the communities in the Niger Delta. Their unrestrained operations have led to the devastation of farmland and fishing waters.

Embittered by the recklessness of the oil multinationals, the local communities took their complaints to the state. They demanded the protection of the environment and of their physical existence. They demanded guaranteed access to oil based benefits including oil revenues. The state's response has been or is to employ force to repress the agitation of the local communities. This biased mediation of the conflict between the multinational oil companies and the oil producing communities is informed by the near total dependence of the state on the multinational oil companies for its oil revenue. In the face of an enduring threat to their livelihood and security, the local communities in the Niger Delta mobilized stiff opposition against the state and oil companies. This situation has led to a cycle of violence, deaths, the displacement of hundreds of thousands of people, and the destruction of whole communities and valuable properties.

The high level of adverse public opinion and action generated both locally and internationally, by the state motivated violence in the Niger Delta, however has forced the Nigerian State to shift its policy towards the region. The state is presently intensifying efforts to pacify the communities in the Niger Delta. The pacification policy seeks to soothe disaffection held by the oil producing communities against the multinational oil companies and the state over long years of neglect, deprivation and environmental abuse. This paper provides an overview of state policy towards the Niger Delta. It attempts a critique of state policy and highlights ways in which state policy aggravates conflict in the Niger Delta. Finally, it assesses the prospects of the pacification policy in the Niger Delta.

6.2. Niger delta crisis: the fundamental issues

The basis of the conflict in the Niger Delta can be found in the exploitative relationship between the oil producing communities on one hand and the multinational oil companies and the Nigerian State on the other. The confrontation between the two parties started when the oil producing communities began to express dissatisfaction over long years of neglect, deprivation and environmental abuse.

In these communities, oil spillage arising from the reckless operations of the multinational oil companies pollute the environment, destroy aquatic life and crops, and make water unsuitable for drinking and fishing (Ihonvbere and Shaw, 1998:237). In addition to oil spillage there is the problem of gas flaring. About 76 percent of the natural gas produced in the process of petroleum production is simply flared. Gas flaring poses a serious health hazard because at the temperature of 1300 to 1400 degree centigrade, the flares in the Niger Delta heat up the environment and also cause noise pollution (Ihonvbere and Shaw, 1998). Gas flaring in the vicinity of human dwellings gives rise to acid rain, deforestation and destruction of wildlife. Similarly, the dumping of toxic, non-biodegradable by-products of oil refining in many communities in the Niger Delta is dangerous to both fauna and flora, including man (Ibeanu, 1999).

Apart from the above preventable environmental hazards, communities in the Niger Delta were also exposed to severe socio-economic deprivations. There were growing cases of deepening socio-economic dislocations, leading a well known writer and activist Saro- Wiwa (1992:7) to lament:

"I have watched helplessly as they (the inhabitants of Niger Delta) have been gradually ground to dust by the combined efforts of multinational oil companies, the murderous ethnic majority in Nigeria, and the country's military dictatorship. Not the pleas, not the writings over the years have convinced the Nigerian elite that something special ought to be done to relieve the distress..."

The socio-economic and environmental situation in the Niger Delta has continued to degenerate. A two-volume study, "Defining an Environmental Development Strategy for the Niger Delta" (1995) conducted by the World Bank described the social and economic conditions in the Niger Delta in these words:

...despite its vast oil reserves, the region remains poor. GNP per capita is below the national average of 280 dollars. Education levels are below the national average and are particularly low for women. In the urban areas of River State, the cost of living index is the highest in Nigeria.

Only one-fifth of rural housing is considered physically sound; water borne diseases are common, electrification (despite the enormous energy resources) is poor, so are water supply and sanitation, with only 12 percent adequate sanitation in River state.

6.3. Local agitation in the niger delta

The impoverishment of the local people in the Niger Delta contrasts with the fabulous fortunes earned by the officials of Nigerian government and oil companies. For more than four decades now, the petroleum prosperity has continued to 'oil' Nigerian economy and the careers of politicians, ambitious soldiers and the staff of oil companies. Being increasingly aware of the contrast in socio-economic conditions between themselves on one hand and government officials and oil company workers on the other hand, the local communities have since begun to hold the oil multinationals and Nigerian government as responsible for their deprivation. Consequently, the communities and their groupings have intensified their agitation for better living conditions and safer environment.

Naanen (2001:221) observed that the agitation of the communities in the Niger Delta have taken three major patterns. The first is civic, involving mainly constitutional steps taken by community leaderships, community – based organizations and interest groups. Such actions include petitions to the political authorities of the day, sponsoring of motions in the legislative houses during civilian administration, and media publicity.

The second approach is the mass movement approach, which is based on the concept of people's power. This approach focuses on mobilizing the oil producing communities to demand for their rights. These rights include local autonomy, adequate representation in government, payment of oil royalties to the communities, protection of the environment, and payment of compensation for land acquisition and environmental damage. The third approach is the radical approach, which is defined by the youth militancy that has dominated Niger Delta in the last decade. The approach is an expression of the frustration engendered by slow response of the state to the situation in the Niger Delta.

Of all the communities in the Niger Delta, the Ogonis made the most outstanding agitation in the early 1990s. Given their history of neglect and deepening material deprivation, Saro-Wiwa (1992:92) reported that:

> In 1990, the Ogonis took stock of their situation in Nigeria and found that for all the wealth of their land, and in spite of the fact that an estimated 100 billion dollars had been taken away from their land in thirty-two years of oil mining, they had no schools, no hospitals and no roads. They found that there was an intense pressure on their land and that they live in a poisoned environment in which even wildlife could not survive.

The leadership of the Ogoni Community realized that given the insensitivity of Nigerian State to their plight, only an internationalization of their claims would force the government to take their agitation seriously. Consequently, they took three major steps. First, the Ogonis carefully studied their problems, evaluated their strengths and weakness, and effectively mapped out a strategy for pursuing their goals. The preparation and presentation of the 'Ogoni Bill of Rights' in October 1990 and the 'Addendum to the Bill of Rights' in August 1991 reflected these preparations. Second, the Ogonis constituted a broad based grassroots organization (the Movement for the Survival of the Ogoni People (MOSOP), which in spite of its leadership problems, fully articulated and represented all Ogoni people in their Six Kingdoms cutting across gender, class, generational, and sub-ethnic levels. Third, MOSOP utilized the opportunity to disseminate its agenda and struggles through the channels and resources of other Non-Governmental Organizations (NGOs) and grassroots organizations within and outside Nigeria (Ihonvbere and Shaw, 1998).

Though the strategy of the Ogonis was largely peaceful; letter writings, demonstrations and rallies, press releases and international campaigns, State officials and the multinational oil companies saw the growing influence of MOSOP and the publicity surrounding the Ogoni struggle as a serious threat, which must be stopped.

6.4. State repression in the niger delta

The main reason for the adoption of repression as the state oil policy is the fear by the government and oil companies that the mounting grievances and struggles in Ogoniland may spread to other parts of Niger Delta (Ibeanu, 1999). The government and multinational oil companies feared that the disruptions of oil production in Ogoniland, which has led to the loss of enormous oil revenue and profit might diffuse, leading to further losses. To check the diffusion, the state embarked on massive repression of the struggles against the state and oil companies in the Niger Delta.

Ibeanu (2001:11) reported that state repression in the Niger Delta have taken four major forms. The first is the harassment of community leaders through surveillance, arrests, detention and execution. From 1991 when the struggle in Ogoniland intensified, several community leaders were arrested and questioned, while many others were detained without trial. In 1994, Ken Saro-Wiwa and eight Ogoni leaders were arrested and arraigned before Justice Ibrahim Ndahi Auta led Civil Disturbances (Special Tribunal) for murder.

Following a clearly flawed trial at which evidence was suppressed, witnesses intimidated or bribed to lie against the accused, and defense counsels were physically abused and intimidated, the nine Ogoni leaders were sentenced to death (Ihonvbere and Shaw, 1998). On November 10, 1995 they were hanged amidst international and local calls for clemency apparently, to deter other community leaders from continuing the struggle (Gwexe, 1999).

The second form of state repression in the Niger Delta involves the encouragement of violent conflicts among neighbouring communities, and using that as a pretext to repress the people. Such conflicts are usually proclaimed to be ethnic clashes by the government. But the frequency of the clashes, the sophistication of weapons employed and the extent of damages convinced many observers that the conflicts might have been induced by the government to derail the agenda of the local communities.

Third, the government induced confrontations among inhabitants of the same local community. The case of the Ogonis is very illustrative. Because of the popularity and effectiveness of MOSOP, the organization became a major target of the government. The government cautiously followed the persistent disagreements among the leadership of the organization, and finally, when internal division within MOSOP led to the death of four prominent Ogoni sons, the government capitalized on that opportunity and dealt conclusively with MOSOP by arresting, trying and executing its leaders for the murder of the four men.

Finally, state repression came in the form of outright violence using the armed forces and police. The deadly activity of the Internal Security Task Force set up by the River State government to oversee the security of Ogoni is an example (Ihonvbere and Shaw, 1998). Another instance is the military invasion of Odi town on the order of President Olusegun Obasanjo in November 1999. This incident led to the dealth of over one hundred inhabitants. Thousands were forced to flee and virtually all the houses in the town were destroyed (Ibeanu, 2001).

The state policy of repression has left a number of negative effects on the oil producing communities, the state and the multinational oil companies as well. On the part of the local communities, the policy has led to unprecedented loss of lives and properties as well as severe humanitarian situation arising from the displacement of people. There have been the extra-judicial killings, torture, rapes, extortion and looting by security forces against the local communities. Also, the inter and intra communal confrontations instigated by the state as part of the policy have caused serious feud and acrimony among neighbouring people and communities who were hitherto living in peace and harmony.

With regard to the State and oil multinationals, the violence precipitated by the state and counter violence by the communities, particularly their youths have led to the disruption of oil production in many parts of Niger Delta. The restiveness, tensions and insecurity that have followed have caused many oil company workers to stay away from their duty posts for fear of their safety. The result is that in 1993 alone, daily production of over 28,000 barrels of oil was halted in Ogoniland, leading the government and multinational oil companies to lose revenues and profits worth over half a million dollars daily (Ibeanu, 1999).

Also, the policy exposed and embarrassed the Nigerian State in the comity of Nations. It portrayed Nigeria as corrupt, callous, repressive and violent. As a result of these impressions, various countries imposed different degrees of sanctions on Nigeria. The Commonwealth of Nations – an organization in which Nigeria has been a very influential member also imposed a suspension on her. The multinational oil companies were further embarrassed by the response generated by state repression. For instance, an international campaign sponsored by the oil producing communities portrayed Shell as the major ally of the State in the pursuit of violent repression. This impression led to widespread condemnation and demonstrations against Shell. In some countries, Shell products were boycotted (Ibeanu, 2001).

The impact of State repression on the conflict in the Niger Delta has also been negative. The policy failed to quell the conflict in the Niger Delta. Rather it fuelled it. State repression jolted the activists, groups and communities and emboldened them. The Ogonis in particular increased their mobilization and media campaign. They also deepened their contacts and politics. MOSOP tried to widen the conflict terrain by sensitizing neighbouring communities and ethnic groups suffering similar conditions as the Ogonis. As a result, the opposition against the State and Oil Companies in the area continued to grow. Therefore, State repression tended to increase the resistance of the local communities in the Niger Delta, resulting to more repression and more resistance. The oil conflict in the Niger Delta has therefore become intractable and difficult to resolve (Ibeanu, 1999).

6.5. Pacifying the niger delta

Following the violent aggression unleashed on the local communities in the Niger Delta by the Nigerian State and the resultant international attention and action the matter generated, the state made a shift in its policy towards the Niger Delta. At this point, the state began to show greater commitment to pacifying the Niger Delta than repressing their agitation. Pacification of the Niger Delta is not a new phenomenon. Successive governments in Nigeria since independence acknowledged the need to address the complaints of the communities in the Niger Delta, but they were unable to do much due to the tenuous relationship between the state and the oil Industry.

In actual sense, the multinational oil companies have a tight grip on the industry, controlling information, technology, expertise and access to global market. The Nigerian state therefore operates the country's oil industry at the mercy of the oil companies. Given the rentier character of the Nigerian State as evidenced in its near total dependence on oil for revenues, the oil industry has come to occupy the most strategic location in its socio-economic, political, and security calculations. Therefore, the state at all times endeavours to maintain cordial and close relationship with the multinational oil companies, even when this relationship may be at the detriment of its citizens. The cordiality that exists between the state and multinational oil companies is driven by mutual quest for revenue and profit. This explains the biased pro-oil multinational oil policy of Nigerian State. It also accounts for the low commitment of successive regimes to measures aimed at propitiating local communities in the Niger Delta.

Generally, attempts by the Nigerian State to pacify the Niger Delta have taken three forms. The first is the introduction and application of the principle of derivation in revenue allocation. Nigeria inherited the principle as a criterion for horizontal revenue allocation from its colonial master. The principle stipulates that each region (now state) should receive revenue from the central government in proportion to its contribution to the centrally collected revenue. Towards the mid-1960s, the continued application of the derivation principle became a matter of serious national debate. A number of writers captured the debate (Teriba, 1966; Philips, 1971; 1975). Following the report of the Aboyade Technical Committee on Revenue Allocation in 1977, the principle was eventually jettisoned.

Since the colonial days, matters relating to revenue allocation in Nigeria have been highly politicized (Obi, 1998; Adesina, 1998) It has remained one of the most keenly contested political issues in the country. As a result, Obi (1998) believes that the key to understanding the politics of revenue allocation in Nigeria lies in a clear knowledge of the balance of social and political forces within the country. Thus, he insists that the reason why the principle of derivation was discarded and replaced with that of equality and population is because the former functioned to the disadvantage of the majority ethnic groups, who incidentally are in control of political and bureaucratic power.

Ibeanu (1999) however, argues that the shift in the mid-1960s from strong regional governments to a strong central government in Nigeria's federal system also explains the discard of derivation principle. These reasons notwithstanding, the jettisoning of derivation principle reflects the inconsistency and weakness of state policy. The halt in the application of derivation principle resulted in the decline of the mining royalties paid to regions (and later to states) from 50 percent in 1963 to 20 percent in 1975. With the promulgation of the Land Use Decree in 1978 and a new constitution in 1979, the states lost such royalties completely to the federal government. The Decree and the constitution vested the ownership and control of all lands and minerals on the federal government. Thus the oil producing communities not only lost the oil on their land, they also lost the land. Various formula for revenue allocation introduced since 1979 in Nigeria have simply disregarded the demand of the oil producing communities for a return to the application of derivation principle, even though several commissions set up to consider the demands have supported the derivation option (Ihonvbere and Shaw, 1998).

However, in a renewed bid to pacify the communities in the Niger Delta, the principle of derivation has been reintroduced following the promulgation of a new constitution in 1999. The constitution stipulates that, "the principle of derivation shall be constantly reflected in any approved revenue formula as being not less than 13 percent of the revenue accruing to the Federation Account directly from any natural resources" (Federal Republic of Nigeria 1999). In the meantime, the half-hearted commitment of the Nigerian State to the pacification policy quickly manifested soon after the 1999 constitution came into force. A debate arose over the ownership of the oil wells off the shore. The federal government, which allocates the national revenue, introduced a dichotomy between the ownership of oil on shore and off shore, claiming that all resources off the shore belong to the federal government. This decision was immediately reflected in the application of the derivation principle.

The second component of the pacification policy is the institution of specific-purpose or unconditional grant. This measure also has a long history. It dates back to the historic London conference of 1958, where it was agreed that, "there should be constitutional provision for the establishment of a Niger Delta Development Board, "which will use a special grant to carry out the physical development of the Niger Delta". Since then, some administrations in Nigeria have embraced this strategy.

In 1992, the federal military government under General Ibrahim Babangida established the Oil Mineral Producing Areas Development Commission (OMPADEC). Not too long ago, the civilian government under President Olusegun Obasanjo abrogated OMPADEC and replaced it with the Niger Delta Development Commission (NDDC). The defunct OMPADEC was accused of corruption, mismanagement and squander of the 3 billion naira take-off grant and 3 percent of the annual national revenue it received. Although OMPADEC can point at some roads, schools water and electrification projects it completed, the widespread feeling in the local communities is that the projects were insufficient, at times misplaced and, more importantly, that what has been given and done did not cushion considerably the excruciating effects of long years of socio-economic deprivation.

Observers still believe that NDDC cannot perform better than what OMPADEC has done. They hinge their opinion on the fact that the two bodies share the same form and character, especially in terms of their structure and subservience to federal manipulation and control. This feeling was particularly heightened when President Obasanjo, insisted on the candidature of Onyema

Ugochukwu, his crony, as the commission's chairman, even when the National Assembly had refused to approve his nomination and Abia, his home state had rejected him as their representative.

Experience over the years has also induced fears over the ability of the entire strategy to succeed in the Niger Delta. The fundamental issue over the years lies in the fact that the people of Niger Delta for whose interest the grant is instituted and the agency set up, are often alienated from the actual management and control of the grant. The result has been that in many cases a wide gulf exists between the local people and the management of the development agency who are usually elites. Consequently, the agency often operate with little or no information, and at times false information about the needs and aspirations of the local communities, leading to situations where projects executed do not have direct bearing on the needs of the people.

6.6. Conclusion

The relationship between the policy of the Nigerian State and the conflict in the Niger Delta is obvious. State oil policy has disfavoured the Niger Delta people, supported the repression of local communities and perpetrated violence. The policy of violent repression gave rise to violent resistance by the local people, and what has followed is a cycle of vicious violence that is unprecedented in Nigeria since the end of the civil war. The persistent violence occasioned by the policy of repression blocks the way for reconciliation, especially as the State, an institution charged with the responsibility of mediating and reconciling parties in the conflict is an ally to one of the parties. Thus, the conflict has become intractable and difficult to resolve.

However, the shift towards the policy of pacification by the state opens new opportunities for the resolution of the oil conflict. Addressing the agitation of the local communities encourages them to return their swords into the scab-bard and to give peace a chance. What is required therefore, is a genuine commitment by the state to the policy of pacification. As at now, this seems to be missing. The components of the pacification policy have remained the same with those of the past era. The present administration has continued to pursue the same strategies, which have failed in the past, without any major amendment. There is a need for the present regime to overhaul the components of its policy towards pacifying the Niger Delta.

The derivation principle and the NDDC, as they are presently constituted appear incapable of propitiating the Niger Delta. This accounts for the recent incidents of anti-state/oil companies demonstrations by women and youths in the Niger Delta.

References

Adesina, O. (1998) "Revenue Allocation Commission and the Contradictions in Nigeria's Federalism" in K. Amuwo et al (ed.s) *Federalism and Political Restriction in Nigeria*. Ibadan: Spectrum Books.

Gwexe, S.G. (1999) "Prospects for African Conflict Resolution in the Next Millennium: South Africa's View" *African Journal on Conflict Resolution, 1(1).*

Ibeanu, O. (2001) Our Mothers' Courage silenced all Guns: Women and Conflicts in the Niger Delta, Paper presented at "Listen to Africa Conference", Chicago, 12-14 September 2001.

Ibeanu, O (1999) "Ogoni-Oil Resource Flow and Conflict in Rural Nigeria", in T.Granfelt (ed.) *Managing the Globalized Environment,* London: Intermediate Technology.

Ihonvbere, J and Shaw, T. (1998) The Illusions of Power: Nigeria in Transition. New Jersey: Africa World Press.

Naanen, B. (2001) "The Niger Delta and the National Question" in E.E Osaghae and E. Onwudiwe (eds.) *The Management of the National Question in Nigeria.* Ibadan: PEFS.

Naanen, B. (1994) "Oil Producing Minorities and the Reconstruction of Nigerian Federalism: The case of the Ogoni People" *The Journal of Commonwealth and Comparative Politics* 33 (1).

Obi, C. (2001) Oil Friendships Are Greasy: A Political Economy of Corruption in Oil Company – Community Relations in the Niger Delta. Paper Present at International Political Science Association (IPSA) RC4mid-term Conference, Abuja, 15-17 October 2001.

Obi, C. (1998) "The Impact of Oil on Nigeria's Revenue Allocation System: Problems and Prospects for National Development" in K. Amuwo et.al (ed.s) *Federalism and-Political Restructuring in Nigeria.* Ibadan: Spectrum.

Ojiako, J. (1981) *Nigeria: Yesterday, Today and....* Onitsha: Africana.

Pal, L.A. (1997) Beyond Policy Analysis, Ontario: ITP Nelson.

Philips, A. (1975) "Revenue Allocation in Nigeria 1970-80" *The Nigerian Journal of Economics and Social Studies,* 17 (1).

Philips, A. (1971) "Nigeria's Federal Financial Experience" *Journal of Modern African Studies,* 9 (3)

Sambo, A (1999) "What is Public Policy" in R. Anifowose and F. Enemuo (eds) *Elements of Politics,* Lagos: Malthouse.

Saro-Wiwa, K. (1992) *Genocide in Nigeria: The Ogoni Story.* Port Harcourt and London: Saros

Teriba, O. (1996) "Nigeria Revenue Allocation Experience" *The Nigerian Journal of Economic and Social Studies* 8 (3)

7. Public policy and the niger delta question: the case of interventionist development agencies

A .A. AGAGU

7.1. Introduction

The Niger Delta area of Nigeria has in the recent past attracted attention across the globe not necessarily because of the abundant natural resources with which it is endowed but rather as a result of the crisis that have greeted the negative effects of the exploitation of these resources on the people. The crisis had emanated as a result of the failure of government to anticipate, formulate and implement appropriate public policies that would ameliorate the negative effects and consequences of oil exploitation on the people and their communities.

Unfortunately, for a very long time, government and the multinational corporations operating in the area appeared not to understand the implications and consequences of their actions or inaction. In fact, the inability to manage adverse situations from which conflicts results such as identity, chauvinism, religious fanaticism, political differences and resource competition have underpinned social disorder, conflicts and wars.

7.2. The problematic

The Niger Delta can be said to occupy a critical position in Nigeria's social economic life. This explains why IDEA (2001:237) argued that the Niger Delta is crucial to Nigeria's future. Indeed the Niger Delta area is enmeshed in a system of paradox and contradictions. It is the main source of Nigeria's wealth and source of happiness and affluence to many Nigerians outside the region. Yet the area and its people are pauperized. IDEA (2001:238) stressed the immense importance of Niger Delta to Nigeria. It noted that:

> As Nigeria,s oil belt, the Niger Delta holds the bulk of the resources that sustain successive military regimes in the past and will sustain a democratic society in future. In fact the much -discussed democratic dividend and sustainable development will draw enormously from the resources of Niger Delta because it provides over 80 percent of government revenue, the bulk of the country,s biodiversity and some of it,s best resources (IDEA 2001).

The institute (IDEA 2001) noted with dismay the many social contradictions within the Niger Delta (which find principal expressions in the internecine conflicts among component ethnic groups, clans, classes and personalities) as well as external manipulation of these differences. These it is argued pose serious threats to democratic consolidation in Nigeria. It finally argued that it is this dialectical, albeit paradoxical, combination of threats and opportunities that makes the Niger Delta the phoenix of Nigerian democracy.

In spite of the great neglect and impoverishment of this area, the government, multinationals and people outside this region more often than not, see the people of the Niger Delta as restive, aggressive, pugnacious and excessive. They often fail to thoroughly and comprehensively analyze the problems confronting the region and its people as well as their implications. The government, oil companies, and Nigerians rarely consider the great implications of any serious and unmanageable escalation of crises in the region. Hence, government fails to anticipate, analyze, formulate and implement adequately public policies that can bring about sustainable development in the area.

The study is meant to achieve the following objectives. One is, to examine the various issues involved in the crises in the Niger Delta region. Two is to examine the adequacy or otherwise of the current government policies put in place to ameliorate the problems in the region. Three is to identify the possible implementation problems in relation to current public policy toward the region. Four recommends how to ensure peace and sustainable growth in the area through proper public policy formulation and implementation.

The paper is divided into six sections. The first section is the introduction, which includes background to the study, the problematic and objectives of the study. Section two deals with conceptual framework. Section three deals with the issues involved. Section four deals with past policies and programmes in the area. Section five examines the current public policy and its implementation. Section six is the conclusion and recommendations.

7.3. Conceptual framework

Dudley (1982) has alerted us that one of the two main handicaps a student of African politics suffers from is conceptual. This emanates from the fact that an individual perceives the world in terms of some conceptual framework, which explains the factors one looks for in the search for explanation. The framework that is relevant for the discussion of this paper is the political economy approach. Richard Harris (1975) argued that any genuine understanding of the problems of development in Africa requires an analytical framework that goes beyond the boundaries of the western social science particularly those that divide the study of economics and political science. Deriving from the above, he identified the political economy approach as a means for thorough understanding of national issues since it locates them within a global perspective while focusing on the interrelationship of external and internal factors in explaining the continued underdevelopment of Africa. Indeed Claude Ake (1982: xiii) described western social science scholarship on developing countries as imperialism. He describe it as "one of the most subtle and pernicious forms of imperialism in the guise of scientific knowledge". Harris (1975) thus argued that it is by studying the political economy, which is "the nexus" that we can understand the causes of under-development and pre-requisite for development. According to him, this approach avoids the pitfalls, which often accompany some western social science inquiry in Africa, which concentrates, on the study of processes or structure of society without any social consideration of their effects on people's lives.

In the same vein, Ihonvbere (2000: 73-109) opined that it is important to understand the dynamics of Nigeria's distinctive political economy and social balances as well as the lasting impact of the social formation's historical experiences in order to fully appreciate the crisis of politics and power. According to him, until we have a dialectical and holistic appreciation of the character and politics of the state, we would be incapable of understanding the reasons why the crisis persists. Ihonvbere (2000) posited further that the character of the Nigerian State continues to be directly responsible for reproducing the country's deepening socio-economic and political contradictions. In fact, the state seems to worsen the country's predicaments with every policy action or inaction it initiates or fails to initiate in the process of trying to consolidate the interests of its custodians.

With specific references to the oil economy, Eson (2000: 106-52) argued that in the colonial era, the development of oil was driven by the interests of the colonial economy and her strategic contestation for the control of the supply sources of oil to the world market. According to him, at independence in 1960, the Nigerian oil industry was firmly under the control of multinational oil corporations. The Nigerian bourgeoisie, which ascended political power, was very weak in terms of material base. Consequently, it lacked the capacity to control the economy.

He noted further that specifically with respect to the oil industry, this meant that the multinational oil corporations continued to set the agenda for the development of the industry. Thus, no national strategy was developed to determine the optimal development and extraction path of Nigerian oil. Yet, given that oil is an exhaustible resource, optimal depletion and extraction paths ought to have been determined within the context of the development strategy of the national economy (Eson, 2000).

With regards to the Niger Delta Area, IDEA (2001:241), rightly observed that:

> The recent history of popular democratic struggles in the Niger Delta cannot be understood outside the structure of the Nigeria oil industry and the political economy that has

arisen from it. The main characteristics of the Nigerian oil industry are extroversion, state dependence and multinational control.

IDEA (2000: 237-280) while examining the state- oil nexus noted that the linkage between oil and the Nigerian State has also received considerable scholarly attention. It noted that a good deal of emphasis has been placed on the rentier context of this linkage in which the state excluded from production is reduced to a mere collector of rent from the multinationals. It noted however that this formulation merely captures the form of the state oil linkage, but fails to address its content. It argued the state- oil nexus fixes the dependence of the Nigerian State integration into global oil relations. It noted further that untold harm has been done to the Niger Delta environment by petrol business. But the Nigerian state has shown itself to be grossly incapable of intervening to protect the Niger Delta environment because of the strong alliance between state officials and petrobusiness.

Lamenting on the terrible situation the Nigerian state has plunged the Niger-Delta region, Iyayi (2000) observed that today, "no week passed without some report on the situation in the Niger Delta; gas flares, spillage and environmental degradation, hostage taking, explosion of pipeline, inter-communal violence, the murder of demonstrators and the squalor and deprivation of whole communities. Reading this reports, it is easy to forget that the situation was not always thus and that for hundreds of years, the people and communities of the Niger Delta had lived in relative peace and mutual coperation. However, it is not difficult to locate the main factor responsible for the change in the situation. It is the nature of exploitation of the oil reserves of the area by various oil companies with the predictive collaboration of the Nigerian state. This political economy has created deprivation, which in turn leads to violence. Violence, which has emanated from deprivations, is derived from frustration-aggression theory. According to Dollard et al (1993) interference with goal directed behaviour creates frustration, which, in turn, leads to aggressive responses usually directed against the frustrating agent.

Public policy as it relates to the Niger Delta question revolves around a number of issues. These issues include exploitation of oil, environmental pollution, impoverization of the host communities and inter-communal and intra-communal crises. Others include relations between the oil companies and the host communities and the relation between government and the oil communities.

The exploitation of crude oil in the Niger Delta area rubs the people of the land for farming and in the process also, oil spillage and pollution kill the fish and aquatic organisms. Hence, oil exploration leads to exploitation and pauperization of the host communities. Amadi and Tamuo (2001:95-101) argued that;

In spite of the fact that the oil industry has remained the leading sector of the Nigerian economy for many decades and that it accounts for over ninety percent of the nation's gross revenue, the people of the community remains impoverished.

According to Amadi and Tamuo, (2001:95-101) the paradox is that in spite of the abundant crude oil resources and the huge petrodollar earnings, the quality of life of the average citizen in the region has remained abysmally low. The multinational oil companies' exploration of crude oil resources have contributed in no small measure to ecological degradation and environmental pollution, which have adversely affected the people's quality of life.

They further observed that;

> Although multinational oil companies operating in the Niger Delta carry out socio-economic development programmes in the area, yet, empirical evidence shows that most of the projects executed are based on the wrong perception of the needs of the host communities— there exists significant relationship between the level of oil exploration activities and the level of underdevelopment of the region (Amadi and Tamuo, 2001).

Obi (1999:15) argued that globally determined relations of production and power in the oil industry prevent the local communities from claming their entitlements, obstructs their direct participation in the actual production of oil and prevent them from getting a fair share of the revenue from oil.

Another pronounced issue in the Niger Delta is that of conflict. Ojo (2002:8) noted that while conflicts are occurring every where in the country, "however, it is in the Niger Delta that they are ubiquitous, one might say, endemic'. He noted that, after the early 1980s, under repressive military regimes and within the context of increasing centralised federalism that has left little resources in the control of state and local authorities, violent conflicts have increasingly expanded in scope, deepened in degree and salience. People are pitched against government, people against oil producing or oil servicing companies, people against people, and community against community (Ojo, 2002: 1)

Ojo(200 1) attributed these conflicts largely to distribution of benefits of oil exploration and exploitation and over the human and environmental costs of these activities. In spite of return to democratic government in 1999, these conflicts continued unabated. Ojo (2001) described the dimensions of the conflict thus:

> They confronted, sometimes even killed, government and oil company security forces that had often brutalised them in the name of legitimate protection of personnel and other interests. The youths attacked their counterparts in neighbouring communities over communal ownership of territories on which oil flow stations are located. And they even attacked one another over who in their own community are the legitimate chiefs to act on behalf of the community (Ojo, 2002:1)

7.4. Nature and characteristics of government policy and programmes in the niger delta

Omoweh (2000: 24-45) suggested that an oil policy should be formulated with its thrust on the material well – being of the people and the sustainability of the environment of the host oil – bearing area in particular, and Nigeria generally.

The Willink Commission is important for being the first to recognise the problem of the Niger Delta as a national problem. (Akinyele 1998:76-91) Thereafter, the Federal Government as direct or indirect response to the minority agitation and development problems devised a number of solutions particularly as it affects the oil communities. For instance, a Niger Delta Development Board (NDDB) was provided for in the Independence Constitution of 1960. The Board essentially was to address the development needs of the delta region. Roberts noted that "the NDDB, although entrenched in the constitution, was easily 'smouldered' by political subterfuge and 'killed' (Roberts 1998: 15) . He noted further that other subsequent measures, which touch on the problem include the creation of (more) states, the institution of the 'federal character' principles, the devolution of powers, financial transfers from the Federation Account and lately, the establishment of OMPADEC. The establishment of the Niger Delta Development Commission (NDDC) is the latest in this area.

From the foregoing, it might not be correct to say that government had not been giving attention to the area in terms of policy. However, the commitment to, and the effectiveness of such policies need to be carefully analysed and evaluated. For the purpose of this paper, the Oil Mineral Producing Areas Development Commission (OMPADEC) and the Nigeria Delta Development Commission (NDDC) would be examined.

7.5. The oil mineral producing areas development commission (ompadec)

The Oil Mineral Producing Areas Development Commission (OMPADEC) was established by the Federal Government by Decree, No 23 of 1992. The commission was inaugurated on October 29, 1992 to be an independent body that would bring about development and manage independently the problems of the oil producing areas. Obi (2001: 14) argued that OMPADEC was set up to supervise the 3 percent derivation fund partly in response to the increased agitation in the Niger Delta as well as the political imperative to distribute federal largesse in the oil producing localities.

Among others, OMPADEC had as part of its tasks, the need to achieve physical and human development of the oil producing areas. It was also expected to compensate these areas materially for such ecological damages and deprivation that might have resulted from the exploitation of the areas. OMPADEC was equally expected to open up the areas economically and also to link them up to the other parts of the country. In addition, OMPADEC was also expected to inspect oil spillage sites so as to be able to proffer solutions either by itself or by getting appropriate agencies to rectify such. To achieve this, OMPADEC was allocated three percent of the Federation Account in line with the ratio of oil produced in each state. A take – off grant of N85 million was provided. At the time of its inauguration in October 1992, OMPADEC inherited a net credit balance of N1, 001, 166, 458.12 from the Task Force, being statutory allocations from the Federation Account received from June, July and August 1992. (Robert, 1998:19).

The main strategy adopted by OMPADEC to achieve its goals was community projects. The major goals included provision of portable water, electricity, all – season network of link roads and canalisation, shore protection and land reclamation and the economic development of the individual and "co – operatives". Others were the provision of health services, communication facilities, markets for conductive business transactions, improvement and support of farming and fishing methods, transportation, encouragement of private enterprenuership, provision of specialized education and (other) educational facilities and the creation of understanding between oil companies and the communities in which they operate (Roberts 1998: 19)

By and large, OMPADEC embarked on some projects and provided some employment opportunities to some indigenes. For instance, it has been noted that by February 1996 when the executive board of OMPADEC was dissolved, the commission was executing no less than 1, 202 development projects. (Roberts 1998: 25).

In spite of all these, however, a number of shortcomings can be identified. First is that there was disharmony between the stated and unstated objectives. What became apparent was the prevalence of and commitment to awarding contracts, rather than to the functioning of projects. To that extent, it could be seen more as a means of serving the interest of the elite as opposed to those of the target groups. Thus, it was merely a means of further extracting the state resources by the elite class. Budget or funds allocated to the commission did not translate to services that changed the people's lives.

Second is that the centralized rather than decentralized nature of the commission did not make it effective in responding adequately to the needs and demands at the grassroots level. In addition, it foisted on the commission and upward – downward rather than downward – upward approach to management. According to Roberts (1998) "OMPADEC's management style was the subject of much criticism. To critics, the commission was run in "secrecy", a pointer to which was the 'refusal' to set up an information unit". He added that;

'OMPADEC authorities in their pursuit of autonomy might have ignored relevant provisions for the commission's interactions with states, local governments and local communities. Indeed, the political setting for its operations was such that the commission held the power of decision, while the role of the oil producing communities was merely advisory (Roberts 1998: 34)'.

Moreover, the operations of the commission were characterized by indiscriminate award of contracts, over – pricing, deficit running, poor monitoring, policy drift, favouritism, politicisation of its operations as well as ineffectiveness and failure in its service delivery.

According to Akinyele (1998:76-91), it is beyond doubt that the activities of OMPADEC were dogged by complaints of fraud and nepotism. In 1994, the representative of the Federal Government in OMPADEC had alleged that the major contracts awarded by the commission did not satisfy the condition laid down by the government. The general complaint is that some of the commissioners... simply inflated contracts, approved sub-standard projects or colluded with contractors to bolt away with mobilization fees. The financial impropriety apparently contributed to the inability of the commission to pay the contractors. At a time the commission's debt to contractors amounted to N2.3 billion.

The failure of OMPADEC to ameliorate the negative effects of oil exploitation, misery and dependency as well as under-employment and unemployment in the region had grave consequences on the various groups as well as the different communities at large. It made the youth to be very restive as they were not sure of their future. This fear of the future and the lopsidedness with which projects are identified provoked " a state of war of all " in the area, hence the area was always boiling and engaged in inter-community and intercommunity wars.

In his criticism of OMPADEC, Obi (1999:14) observed that OMPADEC was established a little too late and excluded the majority of the people in the oil-producing communities from decisions affecting their interest. According to him, more emphasis went into the distribution of federal largesse as reward for loyalty to the government. This took the forms of contracts, appointments, and the location of projects. In the end, OMPADEC was consumed by its own contradictions, which had taken the form of intra-elite wrangling, inter-communal rivalry, and corruption.

The inability of OMPADEC to cope with the multi-dimensional problems confronting the Niger Delta area and the people in spite of the fact that the functions and means of funding were clearly stated and provided respectively brought about the necessity to create a new body to tackle the problems of the area.

7.6. The niger delta development commission

President Olusegun Obasanjo at the inception of his administration highlighted a 12- point programme of action to turn around the fortunes of the Niger – Delta region. They include the following;

- Establishment of the Niger-Delta Development Commission (NDDC)
- Achieving Zero gas flaring by 2001.
- Remediation and rehabilitation of impacted / spilled sites using state of the arts technology
- Conflict resolution / crisis management programme
- Community participation in programme
- Public / private sector partnership programme
- Integrated environmental management and development programme
- Provision of basic infrastructure, such as water, electricity, telephone,e.t.c
- Cottage/small scale industries development / support programme;
- Farms empowerment scheme
- Community health enhancement / promotion programme and
- Environmental programmes. (Vanguard, November 7, 2002)

The President acknowledged that the magnitude and complexity of Niger Delta problems and the geographical size of the region call for well-articulated strategy for ensuring proper implementation of joint differential responsibilities for industrialisation and agrarian revolution among the three tiers of government in the Niger Delta. The President also noted that an enduring federal/state partnership is particularly critical at this time in the light of emerging fundamental issues and the need for participation mechanisms under the new democratic dispensation.

Such partnership could prompt greater understanding among the stakeholders. On the specific role of federal government in the programmes, the President said, government will be keen on the implementation of the following activities on both short-term and long term basis;

- Strict monitoring and evaluation of programmes to be executed by the NDDC.
- Increasing the community development responsibilities of oil companies with a corresponding adjustment of fiscal regime under which they operate;
- Strengthening of the activities of all institutions and organizations concerned with development of Niger Delta
- Creation of a mechanism for sustaining and updating various environmental and socio-economic data bases relevant to the Niger Delta and,
- Professionalism, reformation and empowerment of line agencies/ministries with greater autonomy and the right incentives to implement programmes and assure quality monitoring and enforcement (Vanguard 2000: 1-2 November, 7).

The Niger Delta Development Commission is basically a pet project of president Olusegun Obasanjo's administration. Before and immediately after the February 1999 presidential election, the Niger Delta area had become a boiling point. Various dimensions of conflicts had emerged. Some of these include federal government versus community conflicts, community versus oil company conflicts, inter community conflicts as well as intra community conflicts. As Ojo (2001:1) noted, these conflicts had occurred largely over the distribution of the benefits of oil exploration and exploitation, and over the human and environmental costs of the activities.

Obasanjo had during his presidential campaign in 1998 and early 1999 assured the people of the Niger Delta area that he would provide appropriate solutions to the pervasive problems of the area. Immediately he was sworn in as president, he presented a comprehensive bill on the Niger Delta to the National Assembly for passage into law. The first major problem the bill had to face was the unnecessary delay by the National Assembly before the passage into law. This might have resulted from the petty rivalry among the decision-making elite as well as ethnic and ideological conflicts. Moreover, the National Assembly made a lot of alterations to the bill during its passage to the dissatisfaction of Mr President. At a point President Obasanjo rejected the amendments to the bill (Vanguard 31: 03: 2000).

The preamble to the Niger Delta Development Commission Act 2000 emphasizes that the commission is established with the aim of ensuring more effectiveness in the use of the sums received from the allocation of the federation account for tackling ecological problems, which arise from the exploration of oil minerals in the Niger Delta area. Section 7 of the Act specified the functions and powers of the commission, which include the following; (a) formulate policies and powers and guidelines for the development of the the Niger areas; (b) conceive, plan and implement in accordance with set rules and regulations, projects and programmes for the sustainable development of Niger Delta area in the field of transportation (roads, jetties and waterways), health, education, employment, industrialization, agriculture and fisheries, housing and urban development, water supply, electricity and telecommunications (FGN, 2000).

In addition, the Commission is also expected to survey and provide master plans for the development of the area, identify factors inhibiting the development of the area and deal with issues relating to pollution an environmental degradation. Section 14 of the Act highlights the sources of funds of the commission. These include allocation from the federation account, ecological fund and money from oil companies among others.

Generally, from the provisions of the Niger-Delta Development Commission Act, one could say at the facial level that there is hope for development in the area if these provisions will be implementd. Since inauguration of the Niger-Delta Development Commission (NDDC) a lot of developments have taken place.

The NDDC board was charged to produce a master plan for the development of the region by President Olusegun Obasanjo during its inauguration.

> The major advantage of the master plan is that there would be time frame for the beginning and completion of each project. It is also to harmonize the programmes of development in the Niger Delta to avoid undue duplication (Vanguard, November 7, 2000 1-7)

Besides, it is to ensure the sustainable development of the oil rich region, no matter who is in charge of the NDDC. According to Obasanjo "we can no longer afford the uncoordinated development of the region". Development as envisaged was to be deliberately planned and systematically executed. (Vanguard, 2001 26 June).

In another development, the chairman of the commission, Chief Onyema Ugochukwu, stated in March 2002 that the Niger-Delta Development Commission (NDDC) has planned 641 projects which would cost about N 36 billion for the development of the Niger Delta region (Nigerian Tribune 26 March 2002:6). But "more than a year after the Commission was set up, the Commission could not boast of a master plan for the development of the oil – rich Niger Delta" (Vanguard of 26 June 2001). This position was confirmed by the Managing Director of the commission Engr. Godwin Omeme at a consultative forum, organised by the NDDC for stakeholders in Edo State. This can be seen as one of the early failures of the commission. However, Engr. Omeme said that the NDDC had also embarked on human development programmes in which various skill acquisition schemes have been designed to enhance the ability of the indigenes of the Niger Delta region to be self employed and to get better jobs.

Others activities and contributions of the NNDC include building of model science secondary schools in all the states to enhance their ability to produce large number of engineers, doctors and other scientists in the future as well as building of hostels in the Universities.

7.7. Assessing the interventionist agencies and efforts

In spite of all these efforts, impoverisation, misery, frustration and violence still continue. It needs to be appreciated that public policy is not a casual thing. The stake involved is higher than that of private policy. Hence, the process of policy formulation and implementation must be comprehensive and holistic to achieve sustainable development.

Horn and Meter (1977) provided an eight variable cluster that dispose implementation efforts to achieve local programme performance particularly in a federal system. They opined that these variables would help to explain the factors and events that intervene between circulation of national policy and the results that occur within the state and localities. The variables include; policy resources which include financial and other resources such as personnel, information, among others, the adequacy of these policy resources as well as the time of release of funds, policy standards, which clearly state the necessary requirements on how the stated goals are to be implemented, and communications, which ensures transfer of policy standards with clarity to the implementers.

Others are enforcement, which is the mechanism and procedures put in place by the Federal Government to ensure that the implementers act in conformity with policy standards, the dispositions of implementers, in terms of serious concern and commitment to programmes which positively affects programmes, the characteristics of the implementing agencies, which include the formal and informal attributes of implementing organization as well as nature of staff and structure of the organization and its relations with other sections and officials of government. The other variables are the political environment of the implementing agency and the economic and the social conditions within the community where policy is being implemented. With regards to the government policies in the Niger-Delta, there is little consideration for the above criteria in their formulation and implementation.

In relation to the interventionist development programmes in the Niger Delta area, the fund allocated to OMPADEC was relatively small. For instance Femi Falana wondered why the Petroleum Trust Fund (PTF) received over N 100 billion in three years while the Oil Mineral Producing Communities (OMPADEC) was given only N20billion Naira. Worse still, out of this relatively low amount, he noted that large chunk.... ended up in the pockets of military dictators, traditional rulers and contractors (Falana, 2000: 2 – 5). Information is often distorted in the Niger Delta so as to paint a picture of normalcy. This definitely has negative effects on programme implementation.

It is apparent that for public policies to be successful in the Niger Delta Area, there is the need to thoroughly identify all the stakeholders and design appropriate comprehensive socialization, orientation and training for each group so that each can appreciate its expected roles and genuinely conform. For instance, both government and the multinationals should appreciate the need to genuinely redress the injustice that had been perpetuated on the people of the Niger Delta. The policy makers must be equipped with appropriate policies that can meet the challenges of the area. Such policies must take into consideration appropriate implementation feasibility and appraisal. The local people and the target groups must also realize that peace is necessary for programmes to be successfully implemented. The implementing agencies and official must realize the need for commitment and must avoid the temptation of sacrificing the interest of the community people at the altar of corruption. Of importance here is that government officials and policy makers must not feel they know more than enough and refuse to subject themselves to training and new orientation. Indeed, it is very important for policy makers to understand implementation strategies as well as political feasibility and how to go about it.

It is also essential to sign articles of memorandum that is workable, monitorable and enforceable among government, multinational corporations and the local people. Furthermore, there is the need to periodically review projects, programmes, impacts and effects of oil exploration and exploitation so as to provide necessary and timely adjustments and interventions. It must also be appreciated that the greatest and most important stakeholders here are the host communities and as such communities and the people must be paramount in the whole calculation. It must also be appreciated that justice, democracy and participation are central to satisfaction, peace and strategies of correcting past, present and future injustice.

7.8. Conclusion

Policy neglect and inadequate policy formulation and implementation strategies in all areas of life that affect the people of Niger Delta have led to the past and present on going crises. Therefore, it is only good, comprehensive, anticipatory and well-formulated and implemented policies that will eliminate crises and bring about harmony and sustainable development in the area. For this to take place, government and its agencies as well as their officials must develop new orientations and attitudes toward the Niger-Delta area and its people.

References

Ake, Claude (1982) *Social Science as Imperialism. The Theory of Political Development,* Ibadan: Ibadan University Press

Amadi S.N and Tamuo, Steve O. (2001)" Oil Exploration in Nigeria Its Socio-Economic Impact on Host Communities" *Journal of Environment Studies* vol. 2 No1

Akinyele, RT.(1998)"Institutional Approach to the Environmental Problem of the Niger Delta" in Osuntokun, A.(ed) *Current Issues in Nigeria Environment.* Ibadan: Davidson Press.

Anyaegbunan, E.O. (2000) "Niger Delta: A Case for Regional Contigency Plan" in Osuntokun, A. (ed) *Environmental Problems of the Niger – Delta.* Lagos: Friedrich Ebert Foundation

Dollard, J., Doob, L. Miller, Mowrer O., and Seers, R. (1939) *Frustration and Aggression.* New Haven: University Press.

Dudley, B. (1982) *An Introduction to Nigerian Government and Politics*. London: Heinemann.
Eson, P.O. (2000) "The Political Economy of Oil Extraction in Nigeria" in Raji, w; A.C, and Akinsola E. (eds) *Boiling Point* Lagos CDHR Publication.
Falana F. (2002) "Crises in the Oil Producing Communities in Nigeria: A Welcome Address" in Raji, W. Ale, A and Akinsola, E (eds) *Boiling Point* Lagos: Committee for the Defence of Human Rights [CDHR].
FGN (2000) Niger Delta Development Commission Act 2000
Harris, Richard (1975) *The Political Economy of Africa*. Cambridge: Schenkman Publishing Company.
Ihonvbere, J.(2000)"A Recipe for Perpetual Crises: The State and the Niger- Delta Question" in Raji W; Ale, A and Akinsola, E (eds) *Boiling Point* Lagos: CDHR Publication
Iyayi, F. (2000) "Oil Corporations and the Political of Community Relations in Oil Producing Communities" in Raji, W; Ale, A and Akinsola, E (eds) *Boiling Point*. Lagos: CDHR Publication. Nigerian Tribune, 26 March 2002
Obi C. (1999) *The Crisis of Environmental Governance in the Niger Delta 1985-1996*, Zimbabwe: African Association of Political Science (AAPS) vol. 3 No. 3.
Ojefia, I.A (2000) " The Nigerian State and Environmental Question" in Osuntokun, A (ed.) *Environmental Problem of the Niger-Delta*. Lagos: Friendrich Ebert Foundation
Ojo, O.J.B. (2002) *The Niger Delta Managing Resources and Conflicts*, Ibadan: Development Policy Centre
Olajide, I. Fajonyomi, B.Akhape I. And Mudasiru S.U. (2000) *Nigeria Oil Pollution,Community Dissatisfaction and Threat to National Peace and Security*, Zimbabwe: African Association of Political Science (AAPS) Vol. 4 No. 3
Olokesusi, F. (1998)" Law, Policy and Implementation of Solid Waste Management in Urban Nigeria: An Environmentalist Experiences" in Osuntokun, A (ed) *CurrentIssues in Nigerian Environment*, Ibadan: Davidson Press:
Omotor, D.G. (2000) " Poverty, Environment and Sustainable Development: A Review " *African Journal of Environmental Studies*, Vol. 1
Ondo State Government (2001) Ondo State Oil Producing Areas Development Commission Act 2000
Onosode, G. (2000) "Environmental Management and Sustainable D evelopment in the Niger Delta" in Osuntokun (ed) *Environmental Problems of the Niger – Delta*, Lagos: Friendrich Ebert Foundation.
Osuntokun, A. (2000) " The Niger Delta and Nigeria's Environmental Problems" in Osuntokun, A. (ed) *Environmental Problems of the Niger Delta. Lagos:* Friedrich Ebert Foundation.
Roberts, N. (1999) The State, Accumulation and Violence: The Politics of Environmental Security in Nigeria's Oil Producing Areas. Ibadan: NISER Monograph Series No. 17
Roberts, N. (1998) Managing the Minorities and Development Problem in Nigerian Federalism. The OMPADEC Initiative. Ibadan, NISER Monograph Series No. 11
Ukaogo, V. (2000) "Transnational Business Ethics, Government Policies and the Crisis of Pollution and Underdevelopment in the Niger – Delta" in Osuntokun. A (ed)*Environmental Problems of the Niger – Delta*. Lagos; Friedrich Ebert Foundation
Vanguard November 7, 2000 Lagos

Vanguard 7/11/2000
Vanguard 26 June 2001
Van Horn, C. and Van Meter, D. (1977) "The Implementation of Intergovernmental Policy" in Nagel, Stuart (ed) *Policy Studies Review Annual,* Beverly Hill and London, Sage Publications, Vol. 1

8. The resource curse: oil, communal agitation and state repression in the niger delta

Augustine IKELEGBE

8.1. Introduction

The Niger Delta region has been immersed in struggles and challenges and in a terrain of tension and conflict since the 1950s. Beset by a minority status associated with inferiorization and subordination in relation to majority groups in the Eastern and Western regions, the first struggle was to challenge domination and internal colonization of the majoritan groups and to assert themselves through a struggle for the creation of a region each out of the then Western and Eastern regions. Then there was the total absence of development situated in the neglect and marginalization of the regional and federal governments. The politics of exclusion had meant that the minority areas, which associated with opposition parties in the struggle for separate regions were sidelined in the development programmes of the Eastern and Western regions. The minority groups in the Western region had pitted their tent with the National Council of Nigerian Citizens while those in the Eastern region were in the Action Group in the struggle for a Midwest and Cross and Oil Rivers regions respectively.

The situation of exclusion was particularly compounded by the difficult terrain of the region. This was actually the formal veil behind non development as the governments hid behind the prohibitive high cost of infrastructural development. Outside the regions, the Willinks Commission of 1958, reported that 'the difficulties of this difficult stretch of land were not understood at the headquarters of the government'. Thus effectively the Niger Delta remained shut out of the doors of developmental attention.

The initial agitation failed when the Willinks Commission decided against the creation of new regions. The Commission decided on constitutional guarantees of fundamental rights, which were enshrined in the independence constitution. But this did not guarantee group rights. In fact, given the lawless and normless nature of politics and the primacy of bitter and intense inter group and inter party political relations, human right guarantees were actually no guarantees for minorities. But the Nigerian state, through the Willinks report chose a development path, in declaring the region a special area requiring special attention and consequently establishing the Niger Delta Development Board in 1962 that was charged with the development of the region.

It was in the midst of the early agitation that commercial exploitation of oil for export began in the region. Within a few years, oil began to assume an increasing profile. Oil now entered the equation of not just intra region minority-majority relations but that between the regions and the centre. The struggle for the control of oil had begun and it was important enough to have been a vital factor in the intense hegemonical struggles between the majority groups and the northern and southern sections which finally slid into the Nigerian civil war. But in these struggles, the region comprising mainly minority citizens did not matter. It was therefore not surprising when with the commencement of the civil war, the ethnic minorities in the East pitched their tent with the centre and that support was important in the balance of equation that tilted the victory to the Nigerian side.

But the grievance of the Eastern segment and particularly of the Ijaw region was so intense that agitation went on the militant path by 1966. The persisting exclusion and growing developmental neglect in spite of the regions new oil contribution made a group of youths by the name of Delta Volunteer Force (DVF) and led by Isaac Adaka Boro to embark on the armed, militant and secessionist path of seizing and disrupting oil production facilities at Oloibiri and declaring a Niger Delta Peoples Republic in February 1966. The DVF secessionist bid was crushed and its actors tried for treasonable felony but the message was clear; the grievances and agitation had reached a dangerous point. The intervention of the civil war diverted attention from the struggle. But more importantly, the creation of the Midwest region in 1963 and the 1967 creation of states which affected the Eastern segment doused the agitation for a while. The minorities of the region had now gotten states of their own.

However, the neglect of the region persisted. With military rule, the federal government became an instrument of ethnic and sectional hegemony that was used to centralize national resources, take over oil and gas resources, consistently reduce resource flows to the Niger Delta states by enacting revenue allocation systems that reduced the region's fortunes and to dominate power and representation such that the region was marginalized and subordinated in the national power configuration. The regions development crisis reached unprecedented proportion in the late 1970s, as special attention instituted by the 1960 constitution was abandoned by the military and resource flow and development attention became increasingly poor. It was this situation that set the tone, direction, content and methods of the communal agitation which began in disparate forms in the 1970s. The agitation grew in intensity, breadth, involvement, militancy and volatility until the disparate communal protests were replaced by a region wide protest, peaceful protests by militant and violent agitation, community protests by pan ethnic civil groups and youths and the quest for developmental attention by demands for self determination and regional autonomy through a conference of ethnic nationalities.

The agitation has had considerable consequences on the region particularly because of coercive, violent and repressive responses from the Nigerian state. Oil bearing has thus been hugely perilous. It has been like a resource curse because it has brought poverty, misery, sorrows, deaths, sufferings, devastations, disruptions and dislocations.

The paper examines the challenge of the Nigerian state by the communities and ethnic groups in the Niger Delta since the 1970s and the dominant reaction of repression and violence by the Nigerian state. Besides elaborating on critical concepts and the political economy of oil in Nigeria, the paper examines the emergence of agitation, the issues and methods of agitation and the nature of state responses. We focus on state repression and violence, identify cases or victims of such repression and examine the emergence of counter violence. Finally we look at the consequences of state responses and the future in terms of the possibility of rapprochement. The data is drawn from secondary sources. The time covered is 1970 to 2004.

8.2. Conceptual framework of analysis

Certain concepts are handy in the attempt to explain the predicament of resource rich states and regions.

The politics of hegemony and hegemonic and counter hegemonic struggles characterize much of politics in Africa. The postcolonial state has been characterized as hegemonical as it seeks to establish domination over subnations, identities, social forces and groups. This has made the state more consequential as it has become the object of inter and intra group and factional struggle for power and domination. Once captured, it becomes a vehicle of subordinating and excluding other groups and elite factions. State hegemony has been so exclusionary that it has been meant only for the accumulation and welfare of included groups. As a result, the state project has been characterized with increasing centralization of power and resources to further facilitate accumulation of benefits by incorporated groups.

But the tragedy is that state hegemony is pursued at the political, social and other levels by repressive and violent means (Jackson 2002:46). The state has been brutal, rapacious, suppressive and abusive of its citizens and groupings either in the dynamics of direct domination or in its use by ruling groups to suppress the opposition, rival groups and claimants and other identities. One area where the struggle for hegemony and counter hegemony has been intense is in the control of resources. There is often intense resource competition among groups and ruling factions for the control of resources which are often scarce and yet critical to group welfare and survival. The state has often been used and quite successfully in the appropriation of resource benefits for ruling cliques, identity groups and the incorporated elites. This has generally oiled neopatrimonial networks, corruption, intensified counter hegemonical struggles, and conflicts.

The concept of internal colonialism delineates core and peripheral regions within a state, which is characterised with unequal distribution of power, resources and economic advantages that are in favour of the core region. Often times, the persisting inequalities ossify into stratification systems. When ethnic and cultural differences coincide with the core – periphery stratification, the periphery tends to seek greater autonomy (Stone 1983)

One of the main instruments of internal colonialism, at least in the Nigerian situation is "political penetration, through a skillful pursuit of political control, aided by numerical preponderance of the ethnic majorities (Johnnie, 1998). Political power was then utilized to transfer resources from the numerically weaker oil producing ethnic minorities to develop the dominant areas. This created a powerful core, that is economically advantaged and an impoverished and weak periphery (Naanen, 1995). Naanen (1995) asserts that the internal colonisation of the Niger Delta is hinged on first, the ethnic based political domination by the ethnic majorities of the Hausa / Fulani, Yoruba and Igbo. This enabled the expropriation of the ND oil resources for the benefits of the ethnic majors. Second. is the alliance of the ethnic majority dominant groups, the oil companies and state enterprises. This has enabled the ethnic majors, using their privileged access to 'existing opportunity structures', to achieve dominance in the MNCs, in terms of employment, contracts and business. Consequently the access of the oil producing communities (OPC) to the MNCs and the more 'rewarding sectors of the economy' was restricted. Third, oil exploitation perpetuates environmental degradation which 'undermines the traditional peasant and fishing economies' of the OPC, without the provision of 'viable economic alternatives'. This undermines the economic capacity of the Niger Delta, which in the circumstance of trickle benefits and access to oil wealth has created a crisis indicated by agitation and resistance in favour of regional autonomy and resource control.

The theory of the rentier state postulates that state dependence on mineral rents underpins economic stagnation and political authoritarianism (Yates 1996:6). Because the state is allocative and does not depend on public taxation, it is detached, bloated, un-productive, corrupt, paternalistic, repressive and violent. The rentier state has an enclave and exclusivist nature, in terms of the social structure of accumulation. This concentrates wealth and causes high income inequalities (Mkandawire 2002:196-197). Rentierism engenders a rentier mentality and a rentier class that is complacent, consumptive, corrupt, dependent and profligate (Yates 1996:20-22). The politics of the rentier state is characterized by intensely fierce power struggles, instability, political violence, state violence, poor legitimacy of regimes, kleptocracy, conflicts, undemocraticness, political clientelism and neo patrimony.

8.3. The political economy of oil

The 1970s was the beginning of the oil boom in Nigeria. Oil had become the main revenue and foreign exchange earner. There was a massive in flow of oil resources that began to fuel a massive state expansion, infrastructural development of the cities and corruption, extravagance and ostentation of the ruling class. As resources flowed in, so did the hegemonical nature of the Nigerian state manifest in the centralization of resources in the centre, the appropriation of the resources by majority ethnic groups and the marginalization of minority and subordinated groups.

To denote the centrality of oil in the inter-group and class struggle, the federal military government had begun, even in the midst of the civil war to enact decrees that place ownership of oil and gas resources in the centre rather than in the oil producing states. In fact the northern hegemony through military rule became that of the utilization of state power to appropriate oil resources to the centre which it controlled.

As oil became more important with the oil booms and more central to the economy and the nations existence, the struggle for its control became more intense. The majority-minority struggle began to play out as the former sought to totally disinherit the later of the control and access to oil resources. Several arguments were bandied. Oil it was said was under the soil and not tended or produced by human labour as to warrant derivation based allocation. The land anyway had been taken over by government vide the land use decree and act of 1978 and 1979 and therefore oil resources were Nigeria's common property which does not belong to any group. The Niger Delta minorities argued that they owned the land and whatever was in it and that they deserve benefits for the land and water appropriated for exploitation and the pollution of their environment. Besides they deserve some returns or plough back from the huge resources and revenues extracted from their land. Who won in the struggle for appropriation of resources is indicated that by the early late 1980s, derivation based funds had declined from 50% at the inception of military rule to 1.5%.

8.4. Oil, communities and the emergence of agitation

The oil economy has generated deep grievances, bitterness, frustration and anger even before the civil war. After the civil war, there was no co-ordinated platform and avenue for the expressions of these feelings. Different communities bore particular grievances only in relation to the oil companies, expressed them through petitions and beyond some threshold point, undertook peaceful protests. But as the Enviromental consequences became more grievous and devastating and as the fate of the communities became more similar, a generalized regime of complaints began to emerge. Paradoxically, it was at this point that the benefits from the Nigerian state in terms of allocation of revenues thinned out to a mere trickle which mearnt that there was poor and declining resource capacity to address development. The persisting and in fact deepening economic crisis of the 1980s, which coupled with the brazen hegemony, increasing marginalization of the minority groups of the region and persisting neglect of military rule, accentuated the unease and grievances which began to manifest in social restiveness and tensions.

It was at this point that the glaring neglect, poor development attention and marginalization of the Nigerian state became more visible. The state and the political process began to be seen as inequitable, unfair and unjust. The Nigerian state came to be seen as the main culprit, designer and executor of the regions predicament. The region's elites became more uneasy and began to express individually the predicament and grievances of the region. By the late 1980s, some platforms of concerted agitation and influence seeking in relation to the regions problems had begun to emerge.

An increasing awareness led to greater mobilization and by the early 1990s, the region had become awakened enough for three things to happen. First was the emergence of civil groups which began to articulate demands and to construct platforms of common action. Second was a growing awareness and mobilization, in relation to the oil economy and rising identity mobilization and ethnic nationalism. Third was an increasing expression of the demands and grievances of the region and increasing centrality of the state related demands. By the late 1990s, the demands of the region had become more articulate, complex, political, forceful and volatile.

The early demands and grievances were directed at and expressed towards the oil companies. The state emerged slowly as a target of grievances and an object of demand. It only became a serious object of general agitation in the late 1980s. However, by the early 1990s, the state occupied the central place of the regions grievances as its policies and practices had been identified as the central problem. The Niger Delta problem became essentially an encounter between the oil producing communities and the state. The oil companies became mere business partners of the state who were protected in spite of corporate mis-governance and mistreatment by the state. The oil companies became an object of local agitation for local development. Beside that, they became the guinea pigs, whose facilities and production had to be disrupted to hurt state resource flow and compel its response.

The placement of the state at the centre and the target of agitation has transformed the agitation; its demands, methods, intensity and volatility. It raised the agitation to a political and state related, region wide and well co-ordinated one that was conducted by civil society and the region's elites.

The methods became much more political, militant and economically disruptive and the entire agitation became an intense and volatile regional struggle and challenge of the Nigerian state.

8.5. Issues in the agitation

Several issues can be identified in the Niger Delta region's agitation. The early issues were alleged neglect and marginalization in the developmental attention of the regional governments and the federal government. Then there was marginalization and exclusion in political representation. As derivation based resources declined, the issues of increased oil based resource allocation entered the agitation agenda. With ethnic nationalism and the intensification of the agitation, the issues of the mistreatment by multi-national oil companies, environmental degradation and reparation, undemocratic and disinheriting legislations, self determination, structural imbalances, resource control and conference of ethnic nationalities became more important.

Poor developmental attention

The complaints about neglect and negligence of the region in the federal governments development efforts date back to the 1950s. The indications were the poor infrastructures, facilities, and poverty. By the late 1990s, the region had only 2% of the federal roads and less than 27% of the people and 30% of the households had access to safe drinking water and electricity respectively (NDES 1997). The access to health care and educational facilities in the riverine areas were some of the lowest in the country just as the mean annual income per household is less than the United Nations absolute poverty line of USD1/day. The contradiction between the endemic poverty and infrastructural and service under-development of the region and the fact that its oil generated over 80% of the nations revenues has generated intense frustration and disenchantment.

As Wiwa noted in respect of the Ogoni, "there is no pipe borne water, no electricity, very few roads, ill equipped schools and hospitals and no representation at the federal level" (Ogbodo 1992). Chief M. Ak obo, former minister and foremost leader of the Ijaws has observed that "the agitations will vanish once the problem of development is solved" (Niboro 1998). Chief E. K Clark, another former minister and Ijaw leader has said that the rampaging youths "are only doing whatever they can to make the federal government feel the impact of the negligence... lack of facilities (Niboro 1998).

Marginalization in political representation and resources

The region's leadership feels alienated from the mainstream of Nigerian politics and exploited by the non-oil producing majority ethnic groups who have dominated and dominate the center (Obi 1993). The victory of the erstwhile Northern region in the Nigerian politics of hegemony has created a configuration of a highly centralized federal centre, weak and dependent states and a resource distribution system that have sidelined and disadvantaged the states of the Niger Delta. The regions elites have remained at the margins in terms of political representation and access to state largess and resources. More painful is an increasing narrowing of access and control of the revenues, rents and benefits from the oil economy.

It is therefore not surprising that the region has been a hot bed of the criticism of the present federal structure, the agitation over the national question and the calls for state reforms particularly federal restructuring, regional autonomy and resource control. More recently, the region has been demanding power shift to the region. The civil groups and elites believe that exclusion from the apex leadership of the nation has been the cause of the region's predicament. The region has therefore championed the demands for regionalism, the entitlement of all groups to the presidency, rotational presidency and power shifts. Since the fourth republic, the region's elites and groups have sought and are still seeking strategic political alliances for a power shift arrangement that is favourable to the region.

Disinheritance by federal laws

The region is aggrieved by the dispossession of oil resources by the federal government through several legislations. These legislations include the Petroleum Act of 1969, the Land Use Act of 1978 and other oil exploration laws. These laws have placed the ownership of land and particularly oil and gas on land and territorial waters in the federal government. The region sees this legislation as depriving them of access, rights and freedom to control their resources and to benefit therefore. The laws are therefore regarded as discriminatory, inimical, undemocratic and therefore should be expunged, repealed or abrogated.

More recently there was the issue of dichotomy between on-shore and offshore oil production in the computation of the derivation principle. An earlier decree had put off shore oil production as belonging to the federal government but this was abrogated by the Babangida regime following increasing tensions in the region. But the Obasanjo regime persisted with a dichotomy and sought in 2001 a judgement at the Supreme Court to seek interpretation as to who between the federal government and the littoral states owned off shore oil production. The Supreme Court ruling passed off off-shore oil production to the federal government which meant that a considerable portion of the 13% derivation funds was lost. However, the ensuing tensions in the region were such that the federal government made concessions through the abrogation of dichotomy law that was passed by the National Assembly.

Un-fairness and Inequity of the Revenue Allocation System:

The region sees the resource distribution system as been progressively to her disadvantage. The region's leaders particularly resist the insignificant share of revenues based on derivation, which constituted only 1.5% until 1992, 3% until 1999 and 13% since 1999. The groups believe that the majority ethnic groups and federal government have denied them of fair share from oil receipts.

The region regards the system, pattern, formular and principles of resource distribution in the nations federal system particularly since the advent of military rule as unfair, unjust and inequitable

Loss of control over resources

The region has felt shortchanged and disinherited by the progressive loss of rights and access to oil resources since 1969. By the 1980s, the region had almost totally lost out in terms of rights over land and water, off shore oil resources and revenues from oil. This was in spite of their support for the federal government in the Nigerian civil war and after and for the northern-based political parties and elites in the second republic. This has fueled a feeling of betrayal on the part of the elites and a re-examination of the strategies of accommodation and strategic alliances.

The emergent feeling of loss, disinheritance and dispossession and consequent fustration and disillusionment laid the foundation for the mobilization, radicalism, and the struggles for self-determination and the restructuring of the Nigerian state that began in the early 1990s. More importantly, it transformed the demands from mere increases in revenues and benefits to that of control over resources derived from their land. The region¶s elites and groups claim and seek the actualization and guarantee of the peoples right to the resources on or under their land and waters and self-determination to utilize it in their interests and survival as determined by them.

Several leaders and civil groups of the region in fact insist that resource control by ethnic nationalities and autonomy to utilize their resources in line with aspirations is the only panacea for peaceful co-existence and sustainable peace and security in the region. Further it is seen as non-negotiable for the redress of the neglect and exploitation of the Niger Delta (Akpan 2000; Ehigiator2000; Obari1999A; Onwuemeodo 2000A). Initially only the more militant groups and elites agitated for resource control. By 2000, other groups particularly the pan ethnic groups such as the Union of Niger Delta were in the forefront of the resource control agitation. The resource control agitation was perhaps the most consistent, volatile and high profile demand from the region since 2000.

Marginalization in the oil economy

The region feels that they require special treatment and benefits as the revenue basket of the nation. It also believes that it should play a central role in the state and corporate management of the oil industry and in the employment, contracting and other activities of the oil economy both at the state and multinational levels. But this has not been the case. It was not until the late 1 990s, and arising from immense regional pressures that the region began to gain access to fairly adequate representation in the state management of the petroleum ministry at the level of ministers and top government officials. The local staffing and management of the MNCs is dominated by groups outside the region. The headquarters of the oil companies are also outside the region. Therefore the region has felt that it has been cut off from the oil deals, contracts and largess in the oil industry and denied access to influence and even benefit from the oil bureaucracy and MNCs. In the circumstance, there has been growing alienation, which began to manifest since the 1990s in the popular violence and criminal brigandage against the oil facilities.

The region's demands in this area has been appointment of the indigenes as top state officials at the ministry of petroleum and the Nigerian National Petroleum Company, the employment of indigenes as staff, managers and directors of the MNCs, the relocation of the headquarters of the MNCs to the region and greater benefits in terms of contracts, scholarships and related benefits to the region's citizens.

Mistreatment by the multinational oil companies

The relations of the MNCs with the communities have been a major source of agitation. The MNCs were quite insensitive to the communities in the acquisition of land and water, degradation of the environments, the management of spillages and the management of compensation. Particularly, the MNCs relying on federal laws totally disregarded the region's complaints, protests and demands in relation to unrestrained acquisition of their land and water resources, environmental remediation, compensation and development. Until the 1990s; the MNCs simply called in the vicious mobile police units to brutalize the communities whenever exasperated local people protested. In the 1990s, the MNCs were accused of supporting and providing logistics for state repression of communities and youths.

The MNCs rarely contributed to the development of their host communities until a regime of militant protests overwhelmed them by the late 1990s. The Nigerian state simply left its citizens to the mistreatment of the MNCs without any laws or regulations as to obligations and best practices of environmental and compensation practices. As Mr. Ezekiel A. Tebepina, president of Ogboloma Youths Federation (OYF) said of Shell, "They come to our land with strangers to take away our God-given wealth. When we complain they call us names, and give us peanuts", (Iwori 2002).

Structural imbalances

Federal practice has been characterized by over concentration of powers and resources, which has resulted in disadvantageous structural imbalances, marginalization of minorities and deprivation of resources. The region feels disadvantaged by a federal structure that has deprived them of the control of their resources, have left them at the mercy of a central government in which they are underrepresented and marginalized and which has neglected and cheated them in the distribution and application of their resources. The region demands for a restructured Nigeria where every ethnic group would have the right and say over its own affairs and control over its own resources and development. More specifically, the region seeks the devolution of powers to federating units or confederations.

Most of the civil groups in the region demand for self-determination and regional autonomy within a true, balanced and proper federal arrangement. The more radical elites and groups seek a return to the 1963 regional constitution and re-structuring into six geo-political zones. The main instrument sought for restructuring is the sovereign national conference or a conference of ethnic nationalities. Most of the youth, ethnic and regional groupings seek such conference as a platform to re-negotiate the basis of co-existence particularly in relation to power sharing, political structure and fiscal relations. Many groups believe that the conference is the only platform for the achievement of a nation founded on fairness, equity and justice.

Environmental Remediation and Compensation

The environmental cost and attendant socio-economic disruptions of oil production has been enormous. This has been particularly in the loss of vitally important land and water due to pollution. The initial agitation was undertaken by communities for compensation from the MNCs for land and water polluted by oil spillage. Often times, the communities failed to obtain equitable and prompt compensation. The treatment of the communities in the struggle for compensation was often arrogant, condescending and unequal. However, there has emerged environmental rights activists who have placed the environmental degradation and costs to the region in the front burner and have monitored polluted sites, oil spillage management and compensation practices. These groups have heightened the environmental dimension of the agitation. The agitations for speedy and effective remediation and restoration of polluted land and water have also been heightened since the 1990s with the heightening of the over-all resource struggle.

In line with the environmental component and to alleviate decades of accompanying landlessness, economic disruptions and impoverishment, many groups and leaders in the region seek compensation and in some cases demand for reparation. The demand for compensation has ranged from adequate compensation for specific oil spillages, reparation to the region in the form of varying lumpsum payments from the oil companies, a comprehensive programme of development projects by the state and oil companies and a comprehensive programme of land and water restoration and remediation. For example, the Ogoni Bill of rights, issued in October 1990 demanded for N200 billion from the oil companies for three decades of neglect, environmental degradation, impoverishment and absence of development.

8.6. The methods of agitation

Several methods or instruments have been utilized in the agitation. The initial methods were merely petitions to oil companies. There has since been a transformation towards more direct, engaging and confrontational methods.

Articulation and expression of demands

The expression of the grievances and the articulation of demands began in the early 1950s. Glaring exclusion of the minorities in both the Eastern and Western axis led to early accumulation of grievances which found expression in the flowering of cultural and civil groups and minority ethnic based political parties. The ensuing articulation of the grievances and interests and agitation was centred on the resolution of the problems through special developmental attention and the creation of separate regions in the western and eastern axis. There was a lull in the expression of the grievances and demands of the region during and after the civil war.

A new surge of articulation and expressions only emerged in the late 1970s following increasing decline in the region's resource allocation from the federation account. The demands were centred on increased derivation based allocation from the federal government. This continued into the 1980s and 1990s. This was followed by demands on the oil companies for developmental attention. By the 1990s, the articulation of the region's demands had become so concerted. There was further a high degree of agreement among the region's citizenry and a high level of vocalization. The grievances and demands had become so political and were undertaken by a broad platform of opinion leaders, groups and activists.

The methods of expression were initially political through electoral challenges, opposition politics and the media. It degenerated into an armed rebellion on the Eastern axis in 1966. By the 1970s, community protests and the media assumed centre stage. This continued until the 1990s when a regime of broad and popular protests, popular violence, environmental and human rights activism, political challenges, civil society struggles and solicitation of local and international media coverage of press statements, conferences and paid advertorials became the main methods of expressing the grievances and demands.

Activities of mobilized regional elites

Increasingly, the region has witnessed a convergence of the region's elites on the grievances and demands of the region. This has constructed increasing unity as well as common platform for actions. For the older elites who have been the object of incorporation by the state and MNCs and who led the path of collaboration on the national project, the failure and futility had become obvious by the late 1980s. They felt betrayed by their majority ethnic and state collaborators while at the same time, their people portrayed them as selfish elites who had betrayed the region. Besides, there was the challenge of younger and more radical leaderships who were increasingly more popular and mobilizational. By the 1990's, there had coalesced a wave of such strong feelings about the region's predicament that both the old and young elites were compelled to respond to.

The emerging solidarity and unity began to be noticed in the late 1980s. By the early 1990s, gatherings and groupings had began to emerge as they sought responses to the region's predicament. A major opportunity for construction of platforms for elite unified struggle was the constitutional conference of 1994/1995 under the Abacha administration. The region's elites through several groups successfully constructed a common agenda and successfully pushed for increased derivation based funding at the conference.

The region's elite also began to challenge and respond to national issues that related to the regions interest and predicament. In June 1998, for example, about 100 opinion leaders of the Ijaw, Eket, Annang, Isoko, Urhobo, Itsekiri, Ndokwa, Bini, Ogba and Ogoni challenged through a paid advertorial the assertions of Prof. Sam Aluko about mineral rights (The Guardian 07:06:1998; 34-35). Numerous meetings and conferences of elders, opinion and political leaders, traditional leaders, youths and activists began to take place as platforms for addressing and responding to the region's problems.

Ethnic and minority nationalism

The sense of marginalization and injustice generated recourse to the ethnic group and the region. A strong identity consciousness emerged among the ethnic groups, which began to be directed towards the construction of minority and ethnic nationalism. Ethnic and minority oppression and internal colonization became the explanation for the dispossession of oil resources and the exclusion from state power and resources. Among some of the groups, there was the construction of a group threatened, sidelined and deprived by majority ethnic hegemony. The only path to overturn the status was the construction of platform of strong ethnic solidarity and unity. The past history, glories and resistance of the ethnic groups were reinvented for pride, strength and inspiration in a struggle for the restoration of dignity and equity.

What followed were the awakening of intense identity sentiments, mass identity mobilization, the emergence of communal, ethnic and regional groupings of different hues and of co-ordinated platforms for ethnic and regional action.

Mass mobilization and the activism of civil society

Along with the increasing awareness of grievances and frustration came a proliferation of all types of groupings of communities, ethnic groups, youth, women and traditional, political and opinion leaders. Cultural, pro-democracy, environmentalist groupings and non governmental organizations emerged and became the main instruments of engagement in the region's struggle. The civil groups were able to mobilize the region's citizenry, re-construct the struggle in populist form, create a network of civil groups, raise a high level of solidarity and create common platforms. By 1997, the general awareness about the neglect and treatment of the region was such that 10,000 people gathered from several communities in Rivers, Delta, Edo, Akwa Ibom and Ondo at the launch of a resistance and protest movement, the Chikoko Movement at Aleibiri (Efenakpor 1997). The civil groups heightened the tempo, volatility and intensity of the struggle, reconstructed the demands in more fundamental and comprehensive forms and turned the agitation into that for minority and environmental rights, state reforms and self determination.

The Movement for the Survival of Ogoni People (MOSOP) was the first ethnic civil grouping to be organized in opposition to the MNCs and the state. At a pan Ogoni congress in October 1990, the Ogoni Bill of rights was declared, in which autonomy, resource control, compensation and development were demanded. Through massive mobilization of the Ogonis and persistent non-violent peaceful women and community protests, MOSOP was able to compel the stoppage and in-fact the evacuation of Shell operations in Ogoniland by 1994. MOSOP boycotted the 1993 presidential elections and the constitutional conference elections of 1994. The Ogoni nationality on March 13 1993 organised a 500,000 man vigil in 125 towns and villages to pray for deliverance from the 35 years old ecological war waged by the MNCs and from oppressive rulers (Duru 1997).

MOSOP continues to insist that the government must acknowledge that the Ogonis own their resources and that Shell or any other seeking to operate in Ogoniland must be ready to re-negotiate terms, abide by her conditions and play according to accepted rules (Obari 2000B) As at March 2000, Shell was only allowed limited access to Ogoniland in order to execute projects as a prelude to negotiations towards the resumption of production (Obari 2000C)

Targeting of oil facilities

A major strategy in the struggle has been the targeting of the oil companies as a way of getting at the federal government. This is because the communities and youths believe that the MNCs are in alliance and business with the federal government and that the government depends vitally on the oil produced by them. Therefore disrupting their operations and production would hurt and compell the government to respond speedily. Thus the response to inimical incidences, policies and practices of government is to threaten or actually disrupt oil operations. Since 1997, this strategy has underlined resistance. The MNCs have thus become guinea pigs who are targeted for destruction and disruptions even in matters that sometimes do not relate to them.

The second way in which the MNCs became targets since the 1970s has been in the development of the communities. With government negligence, the communities have looked up to the MNCs as the alternate provider of facilities and social services. The communities have compelled MNCs through protests and production disruptions and agreements and memoranda of understanding to provide facilities and infrastructures. Most particularized community conflicts with MNCs relate to non performance in this area.

Transformation of demands into the political, civil rights and democratic demands

A major development of the 1990s was the transformation of the regions struggle from a mere developmental one to that of self determination, minority rights, justice, equity, fairness, environmental and civil rights. This began with the Ogoni struggle for minority rights and self determination and was reinforced by the youth activists of the Ijaws and other groups in the late 1990s.

The second development was the transformation of the demands for developmental solution to that of political resolution. The state and state reforms through resource control, regional autonomy, true federalism and a conference of ethnic nationalities became the objects of struggle. This further entailed a shift of the focus and object of struggle from the oil companies to the Nigerian state and the entrance into the struggle of top business, opinion and political leaders as well as youth, women and democratic activists. It also meant that a broad spectrum of national and international democratic, human and environmental rights groups and activist became involved in the struggle.

The transformation of the objects and content of the struggle was a key factor in the nationalization and internationalization of the struggle.

Internationalization of the conflict

This strategy commenced in 1990 with the Ogoni mass struggle through the Movement for the Survival of Ogoni people. MOSOP was able to link up with international human rights, environmental and other international civil groups which took interest in the Ogoni and the region as a struggle by minorities threatened by Shell environmental degradation and the Nigerian state's marginalization and repression. International civil interests soon became a massive formation of international pressure comprising environmentalists such as Greenpeace, the Sierra Club, Oil Watch, Friends of the Earth, and civil rights groups such as Africa Fund, Human Rights Watch, Amnesty International, Trans Africa, Unrepresented Nations and Peoples Organization and Project Underground.

In addition church groups such as the World Council of Churches, and the United Church of Christ, the Church of Christ Disciples and United Methodist

Church in the United States, trade unions and other groups and individuals became a very forceful formation that mounted tremendous pressures on their governments and the MNCs particularly in respect of sanctions and boycott. There were delegations and fact finding missions from international civil groups and even the United Nations Organization.

Mass actions and confrontations

The Ogoni took the lead in the ethnic wide organised resistance against oil exploitation and its accompanying effects and in the utilization of peaceful mass actions to prevent and disrupt oil production under the auspices of the MOSOP. Since the late 1990s, oil installations have been sporadically seized, and production disrupted by community members, youths and women as part of the region's struggle. By October 1999, about a third of total crude oil production was cut off through the disruptions of productions by demonstrators in the region. Between January and August 1991 alone, there were over 22 blockades of Shell installations (Ogbodo 1992).

Youths of the Ijaw, Isoko, Egi, Obagi and others have at various times since 1998 shut down, seized, occupied and disrupted oil production facilities. In 1999, Ijaw youth activists shut down virtually all of Texaco's production operations in Delta and Bayelsa states and a considerable part of oil production facilities of other companies as part of the ijaw Youth Council (IYC) campaign to enforce a ban on oil production in the Operation Climate Change phase of the Kaiama Declaration. The Ijaw Youth Council (IYC) according to its president had stated that they believe in non-violent direct action, which includes closure of oil flow stations, but not kidnapping of oil workers. Isoko youths under the auspices of the Isoko National Youth Movement (INYM) seized and shut down the flow stations operated by Shell Petroleum Development Company (SPDC) in December 1998, January 19 1999 and 2000 (Ogoigbe 1998:1-2)

By late 1999, the Nigerian Navy was escorting oil tankers and merchant ships entering or leaving Bonny, Okrika, Onne and Port Harcourt harbors as well monitoring and patrolling oil terminals, oil pipelines, gas flow stations and oil rigs to protect them against sabotage and damage by violent youths 2. President Obasanjo had stated in an address read on his behalf by the managing director of the Nigerian National Petroleum Company (NNPC) at the 17th conference of the Nigerian Association of Petroleum Explorationists (NAPE) that;

> The operating companies and their contractors are continuously embattled by militant groups that have taken over the Niger Delta to the extent that it would not be an exaggeration to say that the Niger Delta is in a state of war.

8.7. State responses to the niger delta agitation

The initial reaction of the Nigerian state to the region's agitation for separate region and developmental attention was to consider it as part of the majority-minority, intra-regional and inter -region and party struggles for power. Thereafter, it recognized the geographical difficulties and sought a special developmental attention. Later responses ranged from ignoring it, making symbolic promises and appeals to that of active engagement through developmental attention, militarization and repression.

Ignoring the agitation

The agitation in the military era was largely ignored. The military governments in the period before the 1990s ignored the agitation because they were not concerted and were at the best backed by disparate communal protests. Second, the governments assumed that the small size and population of the pockets of resentments and protests was not significant enough to threaten stability and economic development (Asiodu 1980). It was only when the agitation became intense and the incidence of community protests became pervading that the state began to consider response.

Acknowledgement of grievances

The governments since the early 1990s have sporadically admitted that a crisis has arisen over prolonged neglect of the region. The Babangida and Abubakar administrations acknowledged the problems of the region and either set up an interventionist agency or attempted a development programme response respectively. The Obasanjo Administration attributes the neglect to prolonged military rule, previous governments and the oil companies. The governments since the Abubakar administration have acknowledged that the grievances are genuine but have quarelled with assumed excesses in the conduct of protests.

Symbolic statements and promises

Symbolic statements and promises that were not intended to be backed by actions have also been a strategy since the 1980s. The essence has been to give impressions of concern and sensitivity to the region's problems. Most governments since the late 1970s have made promises which turned out to be empty and were merely intended to create positive impressions in the region. Several times, hopes were raised through promises, official statements, committees and proposed projects.

The non actualization and emptiness of these promises have been part of the reasons for the distrust and aggression in the region.

For example, in January 1994, following increased agitation in the MDs, the federal government set up a high powerful ministerial fact finding team (MFFT) to the region. The team comprised Don Etiebet, minister of petroleum, Alex Ibru, minister of Internal Affairs, Melford Okilo, minister of Commerce and Tourism. The team was to carry out a comprehensive tour of the region and ascertain their grievances and development on the ground. The committee received oral presentations and 136 memoranda (Ofouku 1999). Nothing followed from the committees' recommendations throughout the Abacha's administration.

The Abubakar Administration raised much hopes because of the development blueprints and the promise to spend over N15.7 billion in a special development programme (HRW 1999). Nothing again emerged. Even the Obasanjo Administration promised a special developmental attention through immediate and special efforts. The government promised increased development with the implementation of the 2000 budget and subsequent budgets (Adigun & Lawal 1999). However, apart from the establishment of the Niger Delta Development Commission (NDDC), nothing much has happened.

Appeals for peace and dialogue

Several governments have called for understanding and restraint from the region and urged peace and dialogue rather than aggression and confrontations. The Abubakar Administration and now the Obasanjo administration have been particularly active in this endeavour. The Abubakar Administration set up a Peace Committee chaired by Rear Admiral Victor Ombu to the oil producing states of Delta, Akwa Ibom, Rivers and Bayelsa in 1998. The committee held several peace meetings with stakeholders. The key message was that "vandalizing oil companies is not the way forward'... ... Restraint is the cornerstone to progress" (Obari 1998)

President Obasanjo visited the region within two weeks of his inauguration in 1999. The Obasanjo's presidency has been more sensitive and talking. President Obasanjo has met several times with representatives of the oil producing communities even at the presidential villa (Onwnemecdo 2000B). Meetings have been held frequently between government agencies and officials and representatives of communities and groups in the region. The government has solicited patience, peace and understanding as a platform for the attainment of the objectives of accelerated development (Adigun & Lawal 1999)

The problem however is that government has not been ready to engage in meaningful and substantive dialogue and negotiation. The government has preferred to force developmental strategies on the people and have avoided the political dimensions of the agitation, which certainly requires fundamental changes and reforms. The state has preferred to use "military force to crush peaceful protests rather than reek to address the issues being protested"(HRW 1999). The Ijaw National Congress blamed the violence following the Kaiama Declaration on "official unwillingness to dialogue with the Ijaw people" (Obari 1999B). The government approach and conception of dialogue has been a top down monologue, in which the people acquiesce rather than actively participate.

Peace too is conceived as stoppage of direct and disruptive actions in spite of continued neglect or even provocation. In both cases, it is the people that are expected to comply while the state and MNCs continued with their character and tendencies in relation to the region.

The state has also ensured that it sidelines the groups and activists involved in the struggle in the meetings held. For example, in the Abubakar administration meetings, Niger Delta prominent citizens were invited as "individuals rather than as representatives of organizations". The Ijaw Youth Council, a major actor in the conflicts between 1998 and 1999 was not invited.

Intervention of state governments

The state governments have tended to intervene to broker ceasefire, settlement of disputes and ensure peace in the relations between the communities, the federal government and the oil companies. The federal government places primacy in the maintainance of law and order and non-disruption of production. The oil companies too are only concerned with the resumption of production. But the state governments have tended to engage stakeholders more broadly to resolve the problems, prevent excesses on the part of all parties and particularly reduce the possibility of the deployment of excessive force by the federal government. State intervention has been underlined by first, the prevention of the breakdown of law and order. Second, there is the fear that such breakdowns may lead to sanctions against military administrators during the military regimes and threats of declaration of emergency during civil democratic dispensations. This is particularly because state governments have been saddled with maintaining security even though they do not control the means of doing so. Third, state governments often intervene to protect the interests and integrity of their communities from mistreatment by oil companies. Fourth, the state governments often utilize the opportunity of conflicts to win some concessions for the communities from the oil companies and to underscore the need for federal attention in the resolution of the resource conflicts.

The state governments have been more effective because they tend to have the confidence of the communities. The latter have tended to heed appeals and attend meetings arranged by the state governments. The state governments have been quick to respond to conflict situations through visits and meetings with concerned parties. This has been particularly so since the current democratic dispensation and particularly since 2000, when state governors and political leaders became actively involved in the struggle for resource control. There have been more state government interests in the issues of environmental degradation and the reaching of memoranda of understanding between the communities and the oil companies. In fact, the state governments have recently being walking on a tight rope. While being part of the struggle, they have struggled to prevent excesses in popular reactions and maintain peace, law and order.

Interventionist efforts and agencies

Two major developmental strategies have been utilized by the federal government as a response to the region's agitation. The first is the promise of increased and accelerated development efforts while the second is through developmental agencies.

The promise or actual commitment of resources for the development of the region and the effort to undertake special projects became significant only since 1998. The Abubakar administration unfolded a development package in November 1998. The package included priority projects such as dualisation of roads, they were to be undertaken and completed before May29, 1999. (Ikwunze 1998). General Abubakar appointed a committee to consider the needs of the Niger Delta. The committee recommended the immediate expenditure of N15.7 billion on development projects (HRW 1999) . The 22 member committee was chaired by Major General Oladayo Popoola and comprised military administrators from the south south zone and representatives of Oil Minerals Producing Areas Development Commission (OMPADEC), the Petroleum (Special) Trust Fund (PTF) and other government agencies.

The Obasanjo administration has promised at various times, to make more resources available, accelerate infrastructural development, embark on instant development and initiate programmes to revert the previous neglect and to address the grievances of the region (Ugwunebo 2000; IRIN 2000) . The administration gave an under taking to spend about N 76 billion annually in the region (The Guardian 24:11:1999). In November 1999, for example, the federal government approved projects worth about N5 billion for infrastructures and other projects in Bayelsa (IRIN 2000).

The federal government has intervened four times in the region through special development agencies. First was the establishment of the Niger Delta Development Board in 1962. Then there was the establishment of the Oil Mineral Areas Development Committees in the Second Republic. But this was merely a reaction to litigation and loud grievances following the huge decline in the derivation-based fund in the then Revenue Allocation Act passed by the National Assembly. The third was in the establishment of the Oil Mineral Producing Areas Development Commission in 1992 by the Babangida Administration. Currently, there is the Niger Delta Development Commission established by the Obasanjo Administration.

The promise of special development funding and projects has often been weak in terms of execution. The Abubakar Administration promises remained so at the end of his tenure. The Obasanjo Administration has been more loud in talking about special attention than in executing special projects in the region. The leaders of the region, are still alleging neglect and marginalization even in this dispensation of government acknowledgement of the crisis. The South-South parliamentary caucus in the House of Representatives alleged in 2000, that the south south zone is still being neglected, discriminated against and obtains a disproportionate share of distribution of projects among the geo-political zones even in the Obasanjo administration (Adamu 2000).

The special intervention agencies have not made tremendous contributions to the development of the region. The NDDB was plagued by weak funding and efforts. The OMPADEC was plagued by excessive federal control, corruption and utilization to incorporate and compromise the region's leaders. Incomplete, abandoned and non functioning projects characterized the commission's output. The NDDC is also plagued by excessive federal control, duplication and overlapping project outlay with the governments of the region, loud public relations and poor performance.

Protection of oil interests rather than the citizenry

The state and MNCs for a while, in response to agitations, created a security veil, operating behind military and police protection. The oil companies have had prime access to security agencies for the protection of oil facilities from the popular protests and anger of the youths and other community members. For this purpose, they provided logistics and related support to the security agencies. By the late 1990s, there were favoured deployments of mobile police units which were equipped, funded and accommodated by the oil companies. Both involvements of the security agencies have continued till today.

The central function of these security agencies was the protection of the oil facilities from the citizens of the oil producing areas. Essentially, they were deployed to chase, brutalize and Harass the people and prevent them from popular actions against the oil facilities. The state coercive agencies were used to crush even the most peaceful opposition. The MNCs have tended to use security forces to attack individuals, groups and communities that protest against them and to acquiesce to and justify such actions as " appropriate responses to threats to company personnel or property" (Human Rights Watch 1999). State brutalities and threats are sometimes perpetrated against communities under the pressure of MNC officials.

Incidences of the involvements of state security agencies in brutal suppression of the local people with MNC logistic support abound. Numerous community members have lost their lives and considerable properties have been destroyed as forces have used excessive force to dislodge community protests and peaceful occupation of oil facilities. In January 1999, Chevron ferried and flew in military personnel by boats and helicopters to "attack a community opposed to its presence on their land. About a dozen people died'(The African Fund 1999). As the Tempo magazine (1997) remarked "Everyone seems to recoil in horror at the memory of Nigerian policemen, sacking a town, killing and injuring its inhabitants, at the behest of an oil company'.

In fact, state repression in the Niger Delta has essentially been oil motivated. It is always to assuage the threats to oil production, and to ensure further investments in the oil sector. The economic interest is clear from public statements of top government officials. In a meeting with MNC officials, state governors from region and government and party officials in November 1999, president Obasanjo said, "the federal government is very much aware of the concerns of the oil producing companies for law and order in the Niger Delta" (The Guardian 23:11:99). The key issue to the Federal government is a guaranteeing peaceful atmosphere for the oil and gas industry.

As a matter of fact, numerous community members have suffered detention and brutalities for raising grievances, raising and participating in delegations and meetings with oil companies, seeking compensation and for participating in protests against oil companies. As Human Rights Watch (1999) has noted, many of the abuses and repression have been known "to take place on, or adjacent to company property or in the immediate aftermath of meetings between company officials and individual claimants or community representatives".

Pervasive repression and military actions

The response of the state to militant civil groups has been high handed, suppressive, repressive, brutal, coercive and directed not at resolution but submission The military governments of Babanginda, Abacha and Abubakar believed that the agitation were excessive and provocative (Idowu1993). As the Chief of Defense Staff, Air Marshall Al Amin Daggash warned, in 1999, "any further protests of any kind would be decisively dealt with" (Vanguard1999). The tone of state response was further reinforced by Gen. A. Abubakar's budget speech on January 1, 1999 when he stated that;

> We cannot allow the continued reckless expression of these feelings.-seizures of oil wells, rigs and platforms as well as hostage taking in the name of expressing grievances. We will not accept brazen challenge to the state authority under threat of violence.

The position of the Obasanjo government too has been that the youth agitation has assumed a criminal dimension which threatens law and order and is resulting in near anarchy (Obari & Obinor 1999). According to the Vice President, Atiku Abubakar, "no responsible government would condone criminal activities in the name of rights activism"(Obayuwana 1999).

The state has preferred the crushing of even peaceful protests rather than address the issues being protested (Obari 1999C). The military has been utilized to crush protests since 1993 and large-scale troop and military hardware deployment has continued since December 1998 (Obari 1999D). The civil administration has sustained massive military deployments and operations to crush civil and militant protests. In Ogoni as in other ethnic groups and communities, there were large scale military occupation, repression, killings and human rights abuses.

Soldiers and mobile policemen have constantly ransacked protesting towns and villages since 1990, with attendant arrests, harassment, rapes, killings and destruction of properties. The military operations in Choba, Odi and in the Warri region since 1999 have sustained the pattern of massive destruction, human rights abuses and killings. In the case of Umuechem, the head of the police inquiry, then a deputy Inspector General, Parry Osayande declared that "the extent of damage is not commensurate with the threat by the villagers" (African Guardian 1994)

Since 1998, there have been numerous military and security roadblocks and special security task forces of the Navy and Army to protect oil installations and maintain security. In 1999, the federal government initiated operation HAKURI II to protect lives and properties and particularly oil facilities (Onuorah 1999). Currently, there is Operation Restore Hope. Troops and warships have been perennially deployed.

8.8. Cases of state repression and violence

Initial state repression in the region were localized but highly intensive and destructive. They were undertaken as part of the strategy of protecting the oil companies and oil installations from the local people and undertaking decisive actions to deter further protests in the region. The Ogoni incident, when an entire ethnic group and several local governments areas suffered state repression, was the first large scale and consistent case of repression. It preceded the regime of extensive repression of almost the entire region which began from 1997.

The case of Umuechem in the then Etche local government area of Rivers State blazed the trail. Shell had operated about 28 active oil wells in the community with the oldest wells being more than 30 years old. But in October 1990, a dispute arose over the non payment of compensation for oil spillages and the failure of Shell to honour agreements in respect of community development. In reaction, the community members embarked on a demonstration and obstructed oil operations. Shell called in the Mobile Police Division. A resistance by community youths led to the death of one policeman. This led to a major police action in which about 495 persons, including the paramount chief and his deputy were killed, virtually all houses including the royal palace were burnt or destroyed, all motor cars, motor cycles, electric generators and other personal effects were destroyed and even domestic animals killed. One year after the incident, the survivors refused to return to the town. The government response to the high handedness of the police in the incidence was the disbanding and demobilization of the Mobile Police Unit that undertook the operation (Adewale 1993; Onuorah 1994) .

In Ogoniland, initial protest began with petitions, press statements and protest letters to Shell, the state and the public over environmental devastation occasioned by oil exploration and exportation, developmental neglect and political marginalization (Saro-Wiwa 1992). The agitation only became intense following incidents of Harrassments in April 1993, when villagers who attempted to resist the bulldozing of their farms and crops in Biara, an Ogoni village, by Wilbros staff, a pipe laying contract firm for Shell, were manhandled. This sparked off protests by the villagers. The military was called in to forcefully put down the demonstration. In the process, seven people were killed in Biara and three in Tai. The protests became large scale, persistent, defiant and extended to the entire Ogoniland and undertaken not just by villagers but numerous civil groups under the leadership of the Movement for the Survival of Ogoni People (MOSOP). The protests forced Shell to stop operations in 1993.

The state response, following the threat on Shell's production was an excessively high handed military and police action, directed at protecting Shell's installations, equipment and personnel and containing the agitations. The Ogoni region was occupied by the Rivers state Internal Security Task Force led at various times by Major Obi Umahi.and Major Paul Okuntimo, and deployments from the Bori Camp, 2nd Amphibious Brigade, Port Harcourt led by Major Braimoh. The security forces terrorized Ogoni land and their activities led to loss of lives and properties (Daily Sunray 29:07: 1993; Adeleke 1993; Obi 1990).

Ikebiri I & II are communities in the Southern Ijaw local government area of Bayelsa state where Nigerian AGIP Oil Co (NAOC) began operations in 1969. In April 1999, the communities, exasperated by failed efforts to obtain compensation for incessant oil spillages which had occurred in 1972, 1980, 1990, 1992, 1994 and 1999, by dialogue, resorted to the shut downs of oil wells. Initial AGIP response, was the invitation and ferrying of 3 boat loads of the dreaded Rivers State Internal Security Task Force, on 18th April from the Brass Terminal to forcefully re-open the wells. When this failed, a combined team of the Navy and Army, invaded the communities on 20th April. About 8 persons were killed, 15 declared missing, numerous persons arrested and detained and 2 speed boats seized. Choba is an Ikwerre community in Rivers State, where Wilbros Nig. Ltd., a subsidiary of an American contractor in the oil/gas industry had operated for several years. In September 1999, aggrieved over a meagre share of Wilbros staff strength; 35 out of 1,500, Ikwerre youths began a protest against Wilbros. Community members occupied the premises of Wolbros. An initial resolution seemed to had taken place on September 17, following an agreement between the community elders, Wibros and the Rivers state government but this was repudiated by the youths who continued the protests.

Soldiers were drafted to Choba, on 28th October, on the request of the state governor and the management of Wibros to disperse the demonstration outside the premises of Wilbros. In the process, 4 people were killed, several were declared missing, several women were raped and properties were damaged. The village was deserted as military and anti-riot policemen mounted roadblocks and patrolled the town (Ekeinde 1999; Azualalam 1999; Vanguard 10:11: 1999; Onuorah 1999; Obari 1999C).

Beyond these incidents of localized nature in communities and Ogoniland, was a regime of large-scale harassment, provocation and repression that began since 1997. The region was effectively garrisoned, with deployments of the military and mobile police from outside the region who mounted roadblocks and checkpoints. The military and Navy were detailed to guard oil installations. According to the then minister of defence, Gen. T. Y. Danjuma;

> This operation Hakuri was initiated with the mandate of protecting lives and property particularly oil platforms, flow stations, operating rig terminals and pipelines, refineries and power installation in the Niger Delta (Onuorah 1999).

8.9. The emergence of counter violence

The extensive repression of the region rather than deter further protests, created a regime of defiant and militant protests and resistance. The new regime of agitation was led by the youths who instead of waiting to be brutalized and killed, decided to take their destinies in their hands and to pay the supreme price where necessary through undertaking bold, decisive and militant and direct actions against the oil companies and the Nigerian state. The youths embarked on a regime of armed resistance and counter violence. Through pre-emptive actions on oil installations, armed protection of seized facilities and counter attacks of military and police provocations, the youths challenged state violence and provoked further militarization.

The regime of counter violence was quite disparate but widespread. It began with confrontations with the security agencies rather than running away from them. In Obagi, Rivers State, for example, youths confronted the policemen on 13th February 1994 and in anticipation of large-scale state violence, hurriedly evacuated the town (Tempo 7:4: 97: 7). By 1997 and 1998, co-ordinated youth confrontations began to take place. But this was preceded by unprecedented mobilization of the youths characterized by the proliferation of communal youth groups and the emergence of ethnic and regional youth associations and youth activists. Among the Ijaws, the high level of youth mobilization was also indicated by youth processions and protests.

The high point of the new regime was the emergence of community and ethnic youth militias as armed wings of youth groups and communities. The militias are fully armed with modern firearms and had some training. By the late 1990s, militia groups among the Ijaw ethnic group included the Niger Delta Volunteer Force (NDVF), Egbesu Boys of Africa, the Federated Niger Delta Ijaw Communities (FNDIC). The NDVF is regarded as the unofficial armed wing of the youth activists and associations of Ijaw youths which were led by the Ijaw Youth Council (IYC). These militant groups seized, shut down, occupied and barricaded flow stations and other oil installations. The FNDIC for example, seized several oil installations in the Ijaw areas of Delta state between 1997 and 1998, to compell government resolutions of the Warri arises over the location of a local government headquarters. By October 1998, many of the flow stations seized in the crisis were still occupied. The militias engaged the Nigerian security forces in numerous incidents on the roads, rivers and creeks, communities, oil installations and sometimes in entire areas such as the Warri region (Okafor 1998; Amaize1998;Obari 1999C; The News 18:01: 1999; Ezomon & Okafor 1999; Suleiman 1999; Ikwunze & Ogoigbe1999).

The emergence of an armed and militant youth formation as an active component of the popular struggle in the region and the activities of the youths particularly in terms of popular protests provoked violent encounters with the state. The first major incidence was in Bayelsa state between December 1998 and early 1999. This followed a military crackdown on Ijaw youths in Bayelsa and Delta states. The shoot out in November and December 1999 followed heavy deployments of hundred of troops to Bayelsa and shoot out between the Nigeria Army and resistant Ijaw youths. The Egbesu Boys engaged the military in Yenagoa, Kaiama and other towns in Bayelsa state. This led to the declaration of a state of emergency in the state.

A major incidence of violent confrontation between the state and militias took place in Odi, a town on the bank of River Nun and the second biggest town in Bayelsa state, which had been a center of Ijaw agitation. Odi youths had in the agitation for better deals from oil production and consequent encounters with the state, seized the police station in 1999. In November 1999, a militia group called Kuroweikikime, camped in Odi had arrested the Police Area Commander and his men who were on visit to Odi and killed them. The Kuroweikikime group claimed that the killings were in retaliation for the killings of Ijaw youths in the crisis in Yenagoa. A police reinforcement was assaulted and forcibly turned back by the youth militias. A military action was ordered against Odi and neighbouring towns. The government stated that it cannot condone criminal activities in the guise of rights activism. Though the youths of Odi claimed that they were not responsible for the killings, they decided that they would defend their home land (Ogodo 1999).

The state deployment included truckloads of soldiers armed with heavy artillery, grenade launchers, mortar bombs and aircraft. The federal government claimed that the troops were to restore law and order,dislodge perpetrators of violence and apprehend suspected murderers (The Guardian 30:11:99). Odi youths resisted by ambushing and killing 3 soldiers and firing from houses. After two weeks of military occupation,. Odi with a population of 60,000 people and about 1,500 houses was systematically and completely destroyed and deserted. There were littered corpses, destroyed properties, dead livestock and destroyed crops. What was left un-damaged were a church, a bank, a community center and a few stray domestic animals. A numbers of Odi women, children and youths were detained at Elele barracks and Bori Camp in Port Harcourt (Onweomeodo 1999; The Guardian 23:11:1999; Ogunsakin 2000;Onwubiko 2000).

The state governor accused the federal government of callousness. According to him, "a whole community of about 3.5 kilometers was leveled. A destruction of that magnitude was callous. Defenseless civilians were murdered" (Onwubiko 2000). After the valuation of the property destroyed by the military operations in April 2000, the state government claimed that its reconstruction would cost

N25billion (Onwubiko 2000). The president following huge civil rights movements and general protest at the level of state destruction and murders apologized for the invasion. But the state government had cause later on to accuse the federal government of insensitivity to the plight of Odi indigenes (Onwubiko 2000).

The Warri region was engulfed in communal violence, state violence and militia counter violence between 1997 and 2004. Though the violence was actually between the Ijaws and the Itsekiri ethnic groups, the Ijaws targeted oil companies because of alleged federal government and oil company favoritism towards the Itsekiris. The Ijaw militias seized and occupied most of the oil flow stations at different points between 1997 and 2004. For example, in October 1998 Ijaw youths took over 9 oil flow stations (James and Ajankola 1998) Most of the oil companies had to close down operations and evacuate their staff at various times between 1997 and 2004. The Nigerian Army and Navy were massively deployed to stem the violence, restore law and order and restore oil production

As the agitation became more directly disruptive and violent, actions such as sabotage of oil installations, the seizure of oil facilities, ships, helicopters, equipment, oil tankers become more pervasive. The state reacted by deploying the Navy to protect the waterways, escort oil tankers and merchant ship and to momtor and patrol oil terminals, pipelines and flow stations (Onuorah 2000; Onwuemeodo 2000). The state embarked on more military deployments and repression. In November 1999, the minister of Defence stated that the most contentious security issue in Nigeria was the graved phenomenon of terrorism perpetrated by criminal element who are involved in hostage taking of oil workers, kidnappings, hijacking of helicopters, sea piracy and demands for ransom.

8.10. The consequences of state repression

The state repression and counter violence has led to the militarization of the region. Military deployments have been massive and fairly consistent since the Ogoni crisis in 1993. There have been several military operations such as HAKURI 1, HAKURI 2, and Operation Restore Hope in the region, ostensibly to protect oil facilities, lives and properties. Military deployments have been accompanied by military and militia confrontations and increased insecurity. That the military has not been very successful in its operations is indicated by continued militia activities, communal and intra-communal conflicts and popular violence against oil facilities.

Military deployments have been accompanied by state repression. There have been pervasive human rights violations, such as torture, arbitrary arrests, summary executions, extortion, rape, brutalization and arbitrary detentions. Thousands of citizens have been killed in the process of suppressing popular protests and youth militancy (HRW 1999; IRIN 2000; Onwuemeodo 1999). The Ijaw Protection Organization (IPO) has estimated that between 1997 and 1999, about 2,626 Ijaws were killed and 238 women raped by federal security forces (Onwuemeodo 1999). There have been massive devastations and disruptions. Several communities have been attacked and completely or partially destroyed in the encounters between youth militias and the military in Delta, Bayelsa and Rivers states, entire communities members have been forced to flee and to temporarily migrate and reside in near by communities. Internal displacements have therefore been massive. Between 1997 and 1999, the IPO estimated that more than 39 Ijaw communities were partially or completely destroyed (Onwuemeodo 1999).

There have been economic disruptions and crisis in the region. Repeated incidents of insecurity, violence, crime and conflicts have created a regime of fear and terror which have forced several companies to leave the Warri region. This has had serious consequences for employment and incomes. Fishing and Farming; the traditional occupation of the region have been disrupted and largely undermined by state violence and counter violence. These have heightened poverty and immiserated the regions citizenry.

8.11. Towards a rapprochement: problems and prospects

The magnitude of grievance and agitation of the region has been such that at different periods, her citizens have engaged the Nigerian military and police in very devastative encounters. The depth of grievance also manifested in the large participation of the indigenes of the region in the military in the April 1990 aborted coup against the Babangida administration. The coup was championed largely by the southern and middle belt minorities in the Nigeria military and constituted the first significant and violent protest against Hausa / Fulani hegemony.

The agitation has been volatile and violent. In some cases, it has degenerated into criminal acts such as hostage taking, piracy, blockage of the waterways and seizures of equipment, ships, helicopters and staff (IRIN 2000). The toll of community unrest in the ND an national economic life stream is immense. The activities of state security agencies have been characterized with extra judicial executions, arbitrary detentions and mass abuses of civil rights.

The disturbances and fighting in the Niger Delta has been by far, one of the greatest challenges faced by successive governments since the administration of General Abubakar (1998-1999). The crisis has posed one of the greatest challenges to the democratic credentials and specifically human rights credibility of the civilian administration. The Nigerian military has seen the most consistent, extensive and intensive actions in the region since the period of the Nigerian civil war.

The situation that has warranted these broad protests have not considerably changed. Even the new democratic government is being accused of not having made much difference to the development of the region. Besides the attitude of neglect and insensitivity to the plight of the OPCs subsists (Obari 2000). The government has continued to see the increased restiveness in the region as criminal action that has potential to generate disorder and anarchy (Obari 1999E). Rather than accede to the demands of the region, the government, it is determined to put an end to youth restiveness and militia activities. As counter violence and direct disruptive actions increased, the government became more violent and repressive. The region has become militarized since 1993. Armed Troops replaced the regular police in the confrontation with protesting communities since the early 1990s. Security forces have protected oil installations since the mid 1 990s.

The federal government explains the under-development of the region by their difficult geographical terrain (Johnnie 1998). This ordinarily warrants extensive deployment of resources to upgrade development efforts. OMPADEC (1994:10) roughly estimated that it would cost about N 80 – N120 billion to fully develop the region as at the early 1990s. The magnitude of resources required, and the commitment and energy demanded have not been forthcoming.

We should note that the restiveness is attributable to the problem of development. The agitation has from the beginning been about spatial and social access to amenities like roads, housing, electricity, water, educational facilities, communication, health facilities and infrastructures. The demand only became more intense as the awareness of overbearing resource wealth and contribution amidst scanty state attention increased. Hence, the swan song demand became that of caring for the "goose that lays the golden egg' and "the owners of wealth shall enjoy the wealth they own' (The Guardian 07:06:98)

The state needs to move beyond repression and the maintenance of law and order, to seek meaningful dialogue on the one hand and undertake extensive developmental activities on the other. First, the problem being in part a developmental one has to be resolved by a satisfactory and effective programme of development. Though the federal government had at various times established intervention agencies such as NDDB, OMPADEC and NDDC, their impact have been mere ripples in a large ocean of under-development. Second, the agitation is a political one that needs political resolution through genuine dialogue. The issues of resource control, regional autonomy, un-democratic laws and conference of ethnic nationalities can only be resolved through dialogue.

References

Adamu, A. (2000) "South-South Reps Lament zone's alleged neglect,' Lagos: *The Guardian,* Wednesday March 15

Adigun, T. and Lawal, Y (1999) "President deplores awarding in Niger Delta, Firms propose reforms" Wednesday, November 17

Adeleke, M. (1993) "The Shooting Field," Lagos: *The News Magazine,* May Africa Guardian April 18, 1994

Akpan, A (2000) "Mobil suspends operatios as youths invade terminal", Lagos:*The Guardian,* Thursday, April

Amaize, E (1998) "Ijaw youths still barricade oil flow stations", *Vanguard,* Thursday, October 29

Asiodu, P. (1980) cited in Saro Wiwa (1997) *Genocide in Nigeria: The Ogoni Tragedy,* Port Harcourt: Saros Publications

Azualalam, C (1999) "Odili's Burden", Lagos: Newswatch, November 21 Daily Sunray Newspapers (1993), Port Harcourt, July 29,

Duru, V. (1997) The performance of oil Mineral producing Areas development commission (OMPADEC) in the development of Delta" Benin city: department of political science and Public Administration university of Benin unpublished M.SC.

Efenakpor, R (1997) "Niger Delta Youth protest against oil primes, launch new group", Lagos: *The Guardian,* Monday, August 18

Ehigiator, k (2000) Niger Delta Union wants 100% resource control, insists on sovereign confab, vanguard, February 6

Ekeinde, A (1999) "Oil Field, Killing Field", *Tell,* May 17

Ezomon, E & Okafor, C (1999) Youths allege plans by Government, Shell to deal with Ijaws. The Guardian, Thursday, January 28.

Frynas, J. (2000) Oil in Nigeria: Conflict and Litigation between oil companies and village communities, Hamburg: Lit.

Human Rights Watch (HRW) (1999) "Crackdown in the Niger Delta", June. See also Human Rights Watch (HRW) (1999) "The Price of Oil Corporate Responsibility and Human Right Violation in Nigeria's Oil Communities", February, New York.

Idowu ,A.S. (1993) clash as govt. deploys soldiers to Ogoni The Guardian May 4. Ikwunze, C (1998) "FG unfolds development plan for Niger Delta', Lagos: *Vanguard,* Thursday November 19

Ikwunze, C. and Ogoigbe, E. (1999) "Group blames killings in Ijaw on CGS order", *Vanguard,* Thursday, January28

Iwori, J. (2002) "Ijaw youth lay siege to Shell flow station", Lagos: *Thisday,* May 7

Jackson, R. (2002) "Violent Internal Conflict and the African state: Towards A framework of Analysis", *Journal of Contemporary African Studies* 20(2)

James, B. and Ayankola M (1998) "Ijaws takes over shell stations", Wednesday October 7

Johnnie,P.A.E (1998) The Role of OMPAJEC in the Development of Oil Producing Communities : A Case Study of Burutu Local Government Area of Delta State, Benin City: Department of Political Science and Public Administration, University of Benin.

Mkandawire, T. (2000) "The Terrible Toll of Post Colonial Rebel Movement in Africa: Toward an Explanation of the Violence against the Peasantry", *Journal of Modern African Studies* 40(2)

Naanen, B (1995) Oil Producing Minorities and the Restricting of Nigerian Federalism: The case of Ogoni People. *Journal of Commonwealth and Comparative Politics* 33(1)

Niger Delta Environmental Survey (NDES) (1997) Niger Delta Environmental Survey Phase Report Volumes I-IV Lagos.

Niboro, I. (1998) "War by all means", Lagos: *Tell,* November 9

Obari, J.O (1998) "Abubakar urges restraint by oil communities",. Lagos: *The Guardian,* Thursday, September 3

Obari, J.O (1999A) "Sexual assault in Choba threatening UNIPORT'S re-opening", *The Guardian,* Tuesday, November 9

Obari, J.O (1 999 B) "Why Niger Delta Crisis Festers Ijaw Leaders", *The Guardian,* January 28

Obari, J.O. (1999C) "Bayelsa lifts curfew Ijaw youths call truce", *The Guardian,* Tuesday, January 6

Obari, J.O (1999D) "Federal government deploys troops in Bayelsa", *The Guardian,* Monday, November 22

Obari , J.O. (2000) "British minister pledges scholarship for Niger Delta indigenes", *The Guardian,* Wednesday 12, January

Obari , J.O.& Obinor, F. (1999). Fed Govt. deploys troops to Bayelsa. The Guardian November 22.

Obayuwana, O. (1999) "No plan to militarize Bayelsa, says Abubaka", Lagos:*The Guardian,* Wednesday, November 29

Obi, E (1990) "Mobile cops drafted to boiling oil community", *Lagos: Vanguard,* Friday November 9

Obi , C (1993) Oil & the National Question NES Conference.

Ofouku, M. (1999) "The Bomb is Ticking", Lagos: *Newswatch,* May 17

Ogbodo, A (1992) "Up in arms", Lagos: *The African Guardian,* August 17 P 34-36 Ogoigbe, E (1998) "Isoko Youths Seize Oil Flow Station" *Vanguard,* Friday, December 18

Onuorah, M. (1994) "Shell Shocked", *The African Guardian July 25*

Onuorah, M. (2000) "Gwadebe 23 other retired with benefits", *The Guardian,* Tuesday, January 25

Ogunsakin, F (2000) "Odi sires government, demands #1 billion damages", *The Guardian,* Tuesday, February 29

Okafor, C. (1998) "Youths Shut Five Shell Flow station", *The Guardian,* Friday December 18

Oil Mineral Producing Area Development Commmmission (OMPADEC) (1 994)Quarterly Report Port-Harcourt Vol.1, No.2 March.

Onuorah, M. (1999) "Trial of ex officers won't affect military's morale say Danjuma', *The Guardian,* Friday, November 25

Onwuemeodo, S. (1999) "Bayelsa Women Urge F G to rebuild Odi", *Vanguard,* Wednesday, December 1

Onwuemeodo, S. (2000A) Navy arrests eight hijackers of Shell boats. Vanguard, Monday 21 February.
Onwuemeodo, S.(2000B) Ikwerre Youths shut down Elf Flow Station. Vanguard, Wednesday 17 May.
Onwubiko, E (2000) "Bayelsa to spend #25 billion on Odi", *The Guardian,* Monday, April 10
Saro Wiwa, K. B (1992) *Genocide in Nigeria: The Ogoni Tragedy.* Port Harcourt: Saros Publications.
Suleiman , T.(1999) "Drumbeats of War The News, January 18.
Tempo Magazine July7, 1997
The African Fund (1999) Nigeria, Blood and Oil. A special Report by Michael P. Freshmen Rights Co-ordinator *September 17*
The Guardian (1999) "Government explains troops deployment in Bayelsa", Tuesday, November 23
The Guardian (1999) "Government sidelines bill, initiates projects in Niger Delta', Wednesday, November 24
The Guardian (1999) "Abubakar pledges federal aid for Niger Delta", Tuesday November 30
The News January 1 (1999)
Ugwunebo, C. (2000) "Obasanjo pledges to reverse ecological devastation of Niger Delta", *Vanguard,* Thursday, February 29
Vanguard (1999) 10 November
Yates, A.D. (1996) *The Rentier State in Africa: Oil Rent Dependency and Neo-Colonialism in the Republic of Gabon.* Trenton & Asmara: Africa World Press Inc.

9. Oil, youths and urban governance: the case of warri

Anthony I. OGBEMI-IFEDIORA

9.1. Introduction

Warri is an urban center in Nigeria with a relatively short biography. It derives its rise and growth to the incidence of European "legitimate' colonial commercial enterprise. In this regard it has its semblance with a few other Nigerian urban centers notably Aba. In this beginning mention could be made of several agencies of growth but today that planned and organized direction of the town's growth that helped it to expand is absent. Warri has grown to become the second most important centre of the oil industry in Nigeria. It is also the largest and fastest growing urban center in Delta state. It is a sprawling conurbation that has grown to swallow up a number of neighbouring communities. Chief amongst these are Effurun, Ugbuwangue, Edjeba, Enerhen, Ekpan, Ugborikoko, Ugboroke and Ekurede. Even Udu, Orhuwhorun, Delta Steel Company, Ovwian and Aladja to the south of the Warri river are easily accessible by the Udu bridge making them part of this same expanding conurbation. The component sections of the city in many ways share common experiences. The transformations in the physical and social environment being considered here are universal and belong to three Local Government Councils.

As a base to the major stakeholders of the petroleum industry, a very significant percentage of the urban workforce is directly or indirectly employed in that industry. Equally true also is that a large number of unemployed youths belonging to the urban communities and migrating from the surrounding towns and villages have become also a very important segment of the population. In virtually every part of the oil producing states of Nigeria, youths have become a major source of socio-economic pressure that cannot be ignored. For government and oil companies alike, the formation of policies and the establishment of various institutions for handling the issues emanating from youth activities remain a major concern. In the same way the governing authorities of Warri in particular and Delta state in general have tasted of the destabilizing of youth gangs. In this paper we shall examine the nature and character of the Warri urban environment, the impact of the oil economy on Warri and how youths and youth activities affect urban governance in the city.

9.2. The warri urban environment: a historical background

In 1901, the government of the protectorate of Southern Nigeria obtained a parcel of land on which they built barracks, offices and residential quarters. This area was known as 'New Warri' and "this government act of acquisition stamped the name Warri on the township' (Ogbemi-Ifediora, 1997:76). This tiny center of administration began to attract migrants who came to work as clerks in the trading firms, as government workers, policemen in the Consular Force and others. From this beginning, the authority took the initiative and began the layout of the new town in order to make provision for roads and housing. Between 1906 and 1911, the government acquired over 300 hectares of land on which to develop a properly planned township (Ogbemi-Ifediora, 1997:77). The objective of the authorities in these early decades was to prevent the haphazard development of the city. The government set up clear policies of urban development and ensured that those specifications were strictly adhered to. Structures that do not fit into the policy objectives of government were promptly demolished.

It is evident that portions of the land acquisitions were unusable except at great costs of reclamation because of the swampy terrain. Such areas fell into an east to west stretch of swampland that separated the "New Warri' township from the two closest indigenous communities of Okere and Agbassa. This was necessary in keeping with colonial policy of separating indigenous disease prone enclaves from areas occupied by Europeans (Ofonagoro, 1979:235). In the case of "New Warri', the population was predominantly white. It was well planned with well laid out streets and plots of land meant exclusively for the colonial middle class. These were those who could afford the new concept of orderly urban life as introduced by the colonial administration. It eliminated the traditional family household organization in favour of a complex personal arrangement of individuals (Handling, 1973:22). An entirely new pattern for disposing and utilizing space was introduced. Outside the boundaries of the township, the indigenous enclaves saw little change and the traditional uses to which space was put remained mostly unaffected. Even after independence, the pattern of development continued unchanged. The indigenous enclaves lagged behind the inner city.

Coming to the society, four distinct groups were identified in the early period. The first were the Europeans who lived almost entirely in New Warri. The second were the Creoles (Saro) from Sierra Leone and Kroomen from the Gold Coast. They lived in New Warri on quarters built by government for African staff. In the local parlance, these were classified as native foreigners. Nigerians who were not indigenes of the Warri area formed the third class in the society constituting also a second category of "native foreigners' (Ogbemi-Ifediora, 1997). The Yoruba,

Ibo, Igala, Nupe and others belonged to this category. The fourth class was the Itsekiri, Urhobo and Ijo indigenes of the town. They come from the villages around and some of their kith and kin have always lived around Warri. Ever since Warri became an important center of the oil industry, a large number of these have migrated into the town. The obvious consequence of this is that the large numbers of jobless youths in the ranks of such migrants have adopted untoward approaches to urban affairs in order to "take care of their needs'.

The exercise of urban governance for Warri and the provision of social services for her citizens was not an easy task up till 1960. Housing was constantly in short supply and so from early times, slum living, a characteristic legacy of colonial rule in Africa began to develop in Odion, Okere and Agbassa. In them could be found most of the social evils characteristic of the poorer areas of western (European) cities during the industrial period (Southhall, 1971: 248). To date such problems have proved intractable for successive governments.

Since 1912, the European colonialist in West Africa had operated on the premise that no land is without an owner, a practice that encouraged land claims to be tested in court instead of settling disputes before the elders council. In this process in Nigeria and elsewhere, some fairly comprehensive bodies of customary albeit untraditional laws that govern the allocation and use of land evolved (Colson, 1971, Meek, 1949). Where no individual claimed proprietary ownership over a given parcel of land, then the right to it must be vested in the political unit whose members lived in and used the region. In this case, the Itsekiri fitted in and the presence of Chief Dore Numa in the absence of an Olu saved the authority the trouble of applying the draconian step of alienating such lands as they did elsewhere in Africa. (Colson, 1971:196). Clashes and conflicts in urban life were not absent in the history of Warri. Between the three principal ethnic groups namely Ijo, Urhobo and Itsekiri the conflicts have almost always had to do with land. The city is located in a low-lying country where elevated and well drained lands are difficult to come by. In the original geographical setting the area under marshlands is larger than high ground. Reclamation has resulted in urban encroachment upon the swamplands so that the original terrain cannot be easily recognized. Nevertheless, we must mention that the land related conflicts of the past were not a youth affair. By comparison the urban conflicts of today have seen the massive in-road of youths of all three major ethnic groups.

9.3. Conceptual analysis

In a developing economy like Nigeria, urban governance is of paramount importance particularly because several projections indicate that they will become home for increasing numbers in the overall population (The Guardian, 1999). Secondly though only a third of the population of developing economies lives in urban centers, the cities nevertheless account for 60% of the Gross National Product (GNP) (Makinde, 1996). It is only by an effective urban management that the identified problems of their urban societies can be addressed.

Ever before petroleum oil became a significant export in Nigeria, the Warri area of the Niger Delta has distinguished itself as an export outlet for palm oil and kernel. Palm oil and its attendant commercial network was the lure that brought increased intercourse between the ethnic groups in the area. Since the 1970s, the word oil has been used chiefly to refer to petroleum, the mineral export that sustains Nigeria. The importance of petroleum to the Nigerian economy cannot be overemphasized. Along with it, has come so much unrest by oil producing communities clamouring for development projects to compensate for the environmental degradation that exploration and exploitation has wrought on their countryside (Omotor, Ogisi & Pessu, 2000). The sensitivity to the exaggerated views that government is rich with "oil Naira' has led many to adopt unwholesome ways and means to get their share of the much-trumpeted national cake. Our reference to oil here is to all the socio economic conflicts arising in Warri due to youth activities as they seek by illegal means to earn a living (get their share of the petro-naria).

Warri is the western headquarter of Shell Petroleum Development Company of Nigeria Limited (S.P.D.C.), the leading oil company of Nigeria. Other companies are Chevron, Elf, Texaco Overseas Petroleum Company of Nigeria, (TOPCON) etc. There are also seismic survey and other petro-allied multi-national companies like Dowell-Schlumberger, Praeklar Seismos, United Geo-physical etc. Oil in this paper refers to all activities in urban life emanating from the managing of public funds derived majorly from oil and the activities of oil companies who are the source of these resources. In this wise oil is the 'connexion' in this relationship between urban managers and the negative interference of youth activism in urban management.

The paper considers the urban environment as social and physical. The social environment at a first level is the urban populace-its character; but then at a second level merges into the physical environment such as the social services manifested in physical structures like roads, educational facilities and health care delivery provisions. The physical environment refers to the natural topography, physical infrastructures and general physical layout of the city namely roads, housing, recreation, and markets. The management and care of both environments are strictly the responsibility of all three tiers of government. The youth in the context of the paper is not defined in terms of age, freshness and vigour, psychological state of mind, marital status or any other commonly adopted frame of reference (Borjor, 2002: 164). Rather the youth is taken to be those between the ages of 18 and 35 years who are involved in all such activities (individually or as associations and organizations), which impinge negatively on urban governance or development. Population falls into the category of youth. Even in the processes, the youths are important either as officials or as thugs for political violence and rigging (Samuel, 2002:9). A large percentage of Warri's population falls into the category of youth. Youth gangs whether addressed as registered bodies or by the epithet "area boys' driven by the materialism of the Nigerian society engage in a number of illegal activities. All who do not fall into this age bracket but are yet engaged in such activities as are characteristic of these youth groups would pass for youth.

Various categories of youth bodies operate in Warri. Some of them are recognized as affiliates of the National Youth Council of Nigeria. The Uvwie Youth Council and the Itsekiri Youth Council are good examples. There are many others, which are illegal and unregistered. Yet, they are active enough and constitute a source of concern to the authorities. Whether they bear a name or not as long as their representatives go about engaging in activities that run counter to the policies of the Local Government Councils (L.G.C.s) they cannot be ignored. The bulk of the problems that the Uvwie Council is currently facing is caused by the youths (Ogheneovo, 2003).

Basically, youth activism is observable in the political, ethnic, occupational, religious, security and other aspects of life of the city. There are the political youth bodies that can be easily identified as the youth wings of the various political parties and other allied interest groups. There are also youth ethnic associations that have assumed more significance in the atmosphere of the volatile ethnic politics that has engulfed Warri in the last decade. Occupational or trade associations like the local wings of the Nigeria Union of the Road Transport Workers have largely been hijacked by youths. Even the Union of Water Transport Operators, and the unrecognized Okada (motor-bike) Riders Association are largely comprised of youths. The latter was born out of the struggle to resist the imposition by the State government of the planned compulsory wearing of helmets and the payment of a daily levy. Though it is easily the largest, it is a very unwieldy and truculent union. Bloody clashes within its ranks have not been uncommon.

The religious youth groups are perhaps the most numerous, being as many as there are churches, christian fellowships and other religious bodies. They are not directly involved in activities that negatively impinge on the management of the L.G.C.s. Nevertheless, they are coveted by urban managers for their moderating influence in helping to curtail the incidence of violence amongst youths. It is believed that a youth converted is one out of the street gangs and on the way to rehabilitation.

Youth associations working for security are often neighbourhood based. They cut across all the other categories and could even take on an ethnic colouring.

The various neighbourhood civil defence squads are recruited from the ranks of the youths. In the Warri South Council area, they have proved indispensable in warding off the attacks of marauding Ijaw youths. In Uvwie, they have equally been useful in keeping Ijaw/Itsekiri clashes out of their domain.

Youth groups of whatever shade, type, class or category, have one thing in common, namely, their members are actively involved in various bodies at the same time. While some of the associations are recognized officially, being duly registered, others bear no name and are not registered. The latter are only identified by their notorious leaders or by their brazen defiance of official policies.

The effects of youth activism is clearly visible in the areas of urban governance; the city's economic life; and in the area of security of life and property.

9.4. The oil economy and its effect on warri

As the largest urban centre in Delta state and the second important center of the oil industry, the effect of the petroleum industry on Warri may not be so evident to a first time visitor because of the years of neglect. This is particularly because of the poor development of infrastructure. However the petroleum industry has had a remarkable effect in many ways (even if in a negative sense). It is the lure for most of the youths who migrate into the city in search of greener pastures (Ogisi, Omotor & Dada, 1999). The resultant effect of population influx in the areas of crime, housing, health care delivery, urban aesthetics, political affairs, educational services, recreation and welfare, ethnic relations and businesses cannot be recounted in great details here.

Some of the highest paying jobs in Nigeria are known to be in the oil industry, and it is the hope of many a youth that a window of opportunity be opened for them in that sector. Shell Petroleum Development Company (SPDC) and Chevron, the leading operators in that sector together with the Nigeria National Petroleum Corporation (N.N.P.C.) and other multi-national and indigenous petroleum outfits are present in the town. These are undoubtedly the highest wage payers within the economy. A large number of oil industry workers reside in Warri with consequences for commerce, cost of living, social expectations and crime. The commercial life is a busy one because it serves not only the residents of the city but also the southern part of Delta state. The cost of living particularly in the housing sector is very high. That sector left majorly to private enterprise has lagged behind demand.

Apart from people directly employed in the oil industry, there are thousands of others who are employed in oil servicing and other petro-allied businesses. The oil economy has led to a high influx of many Nigerians from outside the state. For instance those Nigerians of the Yoruba and Ibo ethnic group constitute a significant percentage of migrants to Warri. They control and dominate the upper ranks of the staff of the oil companies. This has often been a sore spot in the relations between the oil companies and the various youth bodies representing host communities. Their presence in commerce and the light industries is also remarkable. The net effect of this high level of migration is a heterogeneous multi-ethnic society which but for the local ethnic feuds would have presented a considerable advantage.

The local ethnic feuds are also the result not only of historical antecedents but also of the arrival of the petroleum industry. The peculiar character of violence that has engulfed the town in the last decade or more has emerged from the coveted position Warri occupies in the dreams and aspirations of the Itsekiri, Urhobo and Ijaw alike. The question of the ownership of the town and the attendant violence it engenders cannot be recounted here. It is however pertinent to note that it has led to an unprecedented influx of arms into the city to prosecute ethnic wars. This same trend obtains all over the Niger Delta. In the absence of ethnic wars or political violence the youths have found such weapons useful for armed robbery and other related offences.

A very noteworthy aspect of the effect of petroleum on the economy of Warri is in the housing sector. In the absence of government provision of housing private entrepreneurs rose to the demand by putting up several housing, estates since the 1970s (Darah, 2001). Such housing estates have greatly increased in numbers in the last decade. Some of the oil companies themselves have put up private housing estates to accommodate their staff, particularly the expatriates. Closely tied to this is the provision of house ownership schemes by the major oil companies that have helped many workers in the oil industry to own their own houses. The result is the fast spread of the town into new lands outside traditional urban limits. Such new neighbourhoods have further constituted the city¶s mult-ethnic composition.

An aspect of the urban economy that is fast declining in the present circumstance is the hotel and catering sector. From the 1 970s Warri has had a vibrant nightlife concentrated on the Warri/Sapele road. Hoteliers and caterers contributed immensely in the provision of needed services and entertainment. Since the late 80s and 90s, this sector of the economy has been slowly relocating to the outskirts of town and spreading into areas within easy access of the housing estates of the oil companies. Thus, the Refinery road now boasts of some of the finest hotels in the town. Nevertheless, the high rate of crime and violence (traceable to youths with the desire to get rich quick) has killed nightlife. Necessity has advised her citizens to stay indoors at night.

The markets are another important factor in the urban economy. There are five main markets and many other small ones. They are a principal source of revenue for the local councils and for the traders. Trading is a very important aspect of private entrepreneurship in Warri, being a sector open also to any wage earner who desires to make alternative investments. Youth gangs have often secretly engaged the councils for the control of the Igbudu (Warri) and Effurun markets since the civilian administration.

9.5. Management problems of the warri urban environment

Warri presents a peculiar physical environment that puts successive regimes of urban managers to great tasks. The topography is generally low lying, thus elevated land for property development is limited. Consequently, land and land related matters are a very sensitive issue for the three contending ethnic groups laying claim to the town. In addition to this, land reclamation is often necessary. The net effect of this over the decades has been narrow streets and small plots for property development. This has proved a constraint on subsequent schemes of urban renewal.

Added to the above is the fast spread of the town to several sub-urban communities. This has brought into the city, larger numbers of people who are now very sensitive to the issue of the prolonged neglect of the Niger-Delta. Within the town itself, the high density and resultant overcrowding of many neighborhoods has stretched the available social infrastructure to the limit. This is exemplified in the management and disposal of waste. Traditional methods of disposing waste by incineration can no longer handle the many tonnes of refuse generated daily by the large urban population. The management of markets and other shopping districts and control of property development are other problem areas.

One area that demonstrates the management problems emanating from scarcity of land is the matter of encroachment upon and damage to government land. But the government has been largely silent. An example is the large parcel of land situated on Airport Road. It is separated from Shell Petroleum Company Housing Estate Edjeba by a large pond and was originally designated for a modern specialist hospital project in the old Midwest (later Bendel) State. Backed by youth gangs operating in the apparent confidence of their invisibility, tipper drivers and their diggers converted it to a borrow pit. Today, large sections of it will require sand filling before the government can put it to use. The numerous threats and pronouncements of government did not deter the mindless youths. To arrest this trend, structures for the School of Nursing and Health Technology and the Federal Ministry of Health Vaccine Storage Centre for the Niger Delta were erected there. It is apparent that the grandiose scheme of an Ultra-Modern Specialist Hospital for that site is no longer realizable, or has been abandoned.

Another area of difficulty in urban management has to do with urban spatial disposal. Whatever space is available quickly comes under competing demands for use because of the scarcity of land. Commerce is the chief contender and it has given Warri an appearance of one large market. Cargo shipping containers converted to shops line every major street doing visible damage to the city's aesthetics. Along with it is the refuse generated by these unplanned make shift-shopping areas. Succeeding local government authorities and the state government have passed legislations to outlaw this illegality to no avail. The most recent is the declaration by the James Ibori administration within a few months of coming to office to rid the town of ugly container shops and make shift sheds scattered all over the city. The government gave a period of grace and everyone remained expectant that the end of the ugly trend was in sight. Alas it was never to be for the government relaxed and forgot about what it pronounced. The result is that in the last three years the problem has aggravated. The reason is well known. The sites of the container shops along urban roads are allocated to the illegal squatters by various youth associations who collect revenue from them. Unable to halt the illegality the L.GC.s have themselves started to collect shop levies from the operators of container and other make shift shops.

In the area of road construction, many road projects were commenced by contractors and abandoned or not built to specification because of the activities of youths. One example is Fani-Kayode road linking Airport road to Okumagba Avenue that Monier Construction Company had to reduce in width in order to allow youths of Okere-Urhobo rent the space between it and the NPA Housing Estate fence on one side to container shop owners. The youths collect the revenue.

Experience has shown that road construction in low lying and marshy terrains as is obtained in Warri is more cost intensive than it is in upland areas in Nigeria. Land reclamation and the transportation from afar of needed raw materials usually raise constructions bills. Be that as it may, the summation of all the preceding is that when we add the long period of government neglect as is common to all of the Niger Delta, the picture of the urban terrain of Warri is a dismal one (Omotor, Ogisi & Pessu, 2000). The situation is made worse when in spite of this background of neglect the few projects embarked upon are haphazardly executed for reasons of youth activism.

In no other area is urban governance as problematic as in the area of revenue collection. Over the decades, poor and overcrowded neighbourhoods have continued to spring up to cater for the residential needs of the teeming urban poor. This trend has assumed a greater dimension since the mid "80s. Such neighbourhoods in Iyara, Effurun, Okere, Ugborikoko and Orhuwhorhun that are often not traversed by streets are usually out of the reach of the revenue collection staff of L.G.C.s. By the year 2000, about 21.48% of Warri's residential area was taken up by high density neighbourhoods that are prone to social ills like crime, violence, delinquency and deviant behaviour. Omuta and Ikhuoria (2000) computed that 17.6% of Warri are blighted, i.e. prone to social Move ills up. Large sections of Effurun and Udu/Aladja are also blighted. In such areas, youth activism has assumed more negative proportions. The experience of Uvwie Council where revenue collectors for fear of molestation have abdicated their duties in such neighborhoods to youth gangs operating clandestinely under the umbrella of Uvwie Youth Council readily comes to mind here. (Ogheneovo, 2003)

In matters of urban waste disposal, all three local government councils have adopted the current practice of contracting blocks of the town to various refuse disposal firms. It was hoped that this would check the prevalence of huge refuse dumps that sometimes block streets and natural drainage. But it has not worked. Councils are required by the ethnic youth associations to contract such assignments to indigenes (often their members) who in most cases do not have the resources to carry out the task. No sooner have they started than such services are abandoned and the ugly street dumps reappear.

In all the major markets in the Warri area, street trading has been a major dimension. This is one big problem for urban managers, for while it is an illegality, it is a principal source of revenue for youths who control it. At the Effurun and Igbudu markets, roadside hawkers pay daily levies of N20.00 – N40.00 to agents of youth gangs in order to be permitted to display their wares and foodstuff for sale. Their unsightly crowd congests the market. This is against the policy of the Councils who want to decongest and keep the markets open and neat. The health authorities on their own consider street trading in the markets as being chiefly responsible for their filthy nature.

Perhaps the greatest constraint facing urban governance today in Warri is the absence of any state government policy of checking rural to urban drift. Several studies have shown that rural to urban drift is very high in developing economies of the tropical world. Ever since Warri became a major centre of the oil industry, the search for income earning and investment opportunities; and the hope of socio-economic changes have been the lure attracting most youths from the rural areas to it. Youths constitute 66.6% of migrants to Warri (Ogisi, Omotor & Dada, 1999).

9.6. Youths, urban economy and urban governance

The most virulent area of youth activism in Warri is the challenge that it poses to the L.G.Cs. Because of the background of the history of ethnic relations and of the petroleum industry, youth restiveness has become more pronounced. Many have pointed to the get-rich-quick attitude and the break down of value systems. Others have blamed the political elites, which have employed and armed the youths for political ends. The utilization of the youths for violence and intimidation has provided a virulent training ground. The unemployed youths who have tasted power and easy money are not in a hurry to relinquish it.

Local government authorities all over Nigeria have limited sources of revenue and are constantly in a drive to broaden their revenue base. Some of the popular sources of revenue are rents and levies from market stalls, shop owners, offices, business premises and motor vehicles. Others are tenement rates, service charges for waste disposal; fines for violation of statutory laws like no parking zones; or failure to comply with official regulations. In a bid to generate more revenue, the councils have employed many youths in response to the pressure from various groups. Their frequent use of force in the drive to increase the revenue collected has given the councils a bad image before the public. In addition, several youths capitalize on this to harass and extort money from unsuspecting motorists and other innocent members of the public while posing as council officials. To arrest this development the Uvwie Council (and indeed Warri South Council at one time before it) outlawed all such levies.

Other areas in which youths negatively affect urban governance have been discussed in preceding sections of this paper, still it is needful to recapitulate these here.

Youths constitute the nations mobile labour force and are strong mobilisers in politics whether in urban or rural Nigeria. Without paid employment, deviant youths easily become pawns in the hands of ethnic bigots and politicians. They are useful for the prosecution of the ethnic wars and violent politics that have rocked Warri and environs in the last decade. Political violence provides them an opportunity to obtain quick money. The example of the violence that rocked Effurun in 1999/2000 and 2002 over the local government chairmanship position is a recent one. The escalation in the wars between Ijaw and Itsekiri and between the Urhobo and Itsekiri is in part traceable to youth unemployment (Ogbemi-Ifediora, 2002). This situation has grown worse since 1999.

In making demands on oil companies through threats for financial payments or for the employment of some of their members, the ethnic youth movements have become a worrisome threat to the oil companies. Initially, the motives were genuine and peaceful but recent developments have shown that these unskilled youths are more interested in monetary gratification than acquiring skills and taking up employments. This is evident in their attitude to programmes that were put in place to help them acquire skills. The American aided Nigeria Industrial Opportunities Centre located in Effurun has had very few intakes from the locality. Meanwhile people from other parts of Nigeria and Delta State are availing themselves of the opportunity it provides to acquire some form of formal education and skills that would enable them become self employed. Even the National

Poverty Alleviation Programme training workshops organized for the Uvwie has had limited successes. Exactly the same reasons can be suggested for this trend, the attitude of get-rich-quick and the apparent lack of focus amongst the youths.

There has been a high incidence of crime in Warri and of kidnappings particularly of the expartriate staff of oil companies. Between December 1998 and November 2000 there were ten reported cases of kidnap in Warri involving foreigners with ransom demand ranging from N2m. – N50m. (S.P.D.C, 2000). Although the police has consistently tried to work together with the oil industry to find solutions to the demands of youths, the exercise has only been partially successful (Adetuyi, 2000). The cooperative effort between the Police, the Armed Forces (chiefly the Nigerian Navy) and the leaders of the oil industry notably in the sharing of intelligence information has helped to reduce the incidence of youth unrest in and around Warri (S.P.D.C., 2000). Closely allied to the incidence of crime is the militarization that ethnic clashes introduced. Armed escorts and guards became necessary initially for foreigners and now even for the rich and community leaders. Assassinations and other violent crimes have become common as a result of a culture of violence that is traceable to youths.

The flight of businesses and investments from Warri is also linked to youth activism. Since 1997 when the ethnic crisis and youth unrest in the Warri area assumed a new dimension, many of the oil servicing companies have relocated their operational bases outside Warri, leaving behind skeletal staff. The list includes major outfits like LAMNALCO, United Geo-Physical Surveys, Tidex and DOWELL-SCHLUMBERGER. The many running battles with youth gangs coming to make demands for money payments or the inclusion of some of their members in the company's payroll sometimes resulted in bloody clashes between the youths and the armed guards employed by the companies. To avoid this trend it became necessary to relocate. There is a second side to this flight, which has to do with whole families and private businesses relocating in Sapele, Ughelli, Agbarho and other nearby towns. Youth unrest and the unending demands from youth gangs in some neighbourhoods support such relocation. It is pertinent to point out that relocation within Warri itself is also resulting in the growth of new residential areas outside the zones of conflict.

9.7. Accounting for the youth problematic

For a very long time youths have played significant roles in the history of Nigeria. From the time, of the nationalist struggle through the Nigerian Youth Movement, the West African Students Union and the Zikist Movement, youths have positively contributed to the building of the Nigerian nation. Such positive roles of youths are not lacking in the period after independence. Nevertheless the recent past has shown youths as destructive elements. It has been argued by some that the youths have turned out to be pawns in the hands of the privileged elite or ethnic bigots capitalizing on the vulnerability of youths and that vulnerability is traceable among other reasons to high levels of unemployment, lack of education, general idleness and ethnic intolerance (Borjor, 2002; Ogbemi-Ifediora, 2002).

In the face of strangulating socio-economic conditions the youths from the ethnic communities of Warri are suddenly a landless and nearly homeless class. The older generation have sold all the land to non–indigenes of the locality (the rich ones) who by reason of employment in the oil industry can afford the high cost of land and property development. As it were, without education the only way to get back at a system that leaves them so disadvantaged is anchored in illegality. Closely tied to the above is the high cost of living (market prices) in Warri. This is underlined by the huge emoluments of workers of the oil industry. How could an unemployed landless youth survive in this environment?

It has also been argued that youths are the principal agents for political violence, rigging and killings that take place in Nigeria (Etumudor and Erhuvwujiefe, 2002). The implication of this is that the governments of the day will have ridden to power on the ticket of youths. It is very unlikely that such a government would have any inclination to deal decisively with difficult youths whose backing is essential for re-election. There is consequently a freedom to operate undeterred.

9.8. The way forward for the youths in urban governance

The efforts of government in the provision of mass education, job creation, poverty alleviation and the empowerment of youths should not end up as mere slogans as is currently the case all over Nigeria. No wonder some are requesting that more efforts be put into building the people before building the nation (The News, 2001). Skill acquisition and public enlightenment should be taken more seriously. It is not just enough for government to proceed to use law enforcement agencies to combat opposition from youths. The use of youth employment promotion programmes and strategies is recommended. Among these are job creations in the public sector; job creation in the private sector, youth training and apprenticeship programmes and self-employment promotion programmes (Okumagba 1999). The government can encourage more Non-Governmental Organizations to do same. The S.P.D.C. and Chevron self-employment promotion programmes are worthy of mention.

As if recognizing the urgency of the task of making Warri governable, the Delta state government set up a think tank committee on Development led by Gemaliel Onosode, a seasoned industrialist. Working with a view to the future, the report of the committee needs close monitoring, supervision and continuity for it to be beneficial. For instance amongst other things, it recommended the modernization and dualisation of several roads in Warri to give easier access to residential areas (Darah, 2001). In pursuing this, the present defacing of the city through container shops and illegal makeshift structures should be reversed. This should be accompanied by back up legislations and close monitoring to prevent a re-occurrence.

In the mounting conflicts and youth activities the oil companies have continued to be targets. The oil companies have taken community relations as an important fall out of the business of oil exploration and exploitation. The many running battles between S.P.D.C. and Edjeba/Ogunu youths; and between Ekpan youths and The Warri Refinery and Petrochemicals Company have led to the development of programmes and policies for handling youth conflicts. The L.G.Cs can also borrow a leaf from them.

There is need for the government to acquire and designate housing estates wherein persons interested in erecting properties could do so free of the harassment of youths whose excuse in many cases has been that such lands are communal lands. There is need for greater intelligence network between the police and the security agents to rid the society of deviant youths. Crime detection and prevention should be ahead of crime.

Another problem in the effort to tackle opposition from youths is in the manner of enacting appropriate legislation to shield youths from undue exploitation and use by political aspirants and ethnic bigots/war lords. There should be well-defined penalties. Even the youths also should profit from public enlightenment campaigns to warn them of the dangers of crime. A closer scrutiny of and monitoring of the revenue collection machinery of the Local Government Councils that opens them to the infiltration of youths is recommended. The L.G.Cs also should be encouraged to have clearly defined policies for youths. To allow the youths to continue the way they presently do is tantamount to backing up illegality or acquiescing in the face of it.

Still a new area to look into should be income differences and wealth and its distribution. This is especially as it affects those from outside the geographical locality and those from within. Housing policies for the lower income brackets and urban renewal packages are ways to accomplish this.

9.9. Conclusion

The dislocation of the machinery for urban governance in the Warri' area brought about by youth activities and government's acquiescence or inability to manage it have been discussed. The latter has made it possible for youth gangs lacking any central organization for coordination to usurp constitutional organs of governance and impose upon the city hitherto unheard of practices. The paper has shown some of the ways in which youths by their activities oppose and truncate government efforts to make the city livable. Some solutions have been proffered.

A more purposeful and result oriented approach is needed to arrest and tackle the decadence and obsolescence that is engulfing Warri. This calls for a complete re-thinking and re-orientation at all levels of government in conjunction with other interest groups interested in the development of the city.

References

Adetuyi, Samuel A. (2000) "The Role of Law Enforcement in Managing Community Activism", *Paper presented at Delta (Oil Producers) Security Seminar* held at Shell Petroleum Development Company, Warri, May 2000.

Borjor, Eba Maxwell (2002) "The Nigerian Youth and National Development: An Appraisal;' Paper presented at the 46th*Congress of the Historical Society of Nigeria,* Benue State University, Makurdi.

Colson, Elizabeth (1971) "The Impact of the Colonial Period on the Definition of Land Rights' in Turner, Victor (ed) (1971) *Colonialism in Africa 1870-1960 Vol. 3 Profiles of Change: African Society and Colonial Rule,* London: Cambridge University Press.

Darah, G.G. (2001) "Warri: "The Making of Delta's Oil City' Lagos: *The Guardian* Monday, October 29.

Etumudor, M. & Erhuvwujefe, A.J. (2002) "Political Violence in the Nigerian Polity' Paper presented at *National Association for the Promotion of Studies in Arts and Social Sciences* (NAPSASS) Conference held at Nwafor Orizu College of Education, Nsugbe Anambra State.

Handlin, Oscar (1973) "The Modern City as a field of Historical Study' in Callow, Alexander B.Jr. (ed.) American *Urban History: An Interpretative Reader with Commentaries;* London: Oxford University Press

Makinde, Remi (1996) "Strategies for Achieving Purposeful Human Settlements' Lagos: *The Guardian,* Monday October 28.

Meek, C.K (1949) *Land Law and Customs,* London: Oxford University Press. Ofonagoro, W. Ibekwe (1979) *Trade and Imperialism in Southern Nigeria 1881-1 929,* Lagos: Nok Publishers International

Ogbemi-Ifediora, A. I. (1997) The Growth and Development of Warri as an Urban Centre 1849-1960: A Historical Survey M.A. Thesis University of Benin, August 1997

Ogbemi-Ifediora, A. I. (2002) "Ethnic Intolerance, Human Rights Violation and Terror in a Nigerian City: Warri 1990-2001' *Paper present at 1st Annual National Conference of the South-East Zone of Historical Society of Nigeria* at Imo State University, Owerri, December

Ogisi, O. D., Omotor, D. G. and Dada (1999) "Determinants of Rural to Urban Migration: A Case Study of Warri, Nigeria" in Egbon, Peter C. & Orubu, Christopher O. (eds.)

Critical Issues in Nigeria's Development Abraka: Faculty of Social Sciences, Delta State University, Abraka.

Ogheneovo, E. (2003) Interview with Personal Assistant, Office of the Chairman Uvwie Local Government Council Secretariat, G.R.A. Effurun, May 5th,2003.

Okumagba, Paul O. (1999) "Strategies for Combating Youth Unemployment: The Nigerian Experience' in Egbon, Peter C. & Orubu, Christopher O. (eds) *Critical Issues in Nigeria's Development* Abraka: Faculty of the Social Sciences Delta State University.

Omotor, Douglason G., Ogisi, Dicta & Pessu, Emmanuel (2000) "Youth Restiveness and the Development of the Niger Delta: An Exploratory Analysis of Selected Communities in Delta State Nigeria" in Orubu, Christopher, (ed.) *Nigeria Petroleum and Energy,* Abraka: Delta State University

Omuta, G. E. D. & Ikhuoria, I. A. (2000) "Problems and Policy implications of Urban Residential Land Use Characteristics in Nigeria" in *Benin Journal of Social Sciences,* Vols 8 & 9, Nos. 1 & 2, 2000.

Samuel, Dare (2000) "Can Youths Reduce Election Rigging' in *The Guardian,* July 25.

Southhall, Aidan (1971) "The Impact of Imperialism upon Urban Development in Africa' in Turner, Victor (ed) (1971) *Colonialism in Africa 1870-1960 Vol. 3*

S.P.D.C. – The Shell Petroleum Development Company Limited "Information Sharing and Proactive Intelligence Security Management", paper presented by the Director, State Security Service Asaba, at the Delta (Oil Producers) Security Seminar held at Shell Petroleum Development Company, Warri, May 2000. Classified records of kidnappings and attempted kidnaps are from Public Affairs Department.

The Guardian "Cities are Key to Planetary Health' Monday August 2, 1999. *The News Magazine* October 3 0th,2001

Section III
crisis of social responsibility

10. The crisis in relations: multinational oil companies and host communities in the niger delta

Augustine IKELEGBE

10.1. Introduction

For over two decades, the relations between the multinational oil companies (MNC) and the Host Communities (HC) in the Niger Delta region of Nigeria has been sour. Since the 1990s, there has been a groundswell of conflicts between the oil producing communities and the oil companies. Between 1990 and 1993, several communities such as Umuechem 1990, Ogbia 1992, and Diebu 1992 in Rivers State and Uzere 1992, Igbide 1992 and Irri 1993 in Delta State, protested against Shell Petroleum Development Company. In the same period, the Izon Youth Vanguard of Burutu and Bomadi protested against Shell in 1992 (Rowell 1994). Other oil producing communities (OPC) such as Ozoro, Ojobo, Kokori, Ogharefe, Escravos in Delta State; Finima, Bonny, Choba, and Eleme in Rivers State and Eket in Akwa Ibom also had conflicts with the multinational oil companies (MNCs). In October 1993, about 500 residents of Obaji in Rivers State, closed down the Elf station while in December 1993, about 300 residents of Brass Town demonstrated outside the AGIP Terminal (Segun & Ayankola, 1998). Other towns in which conflicts took place in the early 1 990s are Ogulagha, Obotobo, Sekebolou, Ogongbene, Enekorogha, Omoku, Ovrode, Araya, Ewu Urhobo (Segun & Ayankola, 1998).

Though incidents of community protests date back to the 1970s, intensive civil activism and extensive, prolonged and consistent community protests began in Ogoni. The Ogoni struggle, though not unique, internationalized the conflict because the key mobilizational civil group, the Movement for the Survival of the Ogoni People (MOSOP) linked up with the international environmental and civil rights activist groups. Since then, oil exploitation in Nigeria, has evoked international attention and condemnation of the extensive scale of environmental degradation and despoliation and the consequent multiple crises that the MNCs have inflicted on the Niger Delta region.

The pervasive conflicts have in several cases been characterized with communal protests, production disruptions, violence, sabotage, the intervention of state security agents, bloody encounters between community members and militias and state security agencies and the militarization of the Niger Delta region. The relations have remained sour. It has continued to underpin the regime of protests and violence in the region. It is indicated in the insecurity and tensions and the continued inclemency of the environment of business for the MNCs. The management of the relations such that a conducive and peaceful environment of business could subsist has remained a major challenge to state and MNC officials, stakeholders, scholars and concerned citizens.

The paper interrogates the crisis in the relations between communities and the MNCs in the Niger Delta region. It begins with some theoretical considerations of MNC relations with indigenous people in resource rich regions. Then we provide a background section on the relations. Thereafter the issues and causes of conflicts in the relations are identified. In the remainder of the paper, the methodology of the conflicts, the MNC responses and strategies, the emergence of violent conflicts and the relations of the Nigerian state with the HCs and MNCs are examined. Thereafter, we conclude the paper and make recommendations.

10.2. Theoretical and literature overview

Resource rich environments are sometimes sites of struggle between forces of capitalist accumulation and profits supported by third world countries in search of foreign investments and rents and indigenous people. They are also sites of claims and counter claims, contestation and struggle between those who gain or lose from resource exploitation and resource based degradation.

Particularly, such struggles pervade the interaction of the multinational businesses with the host nations and their indigenous peoples. They often pertain to the economy of costs and benefits and contestations over power, equity and justice as it relates to ownership, access, entitlements, rights and standards (Redclift 1987; Obi 1999)

Globalization accentuates the struggles by liberalizations or deregulation that increase investments and extraction, and political reforms that raise participation and popular groupings. But the contradiction that arises is that while increased extraction heightens environmental degradations and tensions, and worsens the inequities in the social relations of the resource and environmental economy, the logic of political reforms heightens indigenous and popular resistance (Redclift 1987; Woodhouse 1992; Ake 1993; Boele 1995; Crow 1995; Saurin 1996; Robinson 1997; Ibeanu 1997; Tandon 1998; Obi 1998; 2000).

Consequently, the conflicts between multinationals and indigenous people in host states have been accentuated since the post cold war era by increased political activism, nationalism and self determination struggles of indigenous peoples most of whom are neglected, marginalized or subordinated in the economy of resources by the partnership of MNCs and host states.

Equitable relations, sustainable environmental management, respect for human rights, responsive and effective corporate responsibility, local participation and promotion of good governance ordinarily should underline transnational corporate objectives. MNC – host community relations should also be that of mutual collaboration and support. But some literature may be indicating the contrary.

First, a critical examination of the power relations between the MNCs and resource rich regions reveal a heavy tilt in favor of the former. The immense technological and economic resource of the MNCs is reinforced by political power situated in the joint business with a state that is heavily dependent on resource exploitation and rents. The MNCs are so unequally powerful and superior that even host states and particularly African states have been profoundly incapable of effectively regulating and domesticating them (Obi 2001).

Second, the power relations of the MNCs with their host states is said to be sophisticated, ruthless, hegemonical, secrecive, criminalized, unorthodox, exploitative, treacherous, greedy, opportunistic, violent and corrupt (Watts 1999; Turner 1981; Obi 2001). The MNCs are said to be less altruistic and humanitarian, egocentric and self interested, insensitive and poorly responsive to local and regional dimensions of environmental issues (Warpner 1996) and hostile to civil society (Makumbe 1998). In extreme cases, MNCs have sometimes forged partnerships with dictatorial regimes, compromised state officials and institutions, reinforced and sometimes funded state repression and shortchanged states (Ake 1996)

Third, at the level of the indigenous people or local hosts, MNCs are claimed to sometimes erect a relationship that dispossesses, neglects, subordinates, disillusions, alienates, impoverishes, dislocates, represses, factionalizes, subverts and orchestrates tensions, protests and conflicts (Saro – Wiwa 1992; Robinson 1997; Human Rights Watch 1999; Frynas, J. 2000; Crow 1995; Carr and Onyeagucha 2001; Obi 2001). Obi (2001) claims that MNC relations with the HCs is underlined by corruption, divisiveness, co-optation, exploitation, betrayal and subversion which in the case of the Niger Delta, fuel tensions, conflicts and crises.

What the aforesaid indicates is that trans-national business governance in resource rich regions and with their indigenous people is rarely that of partnership built around sustainable resource management and rural livelihood development. In fact, the objectives of MNCs and therefore the logic of MNC operations, which pertain to control of and uninterrupted production of resources at minimal costs and maximized profits, may be facilitative of peculiar political, economic and social strategies and relations. The logic at a general level cannot and is not disposing and accommodative of sensitive, ethical, supportive, cooperative, participative, mutually reciprocating and altruistic strategies and relations with the host states and communities (Obi 2001). Rather their corporate governance may disarticulate social and cultural cohesion, traditional governance and communal stability just as MNC/ host community relations are being indicated by hostility, conflicts and violence.

10.3. The relations between the multinational oil companies and oil producing communities in the niger delta

Oil was first struck in January 1956 by Shell in Oloibiri (Rivers State) and later in the same year in Afam, Bomi and Ebubu in the Ogoni area of Rivers State. Numerous oil locations followed between 1957 and 1958 in Ughelli and other locations in the present Delta State (Olayiwola 1987; Obi 1993; SPDC 1996). The colonial government granted concessions to Shell in 1938. Such concessions were only extended to other companies as Tennessee Oil, Gulf and AGIP after 1961. Nigeria joined the Organization of Petroleum Exporting Countries (OPEC) in July 1971 and began to take more interests in acquiring participatory interests in the MNCs as well as in oil exploration and exploitation.

There are five major MNCs. These are Shell, Mobil, Chevron-Texaco, AGIP and Elf. In addition, there are other MNCs in the oil exploration service sector among which are include Wilbros, Ashland and Company and Geo – Physical General (CGG). The largest MNC, Shell BP Petroleum entered into partnership with Nigeria in 1973. The Nigerian Government owned 35% while 65% was shared between British Petroleum & Shell. In 1974 government stake went up to 55%. In 1979, government nationalized BP and its stake-holding became 80%. This dropped to 60% in 1990 following a new partnership agreement in which Shell had 30% and AGIP & Elf shared 10%. By the 1994 joint venture agreements, the federal government now owns 55% , Shell 30%, Elf 10% and AGIP 5%. Shell produces about 55% of Nigeria's oil and plays a central role in the Nigerian National Petroleum Company (NNPC) joint venture.

The MNCs are European & American companies operating joint ventures with the Nigerian National Petroleum Company, the state oil company. The state is therefore a formal partner with the MNCs. More recently, the state has production sharing contracts with new MNCs & indigenous companies. The state is therefore in alliance with the MNCs in terms of the interests in the maximum unimpeded exploration and exploitation of oil.

Initially, the oil producing communities (HCs), MNCs and the Nigerian state shared common interests in the discovery and exploitation of oil. The discovery of oil was seen as a blessing. The HCs had hoped that the MNC entry with heavy equipments and activity would bring in a new era of development and related benefits. The initial euphoria and enthusiasm however began to wane when after one decade, then another and a third, the HCs were more impoverished, their ecology and rural economy dislocated and the expected benefits did not materialize.

10.3. The relations between the oil companies and oil producing communities

The relationship between the MNCs and HCs has always been unequal and heavily lobe-sided in favour of the MNCs. When the MNCs came to HC territory in the 1950s and 1960s, they were regarded as agents of the federal government and were seen as potentially of benefit. The community leaders were excited at the paltry sums paid them as compensation. The level of literacy, awareness and resources possessed by the HCs were no march for the MNCs. The HCs did not have the where-withal to verify the MNC valuations and payments for land and damages. In relation to conflicts in the courts, they did not have the resources to contest with the MNCs in terms of the quality of legal personnel and presentation (Lolomari, 1993). Therefore they were disadvantaged in relation to compensation and acquisition of their lands and waters.

It was within this framework that the HCs were docile, tolerant, understanding and even supportive of the MNCs. It was mainly in the 1980s that the relationship of mutual understanding began to turn sour

The government had right from the early days of oil exploration given the MNCs unrestricted freedom and privileges in their operations. As a result, the MNCs dictated terms to the HCs and undertook actions and activities that were to their disadvantage and misfortune. Until lately, the HCs had to fend for themselves against the MNCs. The state did not show concern about pollution and compensations. The MNCs were reluctant and slow to pay adequate compensation. So the HCs with meagre resources had to seek legal redress against the MNCs who have enormous resources and influence. What the HCs got from the court system were prolonged litigations, pitiable awards and the enrichment of their lawyers. Frustrated with the courts and with no state support, the HCs began to take the laws into their hands. The disruption of production activities of MNCs became an instrument of conflict management against the all powerful MNCs.

The alliance of the state and MNCs ignored the stakes, interests and economic rights of the Niger Delta people. The MNCs with state authorization behaved as if, they could acquire communal resources and operate in their territories without seeking the approval and co –operation of the people. The government demonstration of insensitivity and non concern for the interests, welfare and problems of the HCs further reinforced MNC disinheritance, oppression and mistreatment. As Owugah (1993) states,"... today, many oil producing communities wish that oil had never been found in their communities. The discovery of oil is now seen by these communities as a curse rather than blessing".

In the early days of oil exploitation, the HCs being ignorant and weak expressed their grievances through petitions addressed to the Ministry of Mines and Power and top government officials (Lolomari, 1993). Protests only became an instrument when it became evident that help was not coming from anywhere. Even when pockets of reactions became noticeable, the rage of the people and the clamour for rights was not appreciated until it assumed widespread and violent dimensions. The rage of the Niger Delta people became apparent when "an increasing number of communities, began to clamour for their rights," often through violence (Olukoya, 1995 : 9). The HCs decided to take their destiny in their hands as a result of insensitive and alienating state and MNC activities. The result has been the large scale disruptions, violence and insurrection in the region since the 1990s.

Initially, the HC demands for compensation were treated with contempt and disdain (Duru, 1997). Thereafter the MNCs began to placate and subdue agitators by compromising them through the provision of money, employment and contracts. It was only in the late 1990s, when the communities and the entire region had become so aware, angry, mobilized and activated and a regime of broad, extensive, loud, highly vocalized and disruptive protests had emerged that the MNCs began to re-consider and re-configure their erstwhile methods, practices and strategies.

10.4. Issues and causes of conflict in the relations between the oil companies and oil producing communities

There are numerous issues that have been contentious in the relations between the MNCs and the HCs. These range from environmental and compensation practices to the attitude of the MNCs, the level of local people participation and the quantum of benefits.

ENVIRONMENTAL POLLUTION AND SCARCITIES

The HCs have protested environmental damage and the consequent decline of agricultural and primary production, dislocations of the local economy and accentuation of poverty in the region.

Oil exploration is ecologically hazardous and environmentally devastative. A fact-finding team of institutions and agencies from the United States of America in September 1999 found a high level of oil industry related environmental degradation in the region (The Africa Fund 1999; Oluwafemi 1999:21). Serious oil spillages have been reported in Bode town, Rivers State (December 1996); Uzam, Edo State (1986) and Aleibiri, Bayelsa, (June 1996) (Ugbine 1996). In Akwa Ibom state, villages such as Obiamkpe Iwofe, Okopedi, Amanding and Ukpomefe suffered prolonged and unabated oil spillage for four months. It irredeemably destroyed fish ponds, fishing channels, economic trees and farm lands (Ayonote, 1995). Lives have been lost because of the contamination of drinking waters by oil pollution, as the case of Aleibiri¶s five-month oil spillage between June and

October 1996, which spread to Isampon, Ogbobagbene, Tubegbe, Tamgbe, Ekeremor in Bayelsa State and Torugbene and Ojobo in Delta State (Dekama, 1997). In Ijaw land, the Ijaw Protection Organisation (IPO) claims that there were 500 cases of oil spills in Ijaw land between 1997-1999.

Many of the spillages result in major ecological damages, impacting farmland, swamps, lakes, rivers and fish ponds. Oil, grease and oil effluents are released which cover surfaces, destroy farmlands and aquatic life and contaminate groundwater. Farmlands are rendered barren and vegetation and aquatic life are destroyed. Acid rains have been noticed in Port Harcourt and Warri areas and is said to result from gas flaring. Noise and emissions and continuous heating and lighting through gas flares plague neighbouring communities.

As the case of Egbema clan, Rivers State, whose farmland and rivers were polluted by oil spillage in 1976 indicates,

> [3] the scum may be gone, the land is dried, but the impact (remains); the destruction of the community's economy. [3] Fishing was gone, hunting is no more and farming is a waste of time....... Anything planted did not yield a good harvest again' (News watch 1988).

Environmental degradation has created resource scarcities. There is inadequate land for farming because of burrowed pits, oil fields and flow stations. Soil infertility, wasted farmlands and destroyed vegetation have led to poor agricultural productivity, poverty, out-migration and social disruptions. Acid rain and oil spillages and leakages have polluted rivers, lakes, creeks and swamps and destroyed marine and aquatic life. Even sacred and ancestral places have been desecrated. Thus the very root of the primary economic base of the region such as farmlands, swamps, rivers and ponds have been destroyed. The citizenry have been transformed from great fishermen to great consumers of frozen sea fish. (Ogbodo, 1992). The case of Uzere in Delta state is instructive. Oil prospecting has wiped out a hitherto dominant occupation of fishing and groundnut production for which it was famous, thanks to the fatal poisoning of their lakes and land.

The oil companies claim that community members looking for compensation were responsible for most of the spills. They "claim that only 35% of spills are traceable to their operations (Onwuemeodo 1999). However, of the 2,676 recorded spillages between 1970 and 1990, sabotage accounted for only 3% of total spills (Dekama, 1997).

MNC ENVIRONMENTAL PRACTICES

MNCs gas flares and pipelines are located within and close to populated areas thereby increasing the risks of pollution and damage. Pipelines criss- cross many towns such as Otu Jeremi, the headquaters of Ughelli South Local Government Area and gas flares in the Isoko area of Delta state are close to towns and villages. Furthermore pipelines are on the surface in Nigeria. In fact, environmental standards observed in the advanced western countries are outrageously flouted in Nigeria. Some have claimed that the MNCs have merely taken advantage of weak legal framework, weak implementation and a state held in hostage to oil rents (Obi, 1997; MOSOP, 1993).

Old, rusty and leaking pipes running above the ground surface and bending around houses, farmlands and drinking water sources are all over the region (Strudsholm 1999:37). There are hurriedly laid pipelines, aging facilities, faulty equipment as well as the inadequate maintenance, monitoring and replacement of pipelines. The management of spills has also been a problem. In Aleibiri, an oil spill occurred on March 18 1997. By August 1997, the spill has not been cleared because Shell insisted that it was caused by sabotage. In solidarity, the Chikoko Movement was launched there (Efenakpor 1997). It is no wonder that oil spillage has become the most common cause of the environmental damage that has dislocated the economic, social and cultural well being and livelihoods of the HCs.

Selected cases of conflicts over environmental degradation

S/N	*DATE*	*COMMUNITY/ STATE/ OIL COMPANY*	*CAUSES OF CONFLICT*	*Occupation of terminal*
1	August 1993-2004	Ogoni / Rivers State/ Shell (SPDC)	Environmental degradation	Community Youth protests. Shut down of 5 flow stations.
2	October 8, 1993	Obagi/Rivers State/ EIF Petroleum	Non compensation for oil spillage	Peaceful protests, stoppage of production for over a decade.
3	August 12, 1993	Iyokiri /Delta State/ Shell	Compensation	Community protests
4	September 1997	Pere Mbiri / Bayelsa State	Compensation for oil spillage	Closure of Oluasiri Gas Transmission System.
5	October 14, 1997	Gele gele / Dubril Oil	Environmental degradation	Community protests.
6	November 14, 1997	Ekakprame /Delta State/ Shell	Compensation for land	Occupation of terminal
7	January 20, 1998	Rivers State / Shell	Environmental clearing	Community Youth protests. Shut down of 5 flow stations.
8	June 11, 1998	Bayelsa State / AGIP	Oil spillage	Peaceful protests, stoppage of production for over a decade.

9	October 9, 1998	Bayelsa State / MOBIL	Non payment of compensation	Community protests
10	December 9, 1998	Bayelsa State / Shell	Compensation for oil spillage	Closure of Oluasiri Gas Transmission System.
11	March-July 2000	Isokoland / Delta State/ Shell	Alleged dumping of 40 trucks loads of toxic waste at Ozoro	Community protests.
12	1990	Ogoni/Rivers State / Shell	Environmental degradation and demands for compensation	Occupation of terminal
13	1990	Umuechem/Shell	Protests over environmental degradation and demands for compensation	Community Youth protests. Shut down of 5 flow stations.
14	April	Nembe/Shell	Compensation for environment degradation	Peaceful protests, stoppage of production for over a decade.
15	October 1993-Feb 1994	Ogba, Egbema, Ndoni Local Govt./Rivers State/AGRIP and Ponticellie Nigeria Ltd on behalf of ELF	Gas explosion at Obife, Ogba which kills 5 men and injures 15	Community protests

Source: Authors compilation from Newspapers and Magazines. See References.

10.5. Poor compensation practices

Compensation management is a major cause of strained relations between MNCs and HCs. The HCs protest against delayed and inadequate compensation for natural resources destroyed or acquired in the process of oil exploitation and distribution. MNCs often offer excessively low compensation that are usually rejected. Sometimes this results in litigation between MNCs and communities or families. In the 1970s and 1980s, MNCs engaged in numerous contestations of compensation claims that the HCs were often subjected to litigation and appeals from lower to higher courts for sums as low as N25,000 (less than $200), (Adewale, 1993 : 3). The HCs believe that the MNCs use the law and courts to frustrate injured communities that ask for compensation (MOSOP, 1993). Since the early 1990s, beginnings with the Ogonis, the communities have become more assertive in the demand for environment related compensations. In 2000 for example, about 550 communities in Edo, Delta and Ondo states in a communiqué at the end of a meeting of their representatives in Effurun, Delta State, demanded that the MNCs institute negotiation for payment of compensation for the large – scale damage and devastation of their environment or face attacks (Okafor 2000; Uwakue 2000).

Most HCs regard compensation offered by the MNCs for damages to farmlands, economic trees and crops and fishing waters as too low and unfair. As the Rivers State government (1990) has noted; "*the compensations paid for these deprivations are just pittance, meagre pittance, on which people cannot subsist for even six months and they became frustrated with life*'.

In spite of damages, many appellants do not get awarded compensation or get awarded outrageously low compensation in the courts because of the modes of channeling compensation, technicalities, obsolete standards and old rates of compensation for economic crops and trees. Some cases have remained in the courts for several years. Furthermore, because of the process of assessment, claims and compensation in which numerous middlemen such as lawyers, estate valuers and community leaders are involved and the resultant costs, the actual land and crop owners get only a pittance. Sometimes, the compensation gets hijacked in the process. The compensation management system is also plagued by corruption and sharp practices by MNC officials, and HC chiefs, elite and youth leaders.

The MNCs believe that compensation can only be paid in genuine cases and claim that most cases of pollution arising from spillages are exaggerated. Conflict arises because community assessments of damages and compensation usually at their costs are often at variance with those of the MNCs. Besides, while HCs may exaggerate and make spurious claims, the MNCs also try to "minimize the degree of culpability, by underestimating the amount of crude oil spilled and lost in a spillage incident" Lolomari (1993).

Furthermore, the MNCs often act slowly to the chagrin of the HCs in the clearing of spillages, assessment of damages and payment of compensation. Four communities; Obatobo, Sekebolou, Ofongbere and Ekeremor- Zion in Burutu Local Government area of Delta, sued Shell over oil spillage on their lands and waters. The legal tussle lasted for 14 years. In 1997, the High Court in Ughelli found Shell guilty and awarded the communities a mere N 30, 298. 681 (about USD 318.9 as at 1999 rates of $1 to N 95). Shell refused to pay the fine and instead chose to appeal (Peredeke 1999). In Ejamah-Ebubu village in Rivers State, a Shell pipeline burst in the 1960's and polluted their land and waters.

After protracted unsuccessful efforts at compensation, the community took Shell to court in 1983. Shell chose to settle outside the court. By 1992, there was still no settlement. The community went back to court. By 1999, there was yet no compensation or remediation of the land (Strudsholm 1999: 37-39).Even when compensation is accepted by the MNCs, the payment process is sometimes prolonged.

More importantly, the realities of the Nigerian legal system make the task of establishing and enforcing compensation for damages from oil exploitation to be difficult (Human Rights Watch 1999). The Nigerian government entrusts the MNCs to provide the data on environmental impact assessments, assessment of damages, valuation, land claims, terms of compensation for property and labour and assessment of sabotage. Thus the compensation management process is dominated and controlled entirely by the MNCs. They determine the processes of valuation and control the process of negotiation and payment. The agencies and levels of government only play nominal monitoring roles, that is if they are involved at all and therefore do not guarantee due process, fair compensation and the protection of the environment and the accompanying primary occupation and means of livelihoods. For example, the law states that damages and spillages caused by sabotage are not compensated, but the responsibility for the determination of sabotage is that of the MNCs. Thus the HCs are in a weak position in the entire process and often times are compelled to accept what is on offer. Therefore the people do not receive compensation to "the value of the benefit lost" (HRW 1999).

To compound the problems, there have been allegations that MNC officials settle local chiefs and elite rather than pay affected peasants just as evidence exists that some HC leaders appropriate compensations meant for claimants (Obi, 1993). The issue of compensation has bred numerous conflicts. The discontent arising from shortchanging of community members have often been expressed against MNCs officials and HC leaders. Sometimes, there have been MNC / HC conflicts and intra communal conflicts and clashes (Adewale 1989).

Another dimension of the demands for compensation is the issue of reparations. Some of the civil groups and even political society such as some Houses of Assembly in the region have been agitating and have passed resolutions on the payment of reparation to their states for decades of environmental devastation. In 2000, representatives of about 550 communities in Edo, Delta and Ondo states expressed their support for the struggle for reparations and expressed their determination to actualize it (Okafor 2000; Uwakue 2000).

10.6. Inadequate benefits

The HCs expect that the MNCs should be sensitive and responding to their poor existential conditions and problems and therefore should contribute to the development of the communities and the improvement of their living standards (Adewale, 1993). They expect the oil companies to assume some of the responsibilities that government has shirked. Thus they expect roads, electricity, classrooms, pipe borne water, health facilities, scholarships and others, which the governments have failed to provide. In the early years of oil exploitation, the oil companies contributed nothing to the HCs. However by the late 1970s and early 1980s, communities had begun to protest the situation and in extreme cases blocked access roads to MNC facilities to compel attention and contributions. By the 1990s, there was a cocktail of protests and disruptions over MNC non-contributions or insignificant contributions.

The MNCs often claim that these demands are best addressed by governments and that they can only render social, moral and secondary services (Saro-Wiwa, 1993). They claim that it is the federal government to whom they pay taxes that has primary responsibility and should develop the HCs. They believe that they are merely to add "value to the core of government driven development efforts" (Peredeke, 1999). As various senior officials of Shell have pointed out

> it is the responsibility of government to develop the land' .... (and to provide) [3] basic socio-economic infrastructure in the society'......... ours is a business and we cannot take over the responsibility of government......... A situation where citizens look up to the companies to perform the social responsibilities of government is wrong......... [3] These companies are to support government and not to be seen as supplanting it in the discharge of its social responsibilities' (SPDC, 1995; Achebe, 1995; Imomoh, 1998).

However, the MNCs accept some form of assistance to the HCs but not at the scale expected by the communities. As Solomon Oladuni, Executive Vice Chairman, Mobil Producing Nigeria Unlimited, have stated, though the MNCs

"are there to maximise profit from their business, they owe society the unwritten obligation to put something back to the societies within which they operate.

The MNCs have actually been contributing to the development of the HCs. The later have no doubt benefited from the oil economy. The traditional, business and political elites in the region have benefited from contracts, businesses and other spoils of the oil industry. But the majority of the HC people, have rather either been impoverished or remained poor and have not benefited from the oil economy. There have also been inequitable benefits between communities and ethnic groups and sometimes there is deliberate or inadvertent exclusion of some segments of society from such benefits. But the main problem has to do with the degree, extent and magnitude of the MNC contributions relative to the problems and needs, the level of devastation that they have caused and the level of profits that they have appropriated.

Shell claims that it has been running formal community assistance programmes for more than two decades which include road construction, provision of portable water supply and health facilities, assisting with agricultural development and awarding scholarships. (Peredeke, 1999). According to Shell, about 950 community development projects were provided between 1992 and 1997 in the Niger Delta, for which $3.2m (N 2.5 billion) was spent (Bassey 1998). The NNPC Joint Venture with Shell, Mobil, Elf and AGIP has also contributed social amenities ranging from the provision of pipe – borne water and bore – holes, educational facilities, health care services to the construction of bridges and town halls and the provision of agricultural equipment such as cassava grinding machines. Most of the MNCs have at various times provided services in education, health care delivery, water supply, scholarships, infrastructures and agricultural development.

However, critics say that the money spent by the MNCs on community assistance is estimated by some to constitute only "0.000007% of the value of oil extracted" in the past twenty five years (Rowell, 1994). Turshen (2002) has indicated that Mobil spent $5m on community development which amounted to only a fraction of a penny for every dollar it extracted from Nigeria. The HCs while acknowledging some assistance complain of inadequacy of efforts, negligible investment by MNCs in relation to what they get from their soil, insensitivity to their problems and plights, failure to understand and relate to local social and related situations, non consultation of communal development associations, youths and women group leaders and others in the design and implementation of projects, the alignment with and compromise of traditional and rural elites and the prevalence of ad hoc rather than comprehensive programme development.

Since the 1990s, there have been general regimes of community protests by youths, women and entire community members over non-benefits from several decades of oil exploitation on their land. Different communities across the entire region have disrupted MNC operations over disputes relating to the nature and level of contributions to offset the ecological devastations and socio-economic disruptions. Even the state governments through their legislatures have begun to agitate on behalf of the communities. For example, in February, 2000, The Akwa Ibom State House of Assembly, in a motion summoned the managing director of Mobil to tell them what the state had gained from its operations (Bassey, 2000; Konkwo, 2000) .

Since the massive region wide protests and production disruptions of the 1990s, the MNCs have become more forthcoming, active and comprehensive in community development programmes. In fact, several communities have had to force them to reach memoranda of understanding that relate to packages of contributions as a basis for continued operations.

10.7. Inadequate participation

The HCs have severally alleged that they are insignificantly represented in the management cadre, and staffing of the MNCs and the contracting and businesses associated with the oil industry. The educated youths, who are particularly frustrated with non-employment in the oil and related industries in the region, have been very resentful of the MNCs. With the entry of more educated youths in the agitation since the late 1990s, the articulation of grievances relating to under-representation and participation has become more active and forceful.

The inability of the MNCs to absorb unemployed youths is a major cause of protracted communal and regional protests and conflicts. Different youth groups have had to shut down or disrupt oil operations over demands for employment. The Egi Youth Federation shut down operations of the OMC 58 field of Elf Petroleum Nigeria Ltd in July 1999 over demands for mass employment of Egi indigenes and conversion of Egis from casual to permanent employment status (Obari, 1999) Even ethnic associations such as the Egbema National Congress, in "The Egbema Charter' deployed the scandalous employment situation of AGIP and Shell in regard of Egbema indigenes (Vanguard 23:07: 1998).

Most of the HCs and their civil groups agitate for immediate employment of qualified indigenes in middle level and management cadres, and recruitment of low level cadre staff in casual and permanent positions in line with Federal Government policy on employment of junior staff from neighbouring states and communities (Nwankpa 2000). The region¶s leaders and civil groups also agitate for some equity participation and partnership arrangements with the MNCs to give them a sense of belonging and stake holding.

10.7. Inadequate participation

Selected cases of conflict over neglect and non-employment

S/N	*DATE*	*COMMUNITY/STATE / OIL COMPANY*	*CAUSES OF CONFLICT*	*YOUTH PROTESTS*
1	October 10, 1992	Umuechem/Rivers State/ SPDC	Non-employment	Youth protests, seizure of flow station.
2	October 6, 1997	Bayelsa State/SPDC	Employment demands	Hostage taking
3	March 10-20, 1998	Koluama/Rivers State/ Texaco	Demands for employment	Youths attack ELF Base flow station
4	April 28, 1998	Odidi/Egwa/Delta State/ Texaco	Demands for employment	Seizure of oil barge
5	May 25, 1998	Ondo State	Demands for employment	Peaceful protests, compels stoppage of production for over a decade.
6	July 15, 1998	Rivers State/ SPDC	Demands for employment	Prevented/disrupted operation of oil service companies, blocked access road to Mobil jetty, seized vessels
7	August 11, 1998	Nembe/Rivers State/SPDC	Non fulfillment of employment promises	Kidnapping of 19 oil company workers
8	August 29, 1998	Nembe/Rivers State/SPDC	Demands for Employment	Youth protests
9	October 14, 1998	Abiteye/ Alero/ Delta State/ Chevron	Demands for employment	Youth protests, seizure of flow station.
10	December 9, 1998	Warri Region/ Delta State/ Texaco	Demands for employment	Hostage taking
11	October 1993-1994	Obasi, Egi Clan, Ogba-Ndoni, Egbema Local Govt./ Rivers State/ ELF	Neglect/Non benefits	Youths attack ELF Base flow station
12	March 5, 1997	Warri region, Delta State/Chevron	Demands for employment	Seizure of oil barge
13	1990	Ogoni/Rivers State/Shell	Neglect	Peaceful protests, compels stoppage of production for over a decade.
14	April 2000	Ibeno Local Govt. Area/Akwa Ibom State/ Mobil Producing Company	Neglect, Non provision of facilities	Prevented/disrupted operation of oil service companies, blocked access road to Mobil jetty, seized vessels

15	April 2000	Ijaw Youths National Movement, Warri Region/Delta State/Nobel Oil Drilling Company	Non employment	Kidnapping of 19 oil company workers

Source: Authors compilation from Newspapers and Magazines. See References.

Unfulfilled promises and un-implemented memoranda of understanding

The promises by the MNCs to HCs for support or assistance such as the constructions of access roads, building of schools and hospitals, provision of employment and scholarships and even relocation and reconstruction of new communities sometimes go unfulfilled. The Governor of Rivers State have stated that the non fulfillment or outright refusal of MNCs to implement pacts and memorandum of understanding reached with the communities is one of the major causes of conflicts between the MNCs and host communities (The Guardian 09:03: 2005:64). Earlier in February 2000, the then minister of State for Environment, Dr. Imeh Okopido, had advised Mobil Producing Nigeria Unlimited "to minimise crises in its area of operation, by honouring past pledges and agreements (The Guardian 22:02:2000).

In Finima, near Bonny, the Nigerian Liquefied Gas Company project required relocating the people to a new location, called New Finima, 5 kilometres away. The people reluctantly accepted the relocation in 1978 from their ancestral homes because of the promise of modern facilities and infrastructures in the New Finima. But they were asked to relocate in January 1990 to a new town without light, water, roads and other facilities. The people refused. However, when some of the facilities were provided, the people relocated. By 1993, some of the promised facilities were yet to be provided. Ojobo Community's Zuokomor village, a sand fill project promised in 1991 when the community forced Shell to stop production at Benisede for one month was yet to be executed in 1996.

1999, the Bonny community attacked, stopped and delayed the Liquefied Natural Gas (LNG) plant from its scheduled export in September 1999. The community was protesting their disappointment with the noncompliance with the environmental impact assessment, non implementation of a memorandum of understanding signed in October 1998 and failure to fulfill the promise on employment of community members. One year after the MOU, the compensatory development and employment pledges had not been met (Obari 1999: 21). In 2003, forty-nine host communities in the oil rich areas in Ilaje local government of Ondo state gave Chevron Texaco a seven-day ultimatum to honour earlier agreement reached with them or vacate the off shore line of the area (Johnson 2003).

Attitude of mncs

There are numerous complaints by the HCs about the negative and untoward attitudes of the MNCs towards them. They accuse the MNCs of arrogance, refusal to discuss with HC representatives, insensitivity to representations and petitions and disdainful and condescending treatment. The MNCs do not regard or agree to dialogue with HCs except when there were conflicts. There are also allegations of intimidation, threats and propaganda against the HCs such as that made by representatives of about 550 communities in 2000 (Uwakue 2000; Okafor 2000).

10.8. The methodology of conflicts

Several methods or instruments of engagement have been utilized in the conduct of the conflicts with the MNCs by the HCs. The initial method was petitions that were sent to the MNCs and the state. There has since been a transformation towards more direct, engaging and confrontational methods.

Complaints and petitions

This is the oldest form of engagement of the MNCs. Petitions were written to the MNCs and state governments over grievances and claims. However the MNCs often ignored petitions and treated the petitioners with disdain.

Dialogue

The HCs have since the 1970s sought dialogue with the MNCs. Usually they sent delegations and requested meetings to discuss, negotiate and reach agreements on issues of conflict. The demand for dialogue have intensified since the 1 990s because of increasing awareness and the accentuation of oil based environmental and social problems. Apart from the expression of grievances, community and ethnic leaderships and groupings usually seek dialogue with the MNCs as a first and preferred method for peaceful and amicable resolutions of conflicts. Unfortunately, demands for dialogue were sometimes ignored and community representations disregarded. Usually, the MNCs never want to dialogue except their backs are on the wall.

Protests

Protests became the usual form of engagement by the late 1 970s, when it became obvious that petitions and representations were ineffective. Initially, the protests involved the blocking of access roads to the oil locations. By the 1980s, community protesters began peaceful occupation of oil facilities and premises. By the late 1990s, the oil locations were being seized and occupied. Even major installations such as the Shell's Escravos Export Terminal were seized.

Threats of evacuation

Some communities and civil groups have urged the MNCs to leave their land. In December 1999, six communities in Ughelli South in a petition to president Obasanjo over lingering crises with Shell, gave Shell a two week ultimatum to move all their facilities from their land (Uwakue 1999). In August 1999, The Egbema National Congress gave a 14 day notice to AGIP and Shell to redress the neglect of social responsibility to the Egbema people or quit the area (Obari 1999) .

Disruption of production

A major method of conflict engagement since the 1990s has been the disruption of facilities. There were 335 disruptions of Shell facilities between 1988 and 1998. In terms of the distribution by ethnic groups, the Urhobos had 120 disruptions, the Isokos 53, the Ijaws 133, Itsekiris 36, the Binis 9 and the Ibos 1 (Konkwo 2000).

Violence

Since the late 1990s, there has been a regime of violence against MNC facilities, equipment and staff. Community and ethnic youths have intermittently embarked on regimes of boarding oil rigs, holding staff hostage, kidnapping staff and vandalizing facilities (Strudsholm 1999:37)

Sabotage

Another method of engagement has been acts of sabotage against oil production facilities. A particular victim of sabotage are the oil pipelines and intersections. The purpose is often to get compensation for environmental damage or to get contracts for cleaning up the spilled oil. In recent years, sabotage has taken the form of participation or collaboration in the theft and sale of crude and refined oil through breaking into pipelines, pipeline intersections and manifolds.

Though there are numerous other sources of oil spillage such as burst and leaking pipelines due to old, rusty and weak pipes, sabotage accounts for a reasonable number. Between 1976 and 1988, there were 344 reported cases of sabotage with damages to pipelines in 211 cases (Ogbodo 1992). In 1992, there were 2,645 oil pipeline spillages with 1,837 attributed to alleged sabotage. In 1993, there were 3,595 incidents of oil spillage, out of which 1,835 were attributed to alleged sabotage.

The MNCs have in the last decade, constantly invoked the sabotage clause in order to avoid the payment of compensation in oil spillages. This has been a major source of conflict between them and communities. One fall-out of the situation of increasing sabotage is the heightening of the insecurity and the need to protect oil facilities. The government has had to set up special security outfits and strengthen naval patrols in the waterways.

10.9. MNC strategies and responses

The initial response of the MNCs was to disregard community petitions and representations or treat them with levity. Community leaders are treated with disdain and refused access to top officials. The Delta Oil Producing Communities Association (DOPCA) alleged in 1999, that it had for over a year exhausted all possible avenues to meet with Shell Managing Director and that Shell has shown disregard for round table discussion (DOPCA 1999). The MNCs often ignored calls for dialogue by communities on the issues of concern and problems afflicting them. When protests became the major means of engagement, the MNCs did not take them seriously. First they claimed that the issues raised by communities and civil society and what are actually expected from them are the responsibility of the government. Then they sought to stop the protests through the solicitation of the support of the state security apparatus. It was the consistent and accentuated popular violence that compelled the MNCs to seek other strategies.

Pacification

Right from the early period of entry to the communities, the MNCs had given money, drinks and gifts to community leaders to placate, settle and appease the communities and enable continued exploration and exploitation (0rr 1995). The MNCs have continued the system of cash inducements to community leaders and groups. This is like a pacification incentive, a reward for safety of MNC facilities and a settlement or payment for security. Cash incentives however have often underpinned corruption and the non-percolation of such aids to grassroots members and groups. It has also fuelled inter and intra group rivalries and conflicts (The Punch 05:05, 05; 1-2).

Incorporation

The MNCs initial concept of safe working environments was the incorporation of traditional and community leaders and elites through financial and contract inducements. The strategy was a enrichment of strategic elements in the communities as they emerged. Influential elites and traditional rulers have been compromised as they are cheaper to buy than pay the compensations demanded by the communities. But unfortunately, such representatives collect compensation and welfare packages on behalf of the HCs without the materials getting to the people (Uwakue 2000).

The incorporation of traditional governance structures have created numerous problems. First the MNCs disregarded the majority of the community members and governance structures. Second, it has led to the hijack or capture of benefits from the MNCs in the communities and corrupt enrichment. Third it has led to considerable losses of probable community development projects and infrastructural facilities. Fourth, as the evidence of corruption and betrayal by traditional chiefs became increasingly glaring, the traditional governance structures became de-legitimated. Fifth, there emerged intense struggles for access to cheap MNC fall-out among the traditional ruler ship institutions, opinion leaders, and their socio-cultural and interest groups.

The MNCs strategy of incorporation is in part responsible for the youth emergence as dominant but violent players. As Mofia Akobo, eminent senior citizen of the ND and chairman of the Southern Minorities Movement stated in an interview

> [3] now a chief becomes friendly with oil company, he becomes a contractor... and he become more and more ostentatious with wealth. This antagonizes the youth as they feel that he has plundered their inheritance' (Ezaman 1999)

Divide and rule

MNC officials have sometimes preferred to relate to agents, elites and youths rather than genuine community representatives. Oftentimes, the MNCs prefer to allign with and settle those who have greater capability to disrupt their operations. Sometimes the MNC officials have played one group against the other and communities against their leaders such as by revealing details of transactions with one to the other in the bid for safer operations or the management of conflicts. Sometimes the MNCs have preferred some communities to others particularly in the permission of access and the distribution of benefits. These fuel intra and inter group and communal conflicts.

Litigation

In several cases, the MNCs have resorted to litigation. Initial litigations related to issues of compensation. Often times, litigation has been seen as the option when negotiations and settlement fail and particularly when the relations have been stalemated over issues of compensation. Since the emergence of popular violence as an instrument of HC engagement, the MNCs have sometimes resorted to litigation either out of exasperation or as a possible outlet from existing quagmires. In May 2000, for example, Shell sued the Isoko National Youth Movement over the seizure and occupation of its five flow stations in Isokoland (Vanguard 20: 04: 2000).

Harassment and repression

The MNCs have tended to favour a hard stance, sterner measures and violent responses to protests. They regard the issues in conflict and the ensuing protests as that involving security agencies and the maintenance of law and order. Therefore they often seek protection of security agencies through the deployment of military and police personnel on their facilities and trouble spots. The MNCs pay allowances to police, military and other security personnel deployed to their facilities. The incentives that the MNCs provide motivate the security forces to harass and intimidate community members and to apply excessive force whenever they are invited to quell protests. Besides, the MNCs provide logistic support and contribute arms to the security forces (Ake 1996; Doudu, 1996). The Ijaw Youth Council claimed in 1998 that the oil companies invited Nigerian solders to occupy Ijaw land to protect their installations and vowed that such companies would be treated as enemy of the Ijaw people (Kaiama Declaration 1998).

State brutalities and threats are sometimes perpetrated against the communities at the instance of or under the pressure of MNC officials. Abuses and repression have been known "to take place on, or adjacent to company property or in the immediate aftermath of meetings between company officials and individual claimants or community representatives" (Human Rights Watch 1999). At the instance and expenses of Chevron in May 1998, Nigerian security troops were transported to the Parade platform peacefully occupied for about 3 days by between 120-200 unarmed community demonstrators. Chevron had allegedly refused to negotiate. Two men were killed and a third injured. Chevron reportedly provided helicopters and boats in January 1999 to fly soldiers to attack a community that opposed its operations. Over 12 people died (The Africa Fund 1999).

The use of security forces to attack individuals, groups and communities protesting against them have tended to be seen by the MNCs as "appropriate responses to threats to company personnel or property" (Human Rights watch 1999). But the police and the military have to "say the least", been "excessively heavy handed" (Saro – Wiwa, 1993). They have been associated with large scale destruction and considerable human rights violations. Human Rights Watch (1999) found "repeated incidents in which people were brutalized, detained and beaten for attempting to raise grievances, make claims and seek compensation for environmental degradation from the MNCs. There is no evidence that the MNCs have investigated state repression by security forces, protested such abuses to the state, undertaken measures to prevent such abuses or paid adequate compensation to the victims. But quite unfortunately, the resort to state security agencies has achieved only limited success as they have not stopped or prevented community protests and violent skirmishes.

Local and international environmental and civil rights organizations and activists have accused the MNCs of producing oil at "the cost of violations of the rights of those who live in the areas where oil is produced[1]" and of pumping oil out of the Delta without "regard for the area's pollution and desperate poverty. The Human Rights Watch, Project Underground and other groups in the United States have advocated for congressional investigation and testified before the House of Representatives Sub-Committee on Africa about the complicity of American companies such as Chevron in the human rights violations in the Niger Delta. They have also called for public judicial commissions, criminal investigations and compensation for those affected in the state repression of HC agitations (Institute of Public Accuracy 03:08:99).

Community development

The MNCs have also increasingly embarked on community development (CD) projects. Initially, CD was directed at infrastructure development and social services.

More recent efforts have related to vocational training, community based health care delivery system, educational infrastructure in terms of classrooms, books, and teacher sponsorship and agricultural extension Some MNCs such as AGIP have embarked on extensive programmes such as the Green River Project which has prospectively affected the lives of over 10,000 farmers and families. In 2000, SPDC was said to have invested about US$ 60m on community development (CD) programmes.

However, most of the community development programmes have been poorly conceived, prioritized and executed and have had very little positive effects on the pervasive poverty and scanty development condition of the region. The communities have not been active participants in terms of involvement and inputs in the community development projects of the MNCs. The CDs are dictated, imposed and top down directed. It is not surprising therefore that some of CD projects are based on the wrong perceptions of needs of HCs (Amadi & Tamuo 2001)

The community development projects, have witnessed recent reforms. This has been first in terms of shift from mere physical facilities to economic empowerment and vocational training and skills acquisition programmes. Second is a new emphasis on women and youths. Third, this is in terms of greater community input in terms of priorities and choices of projects.

Memoranda of Understanding (MOU)

The HCs have compelled the MNCs to reach MOUs through violent protests and disruptions of production. In fact, the MNCs are now more serious in the negotiations of memoranda of understanding with communities as the basis for peaceful relations. However, in several instances, the MNCs have not adhered to the provisions of MOUs. In recent years, non compliance with MOUs have been a major cause of conflicts in MNC/HC relations

10.10. The intensification of conflicts and the emergence of violent resistance

The opposition to the MNC took on a violent turn in the 1990s following the frustration and hopelessness felt in the region about the prospects of beneficial peaceful change in the status quo of benefits in the context of the existing tendencies of failed negotiations, broken promises and state repression. The youths particularly began to take violent actions in facilitation of change. Particularly, from 1997, the youths became the vanguard of the struggle against the MNCs and the foot soldiers in the militant resistance. Numerous youth movements, groupings and militias emerged at the community, clan, ethnic and regional levels.

Alongside the youth involvement was the emergence of a regime of storming, seizures, occupations, shutdowns and vandalization of oil facilities and the hijack and abductions of their staff across several ethnic groups in the region. Militant groups such as the Ijaw Youth Council, Isoko National Youth Movement, Chikoko Movement, Niger Delta Resistance Movement and Youth Militias such as Niger Delta Volunteer Force, Supreme Egbesu Assembly, Egbesu Boys of Africa and Federated Niger Delta Ijaw Communities (FNDIC) have been at the forefront of the militant and violent operations against the MNCs and the federal government since 1998 (Obari 1998). The FNDIC was responsible for most of the seizures and shut downs in the Warri region during the protracted Warri crisis between 1997 and 2003. The offensives and counter actions of the Nigerian security forces have led to massive militarization, military occupations, violent encounters and insecurity in the region.

Shell (SPDC) has been particularly plagued by communal and youth attacks (The Guardian, 28:01:1999). Shell had to suspend loading of crude oil at its Forcados Terminal in early January 1999 because of low oil output from the Western Divisional operational area arising from persistent community problems (Obari 1999). Shell has lost millions of barrels of oil to community conflicts in Ogoni and the entire region. The company lost 135 man-days in 1992 and 141 in 1993. SPDC has been planning the relocation of its regional headquarters from Warri to Port Harcourt as a result of constant attacks, kidnapping of staff, unending claims by communities for compensation and demands for ransom in the Western Niger Delta. The militant engagements by the HCs have raised operating costs, dwindled fortunes and made the environment inclement for equipment, facilities and staff (Uwaleke 1999). The threat to lives and operations of Shell has become so much that its security staff strength of 2,470 was almost half of the total staff by 1994 (Onwarah 1994).

10.11. The state, mncs & host communities

Initially, the state left the HCs to the superior resources and influence of the MNCs. There was no clear policy as to the obligation of the MNCs to the HCs. The legislative and regulatory framework for MNC environmental behaviour and corporate practices was weak and ineffective. In fact, the MNCs have from the beginning set the agenda and controlled the Nigerian oil industry. The Nigerian state being totally dependent on oil rents and being in alliance and joint ventures with the MNCs have been unable, incapable of and perhaps unwilling to protect the Niger Delta people and environment (IDEA 2001: 237-280) It was only when the HCs became violent and disruptive that the state became interested, more out of survival and maintenance of law and order than of serious interest in the welfare of the HCs.

There are however emerging changes. The federal government has begun to implore MNCs to be more socially responsible to dialogue and to negotiate and reach agreements with their HCs, in order to promote peaceful relations and stability. It has attempted to initiate dialogue and negotiations between MNCs and the HCs (The GUARDIAN 08:01:99:3). Particularly, it has become more assertive and active in the management of the environment, in terms of setting and insisting on deadlines for the stoppage of gas flares and environmental remediation. It has begun to monitor more effectively and to sanction MNC environmental practices. The state governments have often intervened and brokered negotiations and agreements between the MNCs and HCs during conflicts.

Thus, since the late 1990s, governments have been involved in brokering negotiations and resolving conflicts between MNCs and HCs. Such negotiations still involve largely officials of state governments and sometimes the NNPC. Thus in the conflict between the Uzere community and Shell in July 1992, the Delta state government brokered a cease fire and agreement. In the Egi Youth Federation shut down of Elf facilities in Obagi of July 1999; a team from NNPC led by a Group Executive Director, representatives of Rivers State government and Elf were involved in negotiating an agreement. In the Isoko National Youth Movement shut down of Shell facilities in 1999, the Delta State Military Administrator visited the area and appealed to the youths (Ogoigbe 1998).

However, the federal governments have tended to intervene more in protracted conflicts and it¶s reactions have tended to be forceful and suppressive of the communities. The exceptions have been in the cases involving oil spillages and environmental pollution. In the case of Isoko National Youth Movement shut down of Shell oil facilities in Isokoland, Delta state over allegations of the dumping of toxic waste, the Federal ministry of Environment was involved and the INYM only agreed to allow Shell to resume operations, following the intervention of the minister of Women Affairs and Youth Development (Amorighoye 2000; Ezereonwu Jane. 2000).

10.12. The persisting crisis of relations: towards a rapprochement

The people of the Niger Delta were initially happy and enthusiastic about the prospects and opportunities that the oil economy would bring. The initial attitude began to change when the expected benefits were not forthcoming even after two decades. The situation was accentuated by two factors.

First, oil exploitation made the region prostrate as it was plagued by high rates of out-migration and rural-urban migration, unemployment and underemployment, poverty, food insecurity, depleted agricultural resources and threats to it's fora and fauna. Second, the region was piqued by the outright refusal to pay compensation, delayed and insignificant compensation and the tortuous and painful processes of compensation for their appropriated and devastated land and water resources and crops. In addition, the MNCs marginalized the HCs in the oil economy. The employment, contract and business structures and practices of the MNCs were not tailored to give any preference to the region. Rather, the oil economy was dominated by those outside the region. Therefore, the HCs felt dislocated, disinherited, marginalized, alienated, frustrated and angry. It was these conditions that set the stage for the protests and violence against the MNCs.

The MNC governance system contributed to the transformation of the initial peaceful relations into the present cauldron. We noted that as the initial minimum interaction, condescending neglect and deflection of social responsibilities began to produce hostile relations, there began a system of procuring state protection and repression, purchase of local support and some form of community development. It is the persisting conflict and the regime of violent confrontations by the communities that is conducing new strategies of broad based community development by the MNCs and the memoranda of understanding that are being reached between the communities and the MNCs.

Particularly, the cumulative effects of prolonged, disdainful and contemptuous treatment of the HCs, brazen exploitation with scant attention to the environment and the problems of the communities, tedious but pitiable compensations and unorthodox methods of subordination and pacification have been the sour and conflict ridden relations, the unfriendly and insecure environment of business, the huge losses to production disruptions and shortfalls and the threats to MNC staff and facilities.

MNC corporate governance has been severally fingered as fueling conflicts and crisis between the communities and ethnic groups in the region. Shell, the largest oil company in Nigeria has admitted that its manner of corporate governance; contract awards, appropriation of land, relations with community representatives as well corporate and staff actions such as corruption and reliance on state security forces have contributed to conflicts and violence and their escalation in the region (BBC News 2004). Chevron-Texaco, the third largest oil company, has also admitted that its corporate governance methods and activities have rather contributed to corruption by local community leaders, local tension, communal rivalries and divisiveness, communal unrests and violence (The Punch 05:05:05).

The central challenge in the sour MNC/HC relations is how to re-orient MNC governance system and HC attitudes and responses such that there could be cordial relations, partnership and a peaceful and conducive environment of business.

There are several issues involved. First, how can the logic of profit incorporate social responsibility in terms of responsiveness and sensitivity to the problems of the HCs. Second, is how can corporate governance engender sustainable development particularly in terms of the environment and economy of the HCs. Third is how can HCs share in the benefits of resource exploitation such that they can feel that they are part of and therefore committed to the oil economy. More explicitly put, how can the MNCs create a sense of stakeholder ship in the HCs, such that rather than sabotage and destroy MNC facilities, they would protect and guarantee security and safety.

We should acknowledge that there have been changes in the corporate governance and relations with communities since 2000. The MNCs have been calling on the government to increase developmental attention and improve revenue allocation and related benefits that can significantly bring the oil wealth to bear on the Niger Deltans (Uwaleke 1999). Most MNCs have began to relate to multiple interest groups rather than traditional rulers alone (The Nigerian Tribune 22:03:2002). The MNCs are now more serious in the negotiations of memoranda of understanding with communities as the basis for peaceful relations. However, these efforts are undermined by the non-compliance with and actual breaking of the agreements by the MNCs themselves. We noted earlier that this has been a major source of new conflicts.

We should also acknowledge that the MNCs are becoming more environmentally sensitive. Increasing HC awareness and opposition, increasing federal government activity and effectiveness and international pressures are compelling such environmental responsibility.

However, in both areas; community development and environmental responsibility, a lot still needs to be done. The greater challenge seems to be in the generation of sustainable growth in the local economies and the creation of a sense of stake-holdership. Concerning the former, we noted earlier the increasing economic empowering project profiles of the MNCs. There is need to heighten activity here in terms of more youths and women training, skills development and credit provision. More importantly, economically productive and sustainable projects would have to be established and supported. Local content and local contact input into the oil industry would have to be enlarged.

Through these and other sources, employment opportunities would be generated so that more youths would be employed and poverty reduced. The divesting of youth energies and attention into more productive ventures is in part a solution to the pervasive youth restiveness and militancy in the region.

The MNCs would have to devise ways of increasing stake-holdership within the HCs. The civil groups and political society in the region have been agitating for HC participation through equity holding in the NNPC/MNC joint ventures. More conservative opinion tend to recommend a system of MNC rewards of HCs. It is believed that such rewards would motivate the HCs to protect MNC facilities. Whatever the case, given the circumstance of hostile relations and pervasive violent confrontations, there has to be a system of guaranteeing sustained and undisturbed operations by increasing HC participation and benefits from the oil economy. Something substantial has to be undertaken to douse hostility, engender peace and cordiality, improve the environment of business in terms of security and conduciveness, improve the local economies and strengthen sustainable development.

10.13. Conclusion

The paper examined the issue of persisting crisis in MNC and host community relations. The causes of the conflicts, the nature of response from the MNCs, the methods of engagement of the MNCs by the HCs in pursuance of their grievances and the role of governments were examined. We noted that the nature of relations has been such that there has been the intensification of conflicts since the 1990s and the emergence of a popular communal and youth resistance against the MNCs. We stated that the HC method of conducting the conflict has progressively moved from petitions and peaceful protests to production disruptions, sabotage and violence. The MNC responses which up-till the 1990s comprised of disregarding the protests, incorporation, harassment and repression have begun to shift towards the negotiations of MOUs and more effective and broad community development efforts.

The paper recommends that the major strategy of managing the conflict is in increasing the HC stake-holding in the oil economy. The MNCs should undertake programmes and institute benefits that make the HCs grow commitment to sustained and undisturbed oil operations. But this should be in addition to enthroning environmental sustainability and programmes that foster sustainable development.

References

Adewale, O. (1989) Oil compensation claims in Nigeria: Principle, Guidelines and Criteria, *Journal of African Law* 33 (1)

Adewale, O. (1993) Some Legal Aspects of Community Relationship in the Petroleum Industry Port Harcourt, *National Seminar for Government, Oil Companies and Oil Communities* 5-7 April

Ake, C. (1993) What is the problem of ethnicity in Africa? *Transformation* 22

Ake, C. (1996) Is Africa Democratizing? Port Harcourt: Centre for Advanced Social Science. Monograph 5

Amadi, S.N and Tamuo, S.O (2001) Oil Exploration in Nigeria: Its Socio-Economic Impact on Host Communities, *Journal of Environment Studies* 2 (1)

Amorighoye, N (2000) Isoko youths shut down 5 flow stations, *Vanguard,* Monday, 10 April

Ayonote, L (1995) Blood spillages. *TELL,* December 25

BBC News (2004) Shell admits fuelling corruption, Friday, June

Bassey, A. (2000) A Ibom Assembly orders Mobil Officers' arrest, *The Guardian Online* Saturday, February 26

Bassey, U. (1998) Oil Community, *The Post Express* Wednesday, October 14

Boele, R. (1995) Ogoni: Report of the UNPO Mission to investigate the situation of Nigeria, The Hague: Un-represented Nations and People Organization

Carr, S.O.D and Onyeaguch. U (2001) The Ogoni People's Campaign over oil Exploitation in the Niger Delta in A. Thomas, S. Carr and D. Humphrey (eds.) *Environmental Policies and NGO Influence* London: Routledge

Crow, M. (1995) The Ogoni Crisis: A Case Study of Military Repression in South Eastern Nigeria, New York: *Human Right Watch / Africa* No7 (5) July

Dekama, A (1997) Oil became our problem, *Ijaw* New Special Edition 2 (17) September 30

Delta Oil Producing Communities Association DOPCA (1999) The Niger Delta Crisis – Our Stand *The Guardian,* Thursday, January 28

Doudu C. (1996) Shell Admits importing guns for Nigerian Police, *Observer,* London 28 January

Duru, V.O. (1997) The Performance of Oil Mineral Producing Areas Development Commission (OMPADEC) in the Development of Delta State, M.Sc Project (unpublished) University of Benin, Benin City.

Efenakpor, R. (1997) Niger Delta youths protest against oil firms, launch new group, The *Guardian,* Monday, August 18

Ezereon J (2000) Groups grant Shell conditional return to an oilrig, *The Guardian,* Wednesday, May 17

Frynas, J (2000) Oil in Nigeria: Conflict and Village Communities, Hamburg and New Jersey: LIT Verlag and Transaction Publishers

Human Right Watch (1999) The Price of Oil: Corporate Responsibility and Human Rights Violations in Nigeria's Oil Producing Communities New York February

Human Right Watch (1999) Human Rights Watch call for action against abuses in the Niger Delta, New York, December 21

Ibeanu, O. (1997) Oil Conflict and Security in Nigeria: Issues in the Ogoni Crisis *AAPS Occasional paper* series Vol. 1 No. 2

Imomoh, E (1998) As quoted in Oil and Community, *The Post Express,* Wednesday, October 14

Institute of Public Accuracy News Releases August3, (1999)

Johnson, D. (2003) Host Communities give Chevron – Texaco seven –day ultimatum in Ilaje LG, *Vanguard,* Monday, December 15

Kaiama Declaration (1998) is a communiqué issued on December 11 at the end of all Ijaw Youths Conference

Lolomari, O. (1993) Oil Companies and Oil Communities As Partners in Development Paper *Presented at National Seminar for Government, Oil Companies and Oil Communities* Port Harcourt, April 6-7

Makumbe J. (1998)" Is there a Civil Society in Africa" *International Affairs* 74 (2) Newswatch (1988) The Looming war July 18,

Nwankpa, E (2000) President Lists cost of violence, *The Guardian,* Thursday, March Obari, J.O. (1999) Group warns oil firms over alleged neglect, *The Guardian,* Friday, August 13

Obi, I.C. (1993) Oil and the Nigeria State: An Overview *National Seminar on Conflict Containment Strategies for Oil Companies and Oil Communities* Port Harcourt, April 6-7

Obi, C. (1997) Oil, environmental conflict and national security in Nigeria: Ramifications of the Ecology-Security Nexus for sub regional Security, *ACDIS Occasional Paper* January

Obi, I.C. (1998) Environmental Scarcity in Africa: Some Theoretical Concerns and Emerging Issues, *Africa Insight,* Vol. 28, No.1/2

Obi, I.C. (1999) The Crisis of Environmental Governance in the Niger Delta 1985-1996, *AAPS Occasional Paper Series,* Vol.3 No3

Obi, I C. (2000) Globalized Images of Environmental Security in Africa *Review of African Political Economy* No. 83

Obi I. C. (20001) Oil Friendships are Greasy: A Political Economy of Corruption in Oil Company-Community-Relations in the Niger Delta. Paper presented at *the International Political Science Association LC 4 Conference in collaboration with the Nnamdi Azikiwe University, Akwa, and The Africa Centre for Democratic Governance at* Abuja, Nigeria October 15-17

Ogbodo, A. (1992) Up in arms. *The African Guardian,* August 17

Ogoigbe, E (1998) Isoko youth seize oil flow station, *Vanguard,* Friday, December 18 Okafor, C. (2000) communities indicts Shell over spillage compensation, *The Guardian* Wednesday January 5,

Okonkwo, C (2000) communities Threaten Shell. *PM News,* January 24

Olayiwola, P. (1987) Petroleum and structural change in developing country: The case of Nigeria New York: Praeger

Olukoya, S. (1995) OMPADEC Politics: *Newswatch,* May 19

Oluwafemi, B. (1999) U.S. Team indicts oil firms over Niger Delta Crisis Tuesday, *The Guardian,* September 21

Omoweh, D. (1995) Environmental Pollution Culture and Health in Nigeria: The Sad Plight of the Ughelli Oil Communities *Africa Spectrum* 3 (2)

Onosode, G.O (1996) Community Expectations Versus Health Safety and Environmental and Economic Stewardship of Joint Venture Partners: Finding the Right Balance. *Paper Presented at the 8th Edition of the Biennial International Seminar on the Petroleum Industry and the Nigerian Environment.* Port Harcourt November 17-2 1.

Onuorah, M. (1994) Shell Shocked *.The African Guardian* July 25

Onwueneodo, S. (1999) Ikwerre youths shut down EIF flow station, *Vanguard,* Wednesday, May 17

Orr, D. (1995) Haw Shell won over village with case and Liquor. *Tempo,* December 14 Owugah, L. (1993) Oil Companies and Oil Communities *Weekly Sunray,* Sunday, January 24

Peredeke, A.E (1999) The role of OMPADEC in Development of Oil Producing Communities M.Sc Project (Unpublished) University of Benin, Benin City

Redclift, M. (1987) Sustainable Development: Exploring the Contradictions, London: Metheum

Robinson, D. (1997) Ogoni: The Struggle continues. Geneva and Nairobi, World Council of Chambers and All African Council of Churches

Rowell, A. (1994) Shell shocked: The Environmental and social Cots of Living with Shell in Nigeria. Amsterdam: Green peace

Saro Wiwa K. (1992) Genocide in Nigeria: The Ogoni Tragedy, Port Harcourt Lagos: Saros Publication

Saro Wiwa K. (1993) Community Reaction to Oil Exploration. *Paper Presented at the National Seminar on Conflict Containent Strategies for Oil Companies and Oil Communities* Port Harcourt, August 6-7

Saurin, J (1996) International Relation, Social Ecology and Globalization of Environmental change in J. Vogler and M. Imber (eds.) *The Environmental and International Relations,* London: Routledge

Segun, J. and Ayankola, M. (1998) Ijaws take over Shell station, *The Nigerian Tribune* Wednesday October 7

SPDC (1996) Shell in Nigeria: Fact and Figures *SPDC Journal* April 15

Strudsholm, J. (1993) The Oil Field in Nigeria *Africa Insight 29 (1).*

The Guardian (1999) Oil summit shifted to March 8, Friday, January 8 p. 3 The African Fund (1999) Blood and Oil, A Special Report September 17

The Guardian March 9, 2005

The Nigerian Tribune (2002) Oil firms to deal with groups in Niger Delta Friday, March 22

The Punch (2005) Chevron Texaco admits failing oil communities, Thursday, May 5 Tandon, Y. (1998) Globalization and Africa's Option AAPS Newsletter,

Turner T (1981) "Commercial Capitalism in Nigeria: The Pattern of Competition in D.

Cohen and J. Daniel (eds.) *Political Economy of Africa,* London: Long man Turshen, M (2002) Introduction ACAS Bulletin: Its About Oil, *Review of Africa Political Economy* 91, 15 1-162

Ugbine, Z.I (1996) An Evaluation of Oil Mineral Producing Areas Development Commission (OMPADEC) in Isoko LGA MSc. Thesis (Unpublished) University of Benin, Benin City

Uwakue, A (2000) Isoko make case for N1.8 billion derivation fund *Vanguard,* Monday, May 8

Uwaleke, I. (1999) Shell may relocate from Warri . *The Guardian* Thursday, July 15 Vanguard (2000) Shell sues Isoko group communities. Thursday, April 20

Warpner, P. (1996) Environmental Activism and World Civil Politics. Albany (NY):

State University of New York Press

Watts, M. (1999) Petro Violence: Some Thoughts on Community, Extraction, and Political Ecology, Berkeley: *Berkeley Workshops on Environmental Politics Working* Paper No. WP 99-1

Woodhouse, P. (1992) Environmental Degradation and Sustainability in T. Allen and A. Thomas (eds.) *Poverty and Development in the 1990s.* Oxford: Oxford University Press

11. The nexus of corporate irresponsibility and state failure in nigeria: an analytical framework

Remi MEDUPIN

11.1. Introduction: how the market failed

Modern free market doctrine owes much of its intellectual origin to Adam Smith, a British social philosopher who lived in the 18th century when Britain's industrial revolution was maturing (Umoren, 2001). Smith's free market ideas were enunciated in his 1776 book entitled *An Enquiry into the Wealth of Nations* with emphasis on the virtues of free trade/market. In this paradigm, it was assumed that a free, supra-personal market would co-ordinate random individual decisions and actions and channels them into the right direction. The free market philosophy enjoyed absolute loyalty for quite a while. However, this began to change under the impact of several events (Weisskopf, 1971:116-7).

First was the growth of big business, monopolies, market power, cartels, and so forth in the Western economies and the countervailing growth of governmental control and unionism. Second was the disruption of the free market system through World War I, the subsequent abandonment of the gold standard, growing protectionism, the manipulation of monetary systems and of domestic markets through central banking, tariffs, subsidies and regulations. Third was the Great depression of the 1930s, which undermined the belief in the automatic, self-regenerating powers and beneficiality of the free market. Allied to this was the spectacular economic growth recorded by Russia through centralised planning in the aftermath of the 1917 Bolshevik revolution.

In short, these events led to the disintegration of the free market ideology, which dominated the 19th century. Thus, the belief in the invisible hand of the competitive market was gradually abandoned and replaced by the recognition that the visible hand of corporate private and public organisations and their managers allocate resources and determine the economic fate of citizens.

A footnote needs to be entered here, given its direct relevance to the subject of this paper and it is this: the free market ideology excludes and denies the existence of market power. By definition a market is free only if no firm can influence the prices of the things it buys or sells by its own individual actions. This is romanticized as "pure' or "perfect' market even in contemporary texts on Economics.

This introduction is followed by a brief review of the schools of economic thought on the role of the State in the economy. Section three of the paper discusses corporate social responsibility; it contends that companies in Nigeria violate business ethics, which in section four is shown as state failure. The paper concludes with section five.

11.2. Perspectives on the role of the state in the economy

For analytical convenience, we can isolate four contending perspectives on the role of the public sector in economic development (Medupin, 1992). These are:

1. The Classical/Neoclassical School;
2. The Public Choice School;
3. The Keynesian perspective; and
4. The Marxian perspective.

Adam Smith whose robust belief in the supremacy of market forces in resource allocation has been duly highlighted in the opening section of the paper typifies the classical school. To Adam Smith and his fellow classicists, the conduct of economic affairs is best left to private citizens. They reasoned that the pursuit of limited, selfish interest by individuals would end up in accord with societal wish and good, even when there has been no conscious and explicitly formulated social goal (Wilson and Skinner, 1976). Even then, it is apposite to note that Adam Smith saw a wide and elastic range of activity for government, covering defence, justice and public works.

The Public Choice School (PCS) alternatively known as "Economics of Politics' has James Buchanan (1986 Noble Laureate in Economics) as an advocate and operates on the principle that the analytical categories of economics should be applied to the decision-making process of the State. In this framework, political parties and governments are seen as maximisers of political support in the same way as economic agents [consumers and firms] are utility and profit maximisers, respectively. Thus, in its bid to maximise support, government embarks on public spending on the principle of getting most of the votes cast at elections.

Concerning the Keynesian perspective, it is pertinent to note that Keynesianism became the new orthodoxy in economic theory and policy after the Second World War. At the heart of the Keynesian perspective is the observation that left on its own, capitalism is an unstable economic system which cannot provide full employment or even socially adequate capital utilisation without State intervention through appropriate monetary and fiscal policies.

The Marxist School, notwithstanding its obvious and very serious differences with the Keynesian perspective on basic assumptions and consequent analytical procedure, nevertheless shares with the latter, an abiding interest in the nature and impact of State intervention in capitalist economies.

The Marxist conception of the State is both historical and class-focused. According to Marx and Engels (1977:38) in the *Manifesto of the Communist Party*, "the executive of the modern state is but a committee for managing the common affairs of the whole bourgeoisie". There are variants of the Marxian perspective but they can be collapsed. When collapsed, we get the picture of the capitalist state that is very different from the classicists' liberal characterisation in which the State is seen as "an independent institution standing above and outside the society." Instead, we have a State that is simultaneously an instrument of class rule even as it enjoys some relative autonomy from the ruling class. This relative autonomy indeed helps to mask the partisan role of the State. Thus, when State policies are presented as being in the national interest, at bottom, we find instead that they are synonymous with "the well-being of corporations and financial institutions. O il companies typify the Nigerian case.

11.3. Corporate social responsibility and its negation

As Frederick, David and Post (1988:5) rightly highlight, "today's business firms do not operate in a social vacuum." The hallmark of a socially responsible company is ability to interact positively and humanely with people and groups in its social environment. This is without prejudice to the fact that the modern business corporation is one of the most powerful and influential institutions in the world, especially in this era of globalisation. The idea of corporate social responsibility in the United States began in the early part of the twentieth century when corporations were being criticised for being too big, too powerful, and antisocial, and were accused of engaging in anti-competitive practices. As Federick et al (1988:28) recount, "a few farsighted business executives advised the business community to use its power and influence for broad social purposes, rather than solely for making the highest possible profits." This approach appealed to increasing numbers of people in business, and the idea eventually became the concept of corporate social responsibility.

The modern meaning of corporate social responsibility is presumed to rest on two principles, namely; the charity principle; and the stewardship principle.

Corporate philanthropy or gift giving – freely and voluntarily extended to those in society who are in need is the modern expression of the charity principle. A brief survey of the Nigerian business landscape shows that several companies in diverse sectors of the economy appear active on this principle going by their Annual Reports and newspaper publications on award of scholarships and donations to charity. The stewardship principle involves company managers recognising that their decisions affect the lives of many people in society and that they have a responsibility to balance the interests of many groups rather than focus on any one group alone. On this principle, there is evident widespread default by companies doing business in Nigeria. To cite a particular example, Coca-Cola vehicles are parked carelessly by the highway to discharge wares to their retailers and wholesalers with scant regard for other road users. Several other acts of corporate misdemeanour can easily be invoked.

At a more grievous scale, oil companies operating in Nigeria display acts of social irresponsibility. As it is well known, the activities of the oil companies in the riverine areas have created serious environmental and ecological problems for people living in these areas and their sources of living have been seriously eroded. Oil spills and the consequent pollution of land and water resources usually results in the destruction of crops, economic trees and fish with longer term effect being lower productivity of soils and fishing rivers (Adenikinju, 2002). True, pollution and some degree of environmental abuse may be the natural companion of industrial activity. The concern however is how responsive have the operating companies been to the fall-out of their activities. Ibanga and Obi (2001) view the issue from the angle of sustainable development by posing the question, what provision, are the current beneficiaries of oil exploitation making to enhance, to spread and ensure the perpetuity of the relevance of oil to Nigerians?

Pollution, of course, is an example of a negative external effect, posing harmful effects and costs on people other than the polluters (Eskeland and Jimenez, 1992). Unfortunately, the market, if left to itself, is not an effective mechanism for keeping pollution at reasonable levels. This is why an effective and responsible government becomes imperative. However, in desperate assertion of market supremacy, Ronald Coase and others of that philosophical bias, proposed that if those affected by pollution hold the rights to an unpolluted environment – as they should and must – polluters will "bribe" them to allow some level of pollution. However, as Eskeland and Jimenez (1992) observed,

> When polluters and victims are few and the number of beneficiaries from an agreement is fixed, the Coase proposition may indeed be valid: negotiations can provide for the internalisation of externalities .... But, for the Coase proposition to hold, the costs of negotiating and enforcing agreements must be zero or negligible. In practice, the costs will increase with the number of polluters and victims.

The abysmal failure of the compensation principle in the Niger Delta is evident from various reports whose central message is captured in the uncontested observation that the peoples of the Niger Delta region have suffered gross social and economic infrastructure neglect, poverty, frustrations, ecological catastrophe and economic deprivations [Adenikinju, 2002]. This observation has empirical backing in the effort of Orubu [2001], which examined the compensation schemes established by oil companies operating in Nigeria for the use of natural resources. His finding revealed that the rates of compensation are generally less than the market values of such resources. The irresponsibility of the oil companies goes unchecked before the very eyes of the Nigerian State.

11.4. State failure in nigeria

The persistent corporate social irresponsibility observable in Nigeria is a concrete evidence of a failed state. A major indicator of the failed state is the absence of good governance. Broadly defined, good governance refers to the effectiveness with which a government performs its work and promotes the common good (NACCIMA, 2002) Specific components of good governance include the following:

- Availability of opportunities for broad input in governance and decision-making processes for all elements of society;
- Increased governance responsiveness to citizens, including the poor and the disadvantaged;
- Credible institutional mechanisms for monitoring and redressing human rights violations; and
- De-concentration of ownership of capital, commitment to poverty alleviation and eradication of corruption from public life.

In short, good governance demands that government be responsive, responsible and accountable to the citizens.

A very clear, even if saddening, evidence that the Nigerian State has failed to live to the minimum requirements of an effective State is its constant readiness to sacrifice the people's long term interests to the satisfaction of the profit hunger of a few foreign companies. It has even sadistically offered human lives as sacrifice as is evidenced by brutal state repression of oil based agitation in the Niger Delta. Such acts unmask the partisan character of the State. This is in sharp contrast to the morality of the American political leadership that is willing to sacrifice the interests of a few corporations for the larger long term American interest as vividly demonstrated by Professor Emily S. Rosenberg in her book, *Spreading the American Dream*. This is why the Nigerian State must be charged with primary culpability for the crisis ensuing from corporate irresponsibility in the country.

11.5. Conclusion: putting the people at the centre

One consequence of the failure of the State in Nigeria is the practical collapse of the *Iron Law of Responsibility*. The law says that in the long run, those who do not use power in ways that society considers responsible will tend to lose it. Foreign companies ride rough shod over the interests of Nigerian citizens. This is facilitated by the exclusion of the ordinary citizen in the decision-making organs of the State. Florini (1999) probably has a good view of the dilemma in her analysis that

> The µ invisible hand, of the market depends heavily on the support of a thick µ glove, of rules, norms, and institutions, including government. But too often the glove is opaque, obscuring flows of information essential to the efficient and equitable functioning of both markets and the national and international institutions that regulate them.

This explains the clamour for transparency in economic, financial and business operations across the globe. Increasingly, people have come to realise that in politics, elections alone cannot provide accountability; citizens must also be able to monitor the actions of the officials they elect. In the economic sphere, just as in environmental concern, citizens need to press for the disclosure of relevant information by both firms and governments –federal, state and local.

Because increasing transparency generally requires institutions and individuals to release information they are accustomed to withholding, transparency will rarely come about unless pressures can be brought to bear. This is where civil society organisations have a critical role to play. However, as Florini (1999) points out, "demands for transparency make sense only if two conditions are met. First, the targets of the calls for transparency must be able and willing to provide the requisite information. Second, recipients of the information must be able to use it to evaluate the provider of the information according to some accepted standards of behaviours."

Given that transparency is coupled with accountability and helps to improve efficiency and fairness of markets and companies while enhancing the accountability and legitimacy of government, rising to the challenge of enforcing transparency and accountability becomes a historic responsibility in Nigeria¶s Niger Delta and Nigeria at large.

References

Adenikinju, Adeola (2002) "Resource Ownership, Human Development and Economic Growth." Paper delivered at the Conference organised by NINPA on September 18 at the Federal Palace Hotel, Lagos.

Eskeland, Gummar S. and Emmanuel Jimenez (1992) "Policy Instruments for Pollution Control in Developing countries." *World Bank Research Observer.* Vol. 7, No. 2 (July) pp. 145-169.

Frederick, William et. al. (1988) *Business and Society Corporate Strategy, Public Policy, Ethics.* New York: McGraw – Hill Publishing Company.

Florini, Ann M. (1999) "Does the Invisible Hand Need a Transparency?" In Boris Pleskovic and Joseph Stiglitz (eds.) *Annual World Bank Conference on Development Economics.* Washington D.C. Work Bank, pp. 163-1 84.

Galbraith, John Kenneth (1973) *Economics and the Public Purpose.* Boston: Houghton Mifflin Company.

Ibanga, Itoro and B. Obi (2001) " Sustainability and Deplibility of Resource Use in Nigeria: The Case of Crude Oil," *Conference Proceedings of the Nigerian Economic Society,* pp. 35-55.

Marx, Karl and Friederick Engels (1977) *Man ifesto of the Communist Party.* Moscow: Progress Publishers. Second Edn.

Medupin, Aderemi (1992) The Public Sector and Private Capital Accumulation in Nigeria. Unpublished Ph.D. Thesis. Department of Economics, University of Ibadan.

NACCIMA (2002) *Economic Policy Watch* [August]

Orubu, Christopher O. (2001) "Compensation Programme for the Use of Natural Resources by Oil Companies in Nigeria: Implications for Sustainable Development," In: *Conference Proceedings of the Nigerian Economic Society.* P93-116.

Rosemberg, Emily S. (1982) Spreading the American Dream – American Economic and Cultural Expansion, 1890-1945. New York: Hillandwang.

Umoren, Rose (2001) *Economic Reforms and Nigeria's P ol itical Crisis.* Ibadan: Spectrum Books.

Weisskopf, Walter A. (1971) *Alienation and Economics* New York: E:P Dutton & Co. inc. Chp. 5.

Wilson, Thomas and Andrew Skinner, (eds.) (1976) *The Market and the State* – Essays in Honour of Adam Smith, Oxford: Oxford University Press.

12. Oil and the politics of corporate social responsibility in the niger delta

Dauda S. GARUBA

12.1. Introduction

In recent times, the discourse of corporate social responsibility, particularly as it relates to the oil-rich, but poverty stricken, Nigeria's Niger Delta has attracted attention, nationally and internationally. While the growing dissatisfaction about the lamentable socio-economic and environmental conditions of people in the area may be adjudged as fairly documented, the same claim cannot be made about corporate social responsibility discourse, especially as an antidote to the recurring confrontations between oil companies and their host communities.

This paper seeks to examine oil exploitation in the Niger Delta and the political game that corporate social responsibility has become in the resolution of the recurring confrontations in the Niger Delta. Among the fundamental questions to which answers are attempted in the paper are: what is the nature and underlying feature of the Niger Delta crisis? How is it related to the burning issue of corporate social responsibility? To what extent do oil companies consider corporate social responsibility imperative to the business of oil exploration and production? How are their actions responsible for the perennial belligerent behaviour of their host communities? What are the complaints of oil communities against oil companies? How does the character and behaviour of the Nigerian state shape or impact on the interactions between oil companies and their host communities? How does this in turn impinge on Nigeria's economic growth and security profile? To what extent are the corporate social responsibility projects in the various oil communities in the Niger Delta products of genuine consultation with the people? And, above all, what kind of public policy measure is required to bring about lasting and sustainable peace in the Niger Delta region?

The paper is organised into five parts. Background issues in the Niger Delta crisis follow this introduction. While section three considers some conceptual perspectives to corporate social responsibility discourse, part four examines the issues discussed in section three against the backdrop of the politics they portend in the interactions between oil companies and their host communities on the one hand, and between the latter and the government on the other hand. Part five, which is the conclusion, ties up the arguments of the work and proffers solution on how best to apply corporate social responsibility to resolve the perennial conflicts in the Niger Delta.

12.2. Niger delta crisis: background issues

The Niger Delta crisis draws attention to the geography and state of socio-economic conditions of the region upon which oil exploration and production is negatively impacting. The generally acceptable true geographical definition of the region gives its northern limit as Aboh (fairly above the first division of the River Niger into its two main Nun and Forcados rivers), while its west and eastern boundaries are the Benin and Imo river estuaries (Braide 2001). In simple geopolitical term, the Niger Delta includes the states of Bayelsa, Delta, Rivers and Akwa Ibom, with its fringes in Abia, Cross River, Edo, Ondo and Anambra States.

The area consists of several ecological zones. These zones include the sandy coastal ridge barriers with lowland swamp forest, which represents some of the last remaining pristine forest resources of endemism in Africa; the brackish or saline mangrove forest, which is the largest in Africa and third largest in the world; the lowland swamp forest zone with its fresh water and permanent season forest, which is the most extensive in West and Central Africa; and the lowland rain forest, which is usually about 60% flooded annually (Braide, 2001). The difference in the characteristics of the ecological zones in the Niger Delta notwithstanding, the area shares a common trait of vast floodplain of sedimentary derivation with rivers, rivulets, tributaries, creeks and estuaries (Braide 2001; Hutchful 1985:114; Constitutional Right Project 1999:12).

Besides, a hydrology that is determined by a flood regime of the Niger River and a tide of the Atlantic Ocean, the Niger Delta is known to be particularly sensitive to changes in the quality of water which, as earlier noted, shares some characteristics of salinity and pollution. The core Niger Delta's population is estimated between 10 and 12 million, and the settlement pattern is determined by the availability of dry land (Constitutional Right Project 199:7, Aboro 2000:37; Environmental Research Managers 1997:38). Subsistence agriculture still dominates rural land use (though swamp and riverine topography severely constrain land cultivation)), while fishing activities are widely practiced in inland waters, near shore and offshore. Also, because of its rich flora and fauna, the Niger Delta is known to harbour rare species of wild flowers, birds, other aquatic lives and wild animals. It is from the perspective of this ecological wealth in the region that the environmental destruction caused by activities of oil exploration and production (which is the major industry in the region¶s economy), that the corporate social responsibility discourse assumes a critical dimension.

Historically, oil was first struck in commercial quantity at Oloibiri (in present Bayelsa State) in August 1956 by Shell D‘ Arcy, in partnership with British Petroleum (BP). Another discovery in Afam, and Ebubu and Bomu in Ogoniland (now in Rivers State) in the following year vindicated earlier prospects of the Niger Delta as being a major oil region, but it also stimulated further interest by other foreign oil exploration firms, which began to flock to the area.

Among the several oil multinationals, which joined later, were Mobil, Gulf (now Chevron), AGIP, Safrap (now Elf), Tennessee (now Tonneco and Amoseas (now Texaco). Others are Japan Petroleum, America Occidental, Deminex Nig. Ltd., Union Oil, Great Basins, Ashland, Statoil and Dupont. A few indigenous companies such as; Henry Stephen, Delta Oil, Nigeria Petroleum Development Company, Nigeria Oil Resource, among others also operate in the region.

Although some of these companies later quit Nigeria because of unsuccessful exploration activities, while others are presently on merger deals in the light of on-going globalisation forces, there are still over 11 oil multinationals and several other indigenous companies in Nigeria, operating in over 159 oil fields and producing from over 1,481 oil wells; with many fresh bloc licences still being sought and awarded. Of this figure, the Shell Petroleum Development Company (SPDC), operating almost exclusively in the Niger Delta, owns 83 oil fields, 748 oil wells and 931 oil producing wells, linked by a network of 4,780 kilometres of field pipelines and 1,543 kilometres of trunklines to 87 flow stations and 3 oil export terminals (Constitutional Right Project 1999:10; Manby 1997: 14). The company also produces about half of the entire daily oil output in Nigeria, with more than 90% pumped from onshore concessions (Constitutional Right Project 1999:13).

Outside oil, it is claimed that Nigeria has an estimated 104.7 trillion cubic feet (tcf) of proven natural gas reserves, which is the tenth largest reserves in the world. However, little progress has been made in gas exploitation. The capacity of SPDC"s gas plant is 970 million standard cubic feet of gas per day and 13 gas compression stations all of which are in the Niger Delta (Constitutional Rights Project 1999:13).

Since the first oil export from Nigeria in 1958, the Federal Government's control over the exploration of the resource and its sister mineral (gas) has progressed from its initial regulatory role to direct involvement in the upstream and downstream sector, in the form of acquisition of equity interests and joint ventures with several of the oil multinationals.

The significant status that oil (and by extension gas) occupies in Nigeria's political economy is better appreciated by the fact that the mineral resource displaced agriculture from its earlier top role, in barely a decade after independence to the extent that it now accounts for about 85% of national revenue and 90% of foreign exchange earnings. This has not been without devastating side effects, both on the part of the Nigerian State and the Niger Delta region from where the country's oil and gas resources are exploited. Oil has attained such a significant political position in Nigeria, to the extent of being at the root of prebendal behaviour among its military and civilian leaders alike (Omoruyi 2001; Turner 1980; Mkandawire and Bourenane 1999; Joseph 1996). Also, its continued exploitation has not only added to the persisting crisis of socio-economic neglect and the parlous state of underdevelopment which the area now depicts, it has also thrown up worsening environmental insecurity which since the mid-1980s have received increasing global attention (Obi 1999:4). These two problems and the conflicts they evoke have combined to trigger corporate social responsibility discourses in which several individual and interest groups – intellectuals, opinion leaders, ethnic and human rights campaigners, environmentalists and professionals have taken a frontal row.

Whatever the differing views articulated by these individuals and groups, they are nevertheless unison about the lamentable socio-economic plight of the Niger Delta people, which is further worsened by environmental degradation arising from oil exploitation.

Socio-economically, the Niger Delta bespeaks poverty and neglect. Despite its vast oil wealth, the area remains poor with a GNP per capita that is below the estimated national average of US$260 (Manby 1997:46). It plays host to a teeming unemployed local population and intending oil worker' migrants. The poor state of infrastructural development is such that the region suffers the inadequacy of schools for its population density of 100 persons per square kilometer, absence and/or erratic electricity and lack and/or poor medical facilities. Hutchful (1985:115) observes that transportation and communications are major problems in the riverine areas of the region. There is almost total absence of roads particularly in the riverine areas. Because large creeks form the main arteries in the region, long-distance journeys in the Niger Delta are not only slow, but also Hazardous. Potable water is a major problem, especially in salt-water areas where the water table is not only deep but also expensive to tap.

With official statistics that places 72% of households in the Niger Delta within poverty brackets, the story of the region, when viewed against the hardships caused by activities of oil exploration, production and transportation and the widespread social and ecological dislocation, is indicative of a macabre twist of fate. Owing to growing ecological disaster, the Niger Delta is fast losing its capacity to support qualitative life. Explosions from seismic survey, acid rain, pollution from refinery effluents and drilling fluid and widespread disruption of natural terrain from industrial constructions and installations impair the peoples¶ health.

Perhaps, the most controversial of the environmental impacts of oil activities in the Niger Delta remains oil spills and pollution arising from pipeline leaks and blowout, as well as pollution from gas flaring. Industry sources reveal that Nigeria accounts for 78% of all the gas flared in Africa and 25% of the entire world total (Darah 2001). In 1991 alone, the country surpassed by 72%, the world average for natural gas flaring which was 4%.

With the United States' gas flares figure of 0.6%; Holland's 0.0%; Britain's 4.3%; Mexico's 5.0%; and other OPEC'S countries figures of Libyan's 21%; Saudi Arabia's 20%; Iran's 19%; Algeria's % (International IEA 2001:144-145), Nigeria's 76% figure during the period made her the worst environmental polluter through gas flaring.

Table I below shows gas production and flaring in Nigeria from 1970 to 1999. It is evident in the table that as gas production has increased in Nigeria, so has gas flaring. Though there is a decline in gas flared as a percentage of production, there has been a considerable increase in production since the 1 990s and much more gas has been flared

Table I: Gas Production and Flaring in Nigeria 1970-1999

Period	*Production (million cubic metres)*	*Flaring (million cubic metres)*	*Gas Glaring as a % of total Production*
1970-74	17,438	17,172	98.5
1975-79	22,156	21,083	95.2
1980-84	17,699	14,754	83.4
1985-89	19,940	14,904	74.7
1990-94	31,691	24,788	78.2
1995-99	35,780	25,616	71.6

Source: Adewuyi, A.D (2001) as adapted from Central Bank of Nigeria, Statistical Bulletin (December 1998) and Annual Report and Statement of Accoounts (1999).

Gas flaring in Nigeria is so serious and Hazardous because at the temperatures of 1300^0C and 1400^0C, the multitude of flares in the Niger Delta heat up everything, cause noise and pollution, and produce SO2, $_{CO2,}$NOx and particles around the clock. The emission of carbon dioxide from gas flaring in Nigeria releases 35 million tons of methane (CH4) (Agun 1997:57-58), while oil and gas in the Niger Delta mainland alone accounts for 48% of the greenhouse gas emission in Nigeria (Magbagbeda 1999:283). What this point to is that, through its oil and gas activities in the Nigeria Delta, Nigeria may be contributing much to global warming.

Oil spillages abound in the region with considerable negative impacts. Such include the Ebubu Ochame spill in 1970 which covered 10 hectares of land (SPDC, May 1996) and the 1980 Funiwa Well Blowout in which an estimated 400,000 barrels of oil was lost (Adeyemi-Wilson 1995:16; Hutchful 1985:124-

128). In 1984, a spill that wiped out several communities, water and marine lives, as well as economic trees, fishponds and tracks was witnessed in Sakripulo and its suburbs *(Tempo* 29; 01; 1998:8). The Yokri, Sobebolu and Obotobo II areas of Burutu local government in Delta State also experienced 23 spillages between 1991 and 1997, while Alleibiri (an Izon Village) experienced a spill that caused the death of 35 persons in 1997. Other spillages due to pipeline failures, plant upsets and failures, valve leakages and pipeline vandalization, have also been recorded among people from Okrika, Afam, Kalabari, Nembe, Ogoni, Urhobo, Itsekiri, Andoni and Ijaw ethnic regions with considerable devastating environmental impact.

Table II below shows that there were 4,647 oil spills between 1976 and 1996. During the period, the quantity spilled and quantity recovered in barrels were 2,369,470.04 and 549,060.38 respectively, while net volume lost to the environment was 1,820,410.50 barrels.

Table I: Oil Spills in the Niger Delta 1976-1996

Year	*Number of Spill*	*Quantity Spilled (Barrels)*	*Quantity Recovered (Barrels)*	*Net Volume Lost to the Environment (Barrels)*
1976	128	26,157.00	7,135.00	19,021.05
1977	104	32,879.25	1,703.01	31,176.75
1978	154	489,294.75	391,445.00	97,849.75
1979	157	694,117.13	63,481.20	630,635.93
1980	241	600,511.02	42,416.83	558,094.19
1981	238	42,722.50	5,470.20	37,252.30
1982	257	42,841.00	2,171.40	40,669.60
1983	173	48,351.30	6,355.90	41,995.40
1984	151	40,209.00	1,644.80	38,564.20
1985	187	11,876.60	1,719.30	10,157.30
1986	155	12,905.00	552.00	12,358.00
1987	129	31,866.00	6,109.00	25,757.00
1988	108	9,172.00	1,955.00	7,207.00
1989	118	5,956.00	2,153.00	3,803.00
1990	166	14,150.35	2,092.55	12,057.80
1991	258	108,367.01	2,785.96	105,912.05
1992	378	51,187.90	1,476.70	49,711.20
1993	453	8,105.32	2,937.08	6,632.11
1994	495	35,123.71	2,335.93	32,787.78
1995	417	63,677.17	3,110.02	60.568.15
1996	158	399,036.67	11,838.07	387,198.60
TOTAL	*4647*	*2,369,470.04*	*549,060.38*	

Sources: Braide, S.A. (2001) as adapted from the Department of Petroleum Resources (DPR) records.

The overall impact of all these on land, water and air have been soil degradation, water contamination and pollution, deforestation, air and noise pollution, distortion of ecosystem and biodiversity, the virtual loss of traditional fruits, economic trees, farm crops as well as destruction of human and other lives such as fishes, prawns, shrimps, crabs, oysters, manates and crocodiles (Evinson 2000:49-50; Emuakpor 1998:22; Emmanuel 2000:33-34).

Given its huge oil production capacity when compared to other companies, it is obvious that Shell contributes more to environmental destruction in the Niger Delta than any other oil company. For instance, the company was involved in all the incidents of oil spills in 1998 and 1999 (Adewuyi 2001:127; The *Comet,* 05; 06 2000:7).

The ecological and security threats posed by the situation to the socio-economic lives and property of the Niger Delta people is further aggravated by the deliberate efforts of the Nigerian State, under an alleged conspiracy of a politico-military elite to dispossess them of control of their God-given resources. It is in this context that variously enacted laws such as the Petroleum Decree No. 5 of 1969, Off-Shore Oil Revenue Decree No. 9 of 1971, Land use Act of 1978, Sections 40(3), 42(3), 44(3) of 1979, 1989 and 1999 Constitutions respectively, Title Land (Vesting) Decree No. 52 of 1993 and National Inland Waterways Authority Decree No. 13 of 1997, have been explained and challenged by the minority groups in the region (Omoruyi 2001; Sagay 2001).

The deepening antagonism is characterized by a regime of state violence in the forms of harassment of outspoken community and group leaders (who have successfully championed meetings leading to release of historical documents such as Ogoni Bill of Rights, Kaiama Declaration, Oron Bill of Rights, The Ikwerre Charter of Demands Aklaka Declaration of the Egi People, Urhobo Economic Summit Resolution, etc.) through surveillance, arrest and detention. There have been state incitement of "violent conflicts between communities and ethnic groups, the utilization of instigated violence as a pretext for repression", as well as "direct repression by armed forces and police through extra-judicial killings, flogging, torture, rape, looting and extortion". More recently, there has been the military attacks and destruction in the towns of Choba and Odi (International IDEA 2001:248-249). The brutal intervention of the Nigerian State in the conflict has not only eroded the people's confidence in it as a supposed guarantor of their collective interest, but has also re-invigorated agitations in which the demand for corporate social responsibility has become the irreducible minimum.

12.3. Perspectives on corporate social responsibility

The concept of corporate social responsibility is not entirely new especially among scholars and practitioners of management. It is about an "intelligent and objective concern for the welfare of the society that restrains the individual and corporate behaviour from ultimately destructive activities" (Soeze 2001:23). The concept thus refers to an institutional social obligation of a business to the society in which it operates with the sole purpose of creating the overall environment that such business organization requires to survive and realize her objectives outside the organization (Kootnz 1984:74). Our perspective in the discourse of corporate social responsibility centres on Peter Drucker's (1997:568) traditional and modern approaches. The emphasis of the traditional approach is about what people in business should invest and contribute to enhance maximization of profit without regards to the social problem of the business environment. The main thesis of the approach is that the business of business cannot but be business; hence organisations can only make their best to the society if they focus only on profit maximization without involving themselves with social charity that not only obstructs processes of wealth creation for employees and shareholders, but also short-changes them (Friedman 1981; Miller 1981; Lenitt as cited in Elamah 1998:11).

The modern approach to corporate social responsibility, according to Drucker, counters the traditional approach in that it goes beyond what business should do or achieve for those under its sphere of control to incorporate businesses taking responsibility for social problem to the point of becoming the society's conscience and the solver of society's problem (Drucker 1977:261). This notion of having businesses to react to the total environment (the entire society) rather than markets alone helps "to solve virtually numerous problems because people who have a good environment, education and opportunity make better employees, customers and neighbours for companies than those who are poor, ignorant and oppressed" (Kenneth Andrews, as cited in Agun 1996:17).

A middle course or compromise position between both extremes (traditional and modern approaches to corporate social responsibility) is embodied in the work of Adegboyega Ogunsanya (1991). The position is unique in that it argues that even though profit maximization of the traditional approach is the goal of every business, there is the necessity for interest harmonization. Perhaps the uniqueness identified in Ogunsanya¶s thesis is premised on three major stages/developments in the environment. The first is the perception of managers in the classical approach whose primary responsibility was to ensure the efficient allocation of resources to produce goods and services desired by customers at affordable prices.

The second marked a shift in focus when, between 1920s and 1930s, efforts by business executives in large corporations started redefining the system to reflect interpretation of resource – ownership to mean society's resources. In this situation, it was discovered that shareholders of a business gain by contributing to the society from which, and with whom, they do business. The third and final stage perfected the thinking that employees, customers, suppliers and the public have equal interests, and that in the absence of all of them, there can be no performance or result (Ogunsanya 1991:104). This was how the concept of corporate social responsibility shifted its focus entirely from profit maximization to harmonization of interests. Newman and Logan (1991) may have rightly anticipated Ogunsanya's view when they earlier opined that companies that fail to recognize their social responsibilities toward the society, lose ultimately given that whatever affects the society in which a business owner operates equally affects him and his business.

It is in this context that this paper now turns to examine corporate social responsibility and the political game it has been reduced to, in the course of oil exploitation in the Niger Delta.

12.4. The state, oil companies and social responsibility

Niger Delta crises are rooted in the ecological destruction and general dislocation of the social and economic life of the people. Even though there are several perspectives, they nevertheless point to the general crisis of the state, governance and public policy in Nigeria.

Nigeria is a rentier state in view of its status as a mere collector of rents from multinational oil companies (Nkandawire and Bourenane 1999:1-24). The Nigerian state has been a contested terrain and a mediator of conflicts between various social forces such as those in the Niger Delta (International IEDA 2001:234). The weakness and the partisan posture of the Nigeria state with respect to the Niger Delta crisis impinge on its capacity to effectively protect the region's environment and guarantee its development.

It is this constraint and the anger it provokes, in the form of challenges, protests and militia actions by oil communities against oil companies and the repressive Nigerian state, that have triggered contentions about corporate social responsibility in the Niger Delta. The demands for social responsibility by oil communities and sympathizers of their plight have been expressed as "separate from and not alternative to that owed by the government" (Manby 1997:49).

Although it can be argued (as it is often done) from a legal perspective that oil companies can only be responsible to the Federal Government (the constitutional owner of all mineral on and underneath land in Nigeria) by way of prompt payment of tax, and thus should no be under any obligations to provide social or economic amenities to oil communities for operation (Inko-Tariah 1991:1; Uchegbu n.d:42-43), the situation cannot be regarded as the same where activities of the former impact negatively on the natural environment and life of the latter (Adewale n.d:7 1). Indeed, social responsibility under the latter condition is as compelling as it is important. This is all the more essential in view of the attendant difficulty in securing fair and adequate compensation for damages done to property, crops and economic trees, as well as other inestimable valuables such as mangrove trees, rattan cane, firewood, wild food crops, animals and fish. Though the Oil Pipeline Act of 1958 (as amended in 1965), the Petroleum Act of 1969, the Petroleum (Drilling and Production) Regulation of 1969, among others, grant protection rights to oil communities (including payment of fair and adequate compensation for damages), they are hardly enforced, not only because of the inherent ambiguity in ascertaining what constitutes "fair, adequate" compensation, but also because would-be beneficiaries are illiterate peasants without the means (whether individually or collectively) to challenge in court, the cases of refusal by oil companies to pay, which often go beyond the court of first instance (Orubu 2001:99-100; Constitutional Rights Product 1999:31-32). Litigation sometimes arises when oil companies hide under the canopy of sabotage to refuse payment for oil spills and fire damages to property, crops and economic trees.

While oil companies maintain that they need not be socially responsible for development projects in their areas of operation in the Niger Delta, they have made several claims of substantial responses to the social needs of the people as a strategy to douse the recurring cases of unrest and the mounting local, national and international pressures in the region. For instance, Shell claims that its community assistance budget rose from US$330,000 in 1989 to US$7.5 million in 1993 and US$36 million in 1996. Mobil also claims to have spent an average of US$8 million on community development projects between 1994 and 1997. Elf's estimated annual budget on community development is between US$4.5 million and US$5 million, while AGIP's "Green River Project" is allegedly gulping US$2.5 million yearly (Manby 1997:49-50). The oil multinationals claim that their social responsibility projects are based on the needs of the people, and as agreed on by their host communities (Manby 1997:49-50; *The Pointer* 01; 10; 2002: 12). The needs cover areas in education, healthcare, economic development, environmental protection and social amenities.

Through its regularly published pamphlets, bulletin and leaflets, Shell claimed that between 1986 and 1997, it built 269 classroom blocks and provided furniture and science equipment to 250 schools, through contractors and suppliers chosen from local government areas where projects were executed (SPDC, October 1995; February 1997). The company also çlaim that it has worked in collaboration with the National Youth Service Corps (NYSC) to encourage science education by sponsoring teachers in 14 schools in 1995 and 1996. Its concern for economic and social hardship in oil producing areas and the larger Nigeria society is demonstrated in its yearly award of scholarships to secondary schools and tertiary institution students; a scheme which benefited 3,710 students (2,600 secondary school students and 1,110 University undergraduates) between 1995 and 1997 alone (SPDC, February 1997).

Other projects that Shell claim to have undertaken in the areas it is exploiting oil include the construction of 100 roads between 1986 and 1997 (in addition to the 750 kilometres of operational roads to its operating facilities in the Niger Delta); 90 water schemes comprising drilling of boreholes, erecting of tanks and installing of pumps, pipelines and taps between 1986 and 1994; 9 hospitals including Okoroba and Ogulagha, and Terabor General Hospital in Ogoniland; as well as an active role in the national immunisation campaign for children against polio, measles, whooping cough, diphtheria, tuberculosis and tetanus (SPDC, February 1998; February 1999). Shell also claims to have spent N67.3 million on several assistance projects in areas of agriculture and small scale business, including its training of about 145,5 000 new farmers on modern methods of farming for improved yields in 1993 and 1994 alone (SPDC, October 1995).

The Federal Government's efforts at developing the region are largely institutional driven given that special commissions are set up to handled projects (Akinyele 1998). The Oil Minerals Producing Areas Development Commission (OMPADEC) and its recent successor – The Niger Delta Development Commission (NDDC) are the most popular in this wise. Between 1993 and 1997, OMPADEC spent approximately N7.341 billion to execute projects in the areas of water, road, electricity, jetty, building, drainage among others (Elamah 1998:32). Also, a recent revelation by NDDC Spokesman, Anietie Usen indicates that the Commission has done well with 700 projects (among which 160 have been completed) in its barely one and half years of existence. Within an annual budget of N40 billion under which it operates *(ThisDay,* Saturday, November 23, 2002:35), NDDC's mission is "to facilitate the rapid, even and sustainable development of the Niger Delta to a situation that is economically prosperous, socially stable, ecologically regenerative and politically peaceful" *(Insider Weekly11: 11:* 2002; 27).

An assessment of the acclaimed social responsibility projects executed by oil companies and the development efforts by the government in the oil producing areas of the Niger Delta draws attention to major fault lines in the strategies adopted. Apart from the fact that some of the projects "executed" start and end on pages of newspapers, there is a wide gap between community needs and industry perception of what communities should have; not minding the claim by all companies that their community development projects are relevant to the needs of the people that they are meant to serve. For instance, the Shell fish processing plant in Iko,

Akwa Ibom State, is not only far from any potentials for market, but it neither has any qualified local personnel to run it nor electricity to provide cold storage (Manby 1997:50).

Of equal concern is the fact that the huge sum of money often announced by oil companies to have been committed to community development projects in the Niger Delta are a far cry from the actual work done. A case in point was the claim of US$220,000 worth of humanitarian aids granted by SPDC through charitable organizations to cushion the pains induced by the clash between Adoni and Ogoni communities in 1993; an amount critics argued would be enough to transform the two communities for better, if actually such amount was spent by Shell (Elamah 1998:64). Egbuche (1998) also posits that the estimated US$35 – US$40 million claimed to have been budgeted by Chevron and Mobil for community development projects in their areas of operation in the Niger Delta would have made significant impact, if only the oil industry as a sector has a common strategy for community development. There are indications that while it may be true that oil companies budget the huge amount of money often claimed for community development as a form of social responsibility, a large portion of the fund often get diverted into pockets of oil company officials, local contractors, traditional rulers, "community leaders" and anticipated "troublemakers" in oil producing communities. Allegations of corruption and compromise of community leaderships and local elites over compensations and other benefits from the oil companies have led to quarrels, disturbances, court litigations and even killings.

In a nutshell, an analysis of corporate social responsibility efforts in the oil producing areas of the Niger Delta reveals largely dissatisfied tales. Manby (1997:50) recalls the result of a field research undertaken by Shell in its areas of operation where 84% of respondents to administered questionnaires expressed the feeling that activities of the oil company have adversely affected the economies of its host communities, and that 69% felt a high level of deprivation and neglect in spite of the "numerous' community development efforts of the company. What this portends is that even though a lot has been spent on community development, very little success has been attained, given the abundant insincerity and fault-lines in the design and execution of projects.

12.5. Concluding remarks

This paper discussed the politics of corporate social responsibility as it affects the Niger Delta where oil exploration and production activities continue to trigger endemic crisis. The paper acknowledged that the Niger Delta crisis is rooted in the ecological destruction and general dislocation of social and economic life of the people. Added to this is that because of the rentier nature of the Nigerian state, whose state-oil linkage is personified by perpetual weakness for political power among a corrupt politico-military elite, the state has shown a gross inability to protect the environment as well as guarantee the development of the Niger Delta. This weakness and the anger it provokes are manifested in the recurring challenges, protests and militia actions undertaken by oil communities against oil companies and the repressive Nigerian state. How best these conflicts can be resolved to remove obstruction from the country's mono-cultural source of revenue has triggered increased discourse of corporate social responsibility. Though there are perspectives to this discourse, especially in terms of its legal and moral justification, the fact remains that it is seen as a solution.

But because the solution that is offered is characterized by inadequacy, sabotage and insincerity, both in the design and in the execution of community development projects, the Niger Delta crisis remains perennial.

As a suggestion to a meaningful resolution of the crisis therefore, there is the need for government and all multinational oil companies in the Niger Delta to harmonise their strategies for development of the region in order to be able to make meaningful impact. This can be achieved by establishing a body comprising representatives of all stakeholders including Niger Delta communities, which shall be charged with the responsibility of managing the budgeted funds for corporate social responsibility in the region. The new NDDC is a bit far from what is being suggested here because of overbearing government power over it. Another body should be set up along the same line with a responsibility to set priority projects and targets for participating parties, based on wide consultations with the people, to determine what they need most. The processes of these decisions should be regularly publicized in the media for wider public information and endorsement.

It is also the considered opinion here that the efforts of non-governmental organisations (NGOs) should be encouraged as a bridge gap between oil companies and their host communities on the one hand and the latter and the government on the other hand. As institutions that occupy the space between the public realm and private domain, NGOs¶ norm-creating and agenda-setting roles, as well as their patterns of challenge to authoritarian principles which confer some independence and growing world recognition place them at a vantage position to better assess and influence policy issues with a view to promoting the required working partnership among all stakeholders in the Niger Delta. By and large, it is necessary to say that if these suggestions are pursued genuinely, there will obviously be no end to what can be achieved in corporate social responsibility in the Niger Delta.

References

Aboribo, R.I. (2001) "The Political Economy of Environmental Pollution in the Niger Delta: From Compensation to Participation," *African Journal of Environmental Studies,* Vol. 2, No. 1

Aboro, D. E. (2000) "Mr. President, the NDDC Bill and Niger Delta Environment"" Lagos: *The Guardian,* Monday, August 28.

Adewuyi, A.O. (2001) "The Implications of Crude Oil Exploitation and Export on the Environment and Level of Economic Growth and Development in Nigeria" in The Nigeria Economic Society (2001) *Natural Resource Use, the Environment and Sustainable Development: Selected Paper Presented at the 2001 Annual Conference,* Ibadan: The Nigerian Economic Society.

Adeyemi, Babatola (2002) "Corporate Social Responsibility as a Stimulus for Competitive Advantage", Lagos: *The Guardian,* Friday, December 6, 2002.

Agun, E.A. (1997) Social Responsibility of Oil Companies to Communities in their Areas of Operation: A Case Study of Nigeria Petroleum Development company Limited", Unpublished MPA Project, Benin: Department of Political Science and Public Administration, University of Benin.

Ajoje, F.M. (2000) "Community Relation and the Oil Industry", Lagos: *The Guardian on Sunday,* August 20.

Braide, S.A. (2001) "Oil in the Niger Delta: What are the Pollution Issues?" A paper presented in the Niger Delta Environmental Training Programme at Airport Hotel, Port Harcourt.

Constitutional Rights Project (1999) *Land, Oil and Human Rights in Nigeria's Delta Region,* Lagos: Constitutional Rights Projects.

Darah, G.G. (2001) "The Politics of Oil and Resource Control", Asaba: Lecture delivered at the Conference of Delta State House of Assembly.

Drucker, Peter (1997) *Management,* Pan Books Ltd.

Egbuche, Uzo (1998) "Multinational Corporations and Social Responsibility", Lagos: *The Guardian on Sunday,* July 26.

Elamah, A.Y. (1998) "Management of Community Relations in the Nigerian Oil Industry (A Case Study of Shell Petroleum Development Company)", Unpublished MPA Project Lagos: Department of Political Science, Lagos State University.

Emmanuel, Harris-Okon (2002) "Nigeria: An Environment in D ire Need of Redress," *The Comet,* Tuesday, June 6.

Emuakpor, C.O. (1998) "Woes of Oil People", *The Guardian,* Thursday, August 20. Environmental Resources Manager (1997) *Niger Delta Environmental Survey: Report Phase I, Volume IV Abridged Version of Finding and Recommendations of Phase I Report,* Lagos: Environmental Resources Managers Ltd.

Evinson, Ebi (2000) "Politics of Derivation Principle in Delta State", *The Guardian,* Tuesday, September 6, P. 29.

Friedman, Milton (1981) *The Invisible Hand in Economics and Politics,* New York: Institute of South East Asian Studies.

Hutchful, Eboe (1985) "Oil Companies and Environmental Pollution in Nigeria" in

Claude Ake (ed) *Political Economy of Nigeria,* London & Lagos: Longman. *Insider Weekly* (Lagos) (2002) November 11.

International IDEA (2001) *Democracy in Nigeria: Continuing dialogue(s) for Nation – Building,* Stockholm,

Joseph, Richard (1996) *Democracy and Prebendal Politics in Nigeria: The Rise and Fall of the Second Republic,* Ibadan: Spectrum Books.

Kootnz, Harold et al (1984) *Management,* 8th Edition, New York: McGraw-Hill. Magbagbeola, N.O. (1999) "Economic Growth and Environmental Development in Nigeria: Impact of Climate Change," *Nigerian Journal of Economic and Social Studies,* Vol. 41, No. 2, Pp. 275-294.

Manby, Bronwen (1997) *The Price of Oil: Corporate Responsibility and Human Rights Violations in Nigeria's Oil Producing Communities,* New York, Washington, London, Brussels: Human Rights Watch.

Miller, D. (1981) "Configuration of Strategy and Structure," in R.E. Miles and C.C. Snow (eds), *Organisational Strategy, Structure and Process,* New York: McGraw-Hill.

Mkandawire, M. and N. Bourenane (1999) "The State and Agriculture in Africa: Introductory Remarks", in M. Mkandawire and N. Bourenane (eds.) , *The State and Agriculture in Africa,* London: CODESRIA Books.

Newman, W. and Logan J.P. (1991) *Strategy, Policy and Central Management,* Southern Publishing Company.

Obi, C.I. (1999) *The Crisis of Environmental Governance in the Niger Delta 1985-1996,* Monograph, Harare: AAPS Books.

Ogunsanya, Adegboyega (1991) *Strategy and Public Relations,* Richmond Books Ltd.

Omoruyi, Omo (2001) "The Politics of Oil: Who Owns Oil, Nigeria, State or Communities?" Lagos: *The Guardian,* Wednesday, January 24.

Orubu, C.O. (2001) "Compensation Programme for the Use of Natural Resource by Oil Companies in Nigeria: Implications for Sustainable Development", in The Nigerian Economic Society (2001) *Natural Resource Use, The Environment and Sustainable Development: Selected Papers Presented at the 2001 Annual Conference, Ibadan:* The Nigeria Economic Society, Pp. 93-116.

Sagay, Itse (2001) "Nigeria: Federalism, the Constitution and Resource Control", Lagos: Lecture delivered at the Fourth Sensational Programme organised by the Ibori Vanguard at the Lagoon Restaurant.

Soeze, C.I. (2001) "Oil Industry and Corporate Social Responsibility". *ThisDay,* Tuesday, January 16.

The Pointer, (2002) October 1, P. 12.

SPDC (1995), *Nigerian Brief,* Lagos: SPDC, October.

SPDC (1997) *Nigerian Brief Lagos:* SPDC, February,

SPDC (1998) *Nigerian Brief,* Lagos: SPDC, February

Tempo (1998) (Lagos), 29 January.

ThisDay (2002) (Lagos), Saturday, November 23.

Turner, Terisa (1980) "Nigeria: Imperialism, Oil Technology and the Comprador State" in Peter Nore and Terisa Turner (eds.) *Oil and Class Struggle,* London: Zed Press.

Section IV
oil, conflict and national security

13. Conceptions of third world security and the management of the emerging threats in the niger delta

Christopher ISIKE

13.1. Introduction

The resource based agitation and violence in the Niger Delta region is not new neither did it commence with the discovery of oil in Oloibiri, by Shell-BP in May 1956, and its commercial exploration and exploitation in 1958. The problem can actually, in terms of historical origin be traced to 1849 with the imperial activities of John Beecroft, the British Crown's first consul in the Niger-Delta area (Osae & Odunsi, 1982). It was John Beecroft who introduced the concept of 'gunboat diplomacy' in the Niger Delta, through which he forced questionable agreements and treaties upon local rulers in the area, offering them "protection' in exchange for allowing British traders special and unlimited access to trade and economic activities in their domains. In cases where the kings refused to sign or be bound by such treaties, gunboat diplomacy was extended to violence and low-intensity war when such rulers were hounded, murdered and their domains torched or they were dethroned and or exiled, as the case may be. It is in this context that the violence and inhuman treatment visited on King William Koko of Nembe and his people (Akassa massacre), King Dappa Pepple of Bonny, King Jaja of Opobo, Prince Nana Olomu of Itsekiri kingdom and their peoples should be understood. Other imperialists induced attacks include those on Robin Tree and Okokun in 1886, the Obe and Warri creek attacks of 1887, Asaba and Egbedi (1888) and Oguta in 1890. (IDEA; 2000:143) Thus began an imperial enterprise whose sole target was the expropriation of the economic resources (mainly palm oil then) of the Niger-Delta people and the institutionalization of violence and use of coercion as instruments of political control.

This story of imperialism and institutionalized violence through colonialism, continued from John Beecroft to Federick Lugard. Imperialism and colonialism disrupted the traditional indigenous culture and effective political institutions cum administration in the Niger-Delta. Through the mechanism of indirect rule, they imposed puppets who did not emerge from the recognized social bases of political power, and as such were not recognized by the people. Since these puppets were not subject to their people, but to the imperial control of the colonialists, they were not accountable to their people. Thus also, began the problem of political legitimacy and leadership accountability in the Niger-Delta. This has persisted till date, manifesting in the form of unpopular community leaders, traditional rulers, youth leaders and even state Governors forcing themselves on their people and mortgaging their interests for personal monetary gains from the Multinational Oil Corporations (MNOCs) and the Federal Government.

Naturally, the people reacted to this external aggression by British and European traders, which threatened the territorial integrity of the Niger-Delta, through stiff and sometimes violent opposition for which they ended up more battered. The people reacted to the problems of political coercion and control effected through leadership deposition and imposition, through protests and demonstrations, defiance and rebellion. They also people fought against external aggression, the imperial exploitation of their resources, tyranny and political illegitimacy. It is the sum of the peoples' reactions to imperialism and colonialism that underpinned the restiveness of the Niger-Delta people, beginning from 1849 to the colonial period.

With local opposition crushed through brutal force in most cases, as the 1895 Akassa massacre showed, and with illegitimate political control of the people assured, the stage was set for the Royal Niger Company (R.N.C) to expand its trading frontiers beyond the Niger-Delta. This it did northward, following the River Niger down to Idah, in today's Kogi state. This paved the way for the eventual colonization of the territory of what later came to be known as Nigeria.

The point so far is that violence from two sources constitute the Niger Delta problem of today and that these sources date back to our colonial history. First, there is the institutionalized violence from the state aimed at ensuring political control to pave the way for the economic exploitation of palm oil trade then, and secondly, the individual resistance and collective opposition of the people to state repression and violence. This picture plays out in modern Nigeria, with successive federal governments, both military and civilian, including the present one, continuing the repressive state policy of annexing the oil in the Niger-Delta, denying legitimate agitations for resource control, militarizing the area as deterrence measures and the occasional torching and outright sacking of entire Niger-Delta communities. The violence visited on Umuechem, an oil producing community, in 1990, on the orders of the then head of state, Gen. Ibrahim Babangida, which left scores of people dead and hundreds of houses razed, for simply agitating for a better deal from Shell (Rowel; 1996), is an eloquent testimony of state violence unleashed on a people it has a social contractual obligation to protect. A more recent example is the violent sacking of Odi community in 1999 on the orders of a democratically elected president, Olusegun Obasanjo, following the killing of twelve policemen (sent to restore order in the town) by restive youths protesting the environmental despoilation of their community without adequate compensation. Indeed the Niger Delta problem is a story of power and resistance to it; of alien and imposed authority and attempt to indigenize and make it accountable to the people it purports to rule; an epic tale of ordinary people battling against vastly superior forces (beginning from colonial consuls to post colonial Federal Governments) threatening to take the bread from their mouths and feeding them the crumbs therefrom. It is from this background that this paper seeks to locate the clash of two divergent, but mutually reinforcing, conceptions of security – the Traditionalists and Non-traditionalists, held by the two opposing actors in the Niger-Delta crisis, and how they affect the crisis.

The Federal Government and the multinational oil companies (MNOCs) hold the traditionalists conception of security – that the Niger Delta is of strategic interest to their survival and hence must be secured at all cost using coercive and military means. On the other hand, the deprived and displaced people of the area hold the non-traditionalist conception of security, as they see themselves their lives and existence being more threatened by ecological despoilation and injustice which spews poverty and disease, than by the bullets from trigger happy policemen and soldiers.

The remaining part of the paper will explore the constituents of these security conceptions further; how they have shaped the conflict in the Niger-Delta, how they constitute the national security problematique, and the multiplier effects national insecurity (based on human security) has on global peace and security. Grave as the Niger Delta conflict is, it does not defy solutions, therefore prospects for sustainable peace will be explored based on the mutual re-enforcement of both security conceptions, taking local participation into consideration.

13.2. Insecurity in the niger-delta: 'triumph' of the traditionalists

As earlier mentioned, the struggle for emancipation by the peoples of the Niger Delta has a long history which dates back to the 19th century when European traders led chiefly, by John Beecroft, mounted a regime of gunboat diplomacy, coercion and violence, which they used to extract treaties of "protection' from the people, in order to expropriate and exploit their resources and control trade in palm oil as well as in other commodities (Irele; 1993:2). This gave rise to colonialism, which on its own created relationships of inequality between people of the Niger Delta and outsiders, in which the former became increasingly conscious of their loss of fundamental human rights. In the process, colonialism instituted a master-servant relationship between external imperial forces and local people of the Niger Delta. The situation is not different in post-colonial Nigeria where subsequent Federal Governments since independence, in collaboration with MNOCs, have further exploited the resources of the Niger Delta, spoiled the environment and created poverty and disease. Thus the master-servant relationship has been sustained, but in this case, the Federal Governemnt and the MNOCs comprise imperial masters, and the local people of the Niger Delta, the impoverished servants. The people of the Niger Delta have been subjected, not only to neo-colonialism[2] (Nkrumah;1965), but also to *internal colonization*[3] (Isike; 2002). Through the latter, Nigerian leaders have, out of self-serving interests, have mismanaged and embezzled the resources of the state, 80% of which come from the Niger-Delta, while neglecting, marginalizing and subordinating the Niger Delta people.

From General Yakubu Gowon's decree of 1969, which transferred all revenues from oil mined in the Niger Delta to the Federal Government during the civil war, through General Olusegun Obasanjo's Land Use Act of 1978, which converted all lands in the country, including oil minerals obtained from them, to Federal Government's property, to General Ibrahim Babangida's enactment of the Treason and Treasonable offences decree of 1993, and its application by Late General Sani Abacha on Ken Saro-Wiwa and the Ogoni eight in 1995, we see a pattern of state-centric conception of security, where the Nigerian state in its supremacy, deliberately denies the local people of the Niger-Delta access to and control of their resources thus making them the most blessed and yet wretched of Nigeria. Worst of all, the Federal Government sometimes visits violence in its physical, structural and psychological form, on the Niger Delta people in order to perpetuate and legitimize the denial process (Fanon 1968).

State violence in the Niger Delta has taken three forms. The first is the harassment of outspoken community leaders through surveillance, arrests and detention. Second is the encouragement of violent intra-tribal, ethnic and communal clashes, and use of the clashes as a pretext for repression. Third is the direct repression using the armed forces and police to exert physical violence through extra judicial killings, flogging, torture and rape. An example of the first category is the harassment, arrest and detention of MOSOP leaders like Ken Saro-Wiwa, Leton Kobani, Ledum Mitee and others, about three times in 1993 alone. The Ijaw/Itsekiri intra communal clashes since 1997 is an example of the second while the 1995 summary execution of over 70 Ogonis by the internal Security Task Force set up by Governor Col. Dauda Komo of Rivers State and led by Major Paul Okuntimo falls into the third category. All these were informed by the states' traditionalist understanding of "securing' the Niger Delta in the "national interest'.

It is in this state-centric sense of security held by the Federal Government and the MNOCs that we attempt to locate the triumph of the traditionalists' conception of security in the Niger-Delta crisis today. Basically, traditionalists like Stephen Waltz see security from a state-centred perspective and restricts the application of security to threats in the military realm. He equates security with peace and the prevention of conflict through military means like deterrence policies, non-offensive defense through public policy and law (Tarry; 1999:2).

Three realists assumptions figure prominently in Waltz's definition. First, because of the absence of a world government, the international system is characterized by an objectively observable but not socially constructed anarchy. Second, while states are not the only actors in the international system, they are the most important ones. Third, generalization about state behaviour can be developed because the primary objective of all states is survival. As war is assumed to be the most serious hazard to the survival of states, the study of war becomes the focus of realism (Tarry 1999:3). Waltz was not probably thinking of intra-state conflicts and wars when he assembled his thoughts on this issue. However, since intra-state conflicts and wars have become the lot of the Third World, we tend to see the possibility of aligning the traditionalist perspective with national security in Third World States, hence its application here as an analytical tool.

In applying and integrating Waltz traditionalist conception of security to the crisis in the Niger-Delta of Nigeria, we will take each of his three realist assumptions and try to use them as tools of analysis within the Nigerian context. Therefore, Nigeria is viewed as or likened to the international system that Waltz conceptualizes in his analysis. According to him, because there is an absence of a world government, the international system is anarchic. In the Nigerian context, the Niger Delta crisis, the emergence of civil society groups such as the Movement for the Survival of the Ogoni People (MOSOP), the Movement for the Survival of Ijaw Ethnic Nationalities in the Niger-Delta (MOSIEND), Movement for the Survival of Itsekiri Nationality (MOSIEN), Isoko Front, the Urhobo Economic Foundation (UEF), the Ijaw Youth Council (IYC) and other N.G.O.s like Medicin sans frontiers and the Environmental Rights Action and the regime of protests and social tensions have all portrayed the failure of the Nigerian state. These groups, some of which provide services of an alternate government within the state, are occasioned by the failure of the Nigerian Government to provide human security-freedom from hunger, poverty, disease and unemployment, as well as the government's inability to build the confidence of the people, provide for law and order the entrenchment of the rule of law. These are evidences of failed statism and subsequently, *shadow statism*[4] (Clapham; 1996)

Since the Federal Government and *ipso facto,* the Nigerian state has failed, the Niger Deltans do not see the state as providing for their common good any more, rather they conceive the state as an encumbrance and the enemy within which must be resisted. This is the basis for the popularity and legitimacy of locally based groups like MOSOP, MOSIEN, IYC and INC. Thus we see an objectively observable anarchy, as postulated by Waltz, as the local peoples of the Niger Delta, the Federal government, the governing elites and the MNOCs, all pursue different interests (survival) using whatever means they deem effective, based on their own understanding of what security is.

Secondly, Waltz posits that the states are not the only actors, even though they are the main actors. In the Nigerian context, the Federal Government (the state) is not the only actor in the perpetuation of environmental injustice and its attendant consequences on the Niger-Delta. Other actors are the MNOCs and the governing elites. The local people and other civil society groups are also actors, but on the side of the great divide. Finally, in applying Waltz traditionalist conception of security to reality and its triumph in the Niger-Delta crisis in Nigeria, we need to observe that the Niger Delta, because of its oil mineral resources, accounts for over 80% of Nigeria's export earning (IDEA; 2000:142). This has made the region to be of strategic importance to the survival of the nation, its governing elites and the MNOCs which Shell, Chevron, Mobil, AGIP and TotalFinaElf. Since "survival¶ has become the core interests of the Nigerian Federal Government (controlled by local bourgeosie elites) and the MNOCs, any attempt by the local people and civil society groups to challenge the status-quo is seen, not only as a threat to "national survival' and by implication, "national security' but as an act of war. Therefore since war is assumed by traditionalists to be the most serious hazard to the survival of states, techniques of war like deterrence and war itself are among the strategies employed. It is in this light that the militarization of the Niger Delta and torching of numerous communities such as Umuechem and Odi should be viewed and understood. Also, in the last few years, incidents such as federal institution litigations on the ownership of offshore oil, the initial refusal and delay of President Obasanjo to sign the onshore/offshore dychotomy bill, the massive military deployments and the fortification of naval facilities in the Niger Delta, should be viewed in the context of the traditionalists security school triumphing in the Niger-Delta.

The Niger-Delta crisis, from the traditionalist security perspective of the Federal Government, the governing elites and the MNOCs, is seen as a national security problem only in so far as the corporate survival of Nigeria, continuous access to petro-dollars and the reparation of profits respectively, are threatened by the agitations and restiveness of the peoples of the Niger Delta. This is the security conception that has been most dominant in the clash of security conceptions in the Niger Delta. Its devastating effects have been deliberately down played to other Nigerians and the international community. In constrast, the legitimate resistance of the local people is highlighted in terms of "restiveness of irresponsible youths", "rascals", "kidnappers" and "terrorists" who must be dealt with to restore law and order.

However, basic questions remain- who defines security? Who implements it? Whose security? And security from what threats?

13.3. Re-conceptualizing security:

The non-traditionalist conception

So far, this paper has reviewed the traditionalist concept of security as a framework to order, analyse, explain and give meaning to the ecological and social injustice meted out on the peoples of the Niger Delta by the Federal Government of Nigeria with the active connivance of the MNOCs. The inability of the concept to address the basic questions of human security, where individuals, rather than the state, are the referent subjects, has given rise to the need for a broader concept of security. In this light, this paper adopts the *non-traditionalist concept of security, which* is more depicting of the reality in the Third World, as its conceptual framework of analysis.

The position of this paper is that the clash of the Traditionalist and Non-traditionalists concepts of security, as differently held by the main actors, sums up the crisis in the Niger Delta. Having examined the constituents of the traditionalist concept so far, we shall undertake that of the non-traditionalists and its applications to the crisis in the subsequent parts of the paper. The ultimate goal is to explore areas of convergence such that could proffer workable solutions aimed at bringing about sustainable peace and development in the Niger Delta. What then is the non-traditionalist concept of security? What are its underlying assumptions and constituents?

The non-traditionalist concept of security is a paradigm shift from the rather narrow traditional conception to a broader conception of security which recognizes and accommodates a wider variety of issues that constitute threats to human existence, and as such breed insecurity and anarchy. The analytical distinction between traditionalist and the non-traditionalists is not meant to suggest that consensus among the latter has been reached either. Indeed there is disagreement between two sub-groups – the so-called "wideners' and "deepeners' (Tarry; 1999:1). The wideners, like Mohammed Ayoob, argue that a predominantly military definition does not deny that there are other threats to state survival i.e environmental, social and economic, but that these must be sufficiently politicized enough to enter the national security agenda. The deepeners on the other hand ask the question of whose security is being threatened and support the construction of a definition that allows for individuals or structural referent subjects or objects respectively, as opposed to the state.

However, in the context of the Niger-Delta crisis, we will adopt a general framework of assumptions/propositions to explain the non-traditionalist concept of security:

- Individuals and states are the referent subjects in the security calculus.
- Domestic and non-military sources of threats to human existence threaten the survival of states more than military sources. These are in the political, environmental, societal and economic realms (Booth; 1994:3).
- National security is a function of state-building through legitimacy and integration. (Ayoob; 1994:130).
- Human security is the core of national and by implication, transnational security.

While recognizing both the widened and deepened constituents of security as they apply to the Niger Delta crisis, we shall try to be guided by the above general framework of assumptions. However, in order to give this paper a clearer focus and thus an analytical utility, as well as applicability to reality, we shall adopt Ayoob's (1994) widened perspective of security. In the process, we shall incorporate deepened issues of security like environmental despoilation and economic deprivation (the present lot of Niger Delta peoples). These issues have become sufficiently politicized to have the potential to threaten the survival of the state (Nigeria), its boundaries, political institutions and governing regimes.

According to Ayoob, (1995:9):

> Security-insecurity is defined in relation to vulnerabilities, both internal and external, that threaten to have the potential to bring down or weaken state structures, both territorial and institutional, and governing regimes

Though this definition is state-centric, it also recognizes that there are domestic or internal sources of vulnerabilities or threats, which could be hazardous to the survival of the state and its peoples. The only condition being that these internal vulnerabilities should be sufficiently politicized enough to enter the national security agenda. This condition gives Ayoob's definition a leeway that deepens security without taking away its state-centredness. Even then, Ayoob (1997:131) justifies this state centredness by arguing that the state system remains the primary mode of organization for Third World states in their move towards statehood. The more a state or regime- and often it is difficult to disentangle issues of state security from those of regime security in the Third World – leans towards the invulnerable end of the vulnerable-invulnerable continuum, the more secure it will be. Other types of vulnerability, whether economic or ecological, become integral components of our definition of security only if they become acute enough to acquire political dimensions and threaten state boundaries, institutions or regime survival (Ayoob; 1995:9).

Applying this analytical conceptualization to the Niger Delta crisis, we see that oil exploration in the region creates environmental problems through gas flaring, product spills and discharge of refinery effluents into fresh water sources. These all have grave consequences for human lives that is food insecurity, poverty, food poisoning and disease. The green house effect of gas flaring affects not only the humans in the Niger Delta and beyond, but also the wild life in the area. These have casued rural-urban migration and occupational displacement, which escalates the unemployment problem.

Following the states inability to protect the people of the Niger Delta from these hazards due to its rentier status and the failure to provide for the development needs of the area mainly because of corruption and mis-governance, which has in turn informed the states' adoption of a traditionalist concept of security, the Niger Delta people have become increasingly restive. They have engaged in activities that threaten to dislocate state boundaries, institutions and even the survival of the Obasanjo regime, going by the recent agitation for a sovereign Republic of Niger Delta (Thisday 30:12: 2002), culminating in a body, The Front for the Republic of Niger Delta (FROND) declared by a famous human rights lawyer and Niger –Deltan, Festus Keyamo. This is the crescendo of past self-determination agitations like the Ogoni Bill of Rights (OBR) by MOSOP in 1990 and the Kaiama Declaration of 1998 by youths of the Ijaw ethnic group, the fourth largest ethnic nation in Nigeria, in 1998. The crux of these agitations is resource control and devolution of power, which will radically change the present federal structure of government. This no doubt, threatens regime stability because such a fiscal and administrative decentralization will reduce the power and influence of the federal government as well as its officials (governing elites). Officially, at the state level, the power structure will change and at the personal level of the governing elites, political and economic power and relevance will be altered. Either singly or in congruence, the environmental crisis in the Niger Delta has become sufficiently politicized enough to enter the national security calculus.

Unfortunately, the government prefers to see insecurity in the Niger Delta from the traditionalist prism, hence its state-centric response to the crisis. Unless the Federal government appreciates the impact of environmental despoliation occasioned by oil exploration on the people of the Niger Delta, and subsequently responds to security and insecurity from that perspective, the Niger Delta will explode one day in a violent conflagration that will either consume all, or forcefully change the current state structure, political institutions and the social and economic regimes of inequalities in Nigeria.

As it is, it is not the strong presence of security men and armoury that will bring about security in the Niger Delta, rather, it is by inspiring confidence and support of the people through genuine commitment to creating a new political and economic framework, based on the tenets of the non-traditionalist concept of security, a framework that will transfer power and resource control to the local people in the Niger Delta, without necessarily affecting negatively the relevance of the federal government.

Before going ahead to espouse on how this can be achieved, it will be pertinent to explicate the environmental consequences of oil exploration in the Niger Delta. This is to enable us appreciate the need for a paradigm shift, in understanding security in that region.

13.4. Environmental despoilation in the niger delta

Grave harm has been done to the Niger Delta environment by oil exploration and the Nigerian government has shown itself to be grossly incapable of intervening to protect the Niger Delta environment because of the linkage between the governing elites and the oil business, and secondly, because of the rentier nature of the Nigerian state that makes it depend solely on oil and the rents it collects from the MNOCs. As a result, the MNOCs practically get away with all the murderous atrocities they vent on the people of the Niger Delta. For example, according to the IDEA (2000:146) reports, enormous amounts of natural gas are flared daily, usually in the vicinity of human dwellings (IDEA, 2000:146). In 1982, Shell, Gulf, Mobil, AGIP, Texaco, Pan Ocean, Ashland, Philips, Tenneco and Elf flared over 13 billion cubic metres of gas in 145 (oil) communities in Nigeria (IDEA 2000). Of these, Shell was the worst culprit, flaring nearly 7 billion cubic metres of gas that year. In one case alone, in the town of Sapele, Shell burnt over 944 million cubic metres of gas (IDEA 2000). In 1991, Nigeria exceeded the world average for natural gas flaring by 72%. In-that year, while the world average for gas flared, as a percentage of total production was 4%, Nigeria flared 76% of total production. This compared poorly with the OPEC average of 18%. The next worse case, Libya, flared 21% of total production. The report claims that apart from the huge greenhouse gases that gas flaring pumps into the atmosphere, constant flares affect both wildlife and human beings negatively. At temperatures of up to 1,400^{o}C, the "Sauna bath" effects that gas flares produce make living in many communities nearly impossible (IDEA 2000).

Oil spillage and product leakage, which destroy wildlife, farmlands, forests and human lives, are another source of environmental despoliation caused by oil exploration in the region. The MNOCs and the Nigeria National Petroleum Corporation (NNPC) use high pressure pipelines to transfer products from the oil fields and refineries respectively, to various parts of the country. In the process spill and leakages are experienced in devastating magnitudes. The MNOCs, NNPC and even the Federal Government in numerous cases blame such incidences on sabotage by the local people, even though they have not bothered to address the question of what monster called "deprivation', gave rise to the monster called "sabotage'. The truth however is that much of the pipelines and valves in question are very old and therefore prone to failure (IDEA2000: 147).

Another area of environmental despoilation in the Niger Delta, occasioned by oil exploration, is the discharge of refinery effluents (harmful liquid waste that flows out of the refineries during the process of refining crude oil) into fresh water sources and farmlands. This no doubt destroys the environment According to the (IDEA 2000), these effluents contain excessive amounts of very toxic materials like chromium and mercury which can be stored in the brains of fishes for a long time thus causing food poisoning. The hydrology of the area has been affected. This has caused flooding in some areas and water deficiencies in others. Badly constructed canals and causeways also cause or lead to forests atrophy, water hyacinths, and water scarcity and the extinction of many species of plants, animals and fish (IDEA 2000).

These are some of the horrifying environmental consequences of oil exploration that Late Ken Saro-Wiva showed the world before his hanging by the Federal Government in 1995. These are the grave realities facing the Niger Delta people. The threats to their existence that these vulnerabilities pose, happen to be more devastating for them than the threats of physical force and bullets from the federal guns. This surmises their insecurity problem, which of course has been compounded by the Federal Government militarization of the area.

13.5. 'Human security' and the national security problematique in nigeria

According to the Bonn Declaration of 1991, *"human security is the absence of threat to human life, lifestyle and culture through the fulfilment of basic needs"* (Omari; 1995:4). This encapsulates the crux of our insecurity/security continuum in the Niger Delta, which moves from one end to the other at a particular point in time, depending on what an actor in the crisis perceive or conceive as security.

Human security and the organization of a state for the purpose of securing its people did not start with the Bonn declaration in 1991. Indeed, as far back as 1705, the German philosopher Leibniz, expressed the need for a state to provide common security (La securite' commune) to its citizens. This was in reaction to the Clausewitzian and traditionalist conceptualization of security (Ngubane & Solomon; 2002:59). The great French philosopher, Jean Montesquieu, also noted that true political freedom could only occur when people are secured by the state.

Thus we see a link between the state and human security on the one hand, and the state and its own security on the other. This is where the application of Ayoob's state centric but widened (Non traditional) concept of security becomes relevant. To him, security analysis should not undermine the move towards statehood, especially in the third world or the Less Developed Countries (LDCs) (Ayoob; 1997: 131). To Ayoob (1997), the Third World, because of space regime illegitimacy and lack of internal cohesion which breeds domestic conflicts, are more vulnerable to insecurity/threats arising from non traditional sources or better still, from human sources such as poverty, disease, illiteracy and environmental degradation. However, to Ayoob (1997), these issues on their own do not constitute security threats to the state, even if they do to the individuals who make up the state. They must be sufficiently politicized to gain enough prominence to be able to produce political outcomes that can threaten the survival or effectiveness of states and regimes. In other words, non political issues that have the potential to endanger the well being of Third World states and regimes become security problems only when they are able to intrude into the political arena. This is Ayoob (1995: 190) political core of security

In applying this and the Bonn declaration to the Niger Delta crisis, the paper though it recognizes a deepened range of issues (environmental despoilation and its resultant economic deprivation), also recognizes the important role of the state in ensuring human security, hence its analytical application of Ayoob's (1995) widened conception of the non-traditionalists school of security. The main effort in the use of this analytical tool is to prove that the interdependent human problems of environmental despoliation and economic deprivation in the Niger Delta have become prominent enough and sufficiently politicized to enter the national security agenda of Nigeria.

This paper has shown so far, the consequences of oil exploration on the environment of the Niger Delta. It has shown that these environmental consequences in turn result in disease, poverty and death, thus threatening the lives, lifestyle and culture of the people. By implication we see an inability of government (state) to prevent the MNOCs from carrying on with their despoilation of the Niger Delta environment without protecting it, due to the entrenched common imperial interests of both parties. The Nigerian state has not been able to provide for the developmental needs of the Niger-Delta people that is commensurate with what is taken out of their environment, neither has the state being able to cause the MNOCs to do so. Rather we see the state institutionalizing violence and repressing its own people in collaboration with MNOCs, instead of protecting them against external aggression as entrenched in the constitution.

The conclusion in this regard is that the state has failed. It has failed to provide human security for its people in the Niger Delta. As a result, the people feel estranged and alienated hence their resort to violence and the recognition of alternate, locally grown civil structures (MOSOP, MOSIEN, I.Y.C e.t.c.) to direct and guide their fight for survival and emancipation, for a fair deal from their resources, or better still resource control and self determination. These are the issues for the Niger-Delta people, and to the extent that nation building is affected negatively by the ensuing conflict in the Niger-Delta, a national security problem exists. The actions of the Niger Delta people which include move towards self-determination and separation, attacks of state¶s political institutions and the construction informal economy (Kidnapping of MNOCs officials for ransom from the companies) threaten state boundaries, negatively affect state building and portends a crisis of instability. The regular deployment of troops to the Niger Delta and the treatment of the area as a national security concern (traditionalist) by successive governments, are eloquent pointers to the fact that the Niger Delta is a national security problem, which has grave economic and political consequences for the corporate well-being of Nigeria.

Interestingly this "Frankenstein' does not end within the confines of Nigeria. Like foreign policy, which is an extention of domestic policy, the national security problem in Nigeria, as arising from the Niger Delta extends well beyond the shores of Nigeria. Ayoob (1995:195) sums this trend as "the contagion of instability". To him, if the Third World continues to suffer from conflict, insecurity and instability, that situation can harm the international system as a whole. In his opinion, the components of the international system are so closely intertwined today that a Hobbesian periphery (The stateless and anarchic Third World) and a Lockean core (The Organised states of the developed world based on the principle of the common good) cannot co-exist without the periphery exporting some instabilities to the core (Ayoob 1995). The point here is that, the core is too small, compared to the majority periphery that happen to suffer insecurity and instabilities, which must necessarily permeate the core. These instabilities in the Third World, which in some cases result in genocide and civil wars, have become so prominent that they are now the major wars recorded since the post cold war era. For example, Ann Tickner argues that of the approximately 127 "significant wars" which have occurred since 1945, all but two of them, the Gulf War (the only interstate war since then) and the Yugoslavia civil war, were within L.D.Cs (Tarry; 1999:3).

The attendant refugee flows, not only into neigbouring states, but also into developed states gives intra state conflicts in LDCs both a regional and global outlook. This further compounds the global security problem especially as recently redefined following the September 11, 2001 terrorists' attacks on the United States of America (U.S.A). If the Niger Delta problem is not tackled head on by all the actors involved, with a view to marrying their divergent conceptions of security so as to create sustainable peace based on a sustainable management of the environment, it may turn into a conflagration that will be all consuming. All-consuming here implies stretching beyond the shores of Nigeria to the African continent and of course to the developed North thus aggravating social, human and economic insecurity. There is the need for the globe to see itself as stakeholders in the Niger-Delta crises. This is not out of place when one considers that the developed core, such as the U.S., Britain and France are the home countries of the MNOCs, and that it was the exportation of their imperialist capitalist accumulation quest that led to the human insecurity of Niger-Deltans in the first place.

13.6. Towards a convergence of the two security conceptions: prospects for sustainable peace and assurance of human security

Indeed, there is a crisis of instability in the Niger Delta. The violence instituted by the state, with the active collaboration of the MNOCs, and that perpetuated by the Niger Delta people in response to state violence, breeds only more violence, which has very tragic consequences for all the actors.

This paper has summed up the crisis as the result of the clash of two divergent conceptions of security as held by two opposing groups of actors – the traditionalist security conception held by the federal government, the governing elites and the MNOCs on the one hand, and the non-traditionalist conception held by the Niger Delta people, its environmental protection crusaders, civil groups, political activists and its national, regional and international sympathizers, on the other hand. This clash of conception informed the application of Ayoob's (1995) 'wideners' perspective of national security as, an analytical framework to explain the nature of the emerging human insecurity threats in the Niger Delta and their implications for national security. This is with a view to finding a congruence of these conceptions in a manner that will bring about sustainable peace and help nation building.

Ayoob's (1995:4) widened concept of security, which finds meaning especially in the Third World, provides a platform for the convergence we seek in the Niger Delta crisis. This is because his analysis of security in the Third World recognizes the state as the primary mode of organization. The Third World state, in the absence of a developed civil society owing to its infancy in the move towards statehood, remains the mechanism for providing the basic needs of its people and ensuring their security in the process. Also, because of institutional underdevelopment, which is the predominant source of conflict in the Third World, the state remains the main engine room for growth and development. This defines a role of relevance for the governing elites. This value of "stateness' is important here, as the Niger Delta people need to recognize it and strive to struggle for their emancipation within the context of the state system, instead of continually fighting against it a strategy that has not yielded much in terms of emancipation from tyranny, oppression and exploitation.

Second, Ayoob's (1995) analysis also recognizes that domestic vulnerabilities can also threaten state structures and the survival of the state, thus constituting a national security problem. He however sounded a caveat; that these vulnerabilities must be sufficiently politicized to be considered serious enough by the governing elites, to enter the national security agenda, before they can become national security problems. This pre-condition notwithstanding, the point of note is that Ayoob's (1995) analysis recognizes threats to national security outside the military and war realms, threats which can come from the inside (domestic) as a result of economic deprivation, environment despoliation, disease and poverty occasioned by bad leadership.

Since it is the position of this paper that these threats, caused by the activities of the MNOCs, exist in the Niger Delta, and that through the reactions of the people to them (threats), and the federal government's response to the people's reaction in state-centric and military terms, the Niger Delta has indeed entered the national security agenda. These two values of Ayoob's analysis as they relate to the Niger Delta, provide a platform for the convergence of the two divergent conceptions of security. The Nigerian federal government, as the torch bearer of the MNOCs in this clash, and the Niger Delta youths, as torch bearers of the Niger Delta people, must come together under this platform, appreciate each others position and chart a commonly articulated course that will lead to sustainable peace and development, based on an efficient management of the environment.

In the short run, the federal government must as a first step, demilitarize the Niger Delta. If the resistance of the Ogoni and Ijaw youths to the federal might (aided by the MNOCs) has lessons for the government, it is that more armed soldiers and sophisticated weaponry will not solve the Niger Delta security problem and bring about the much desired peace in the area. The people are determined and strong willed, but they want peace and are convinced about the legitimacy of their demands which, according to IDEA research findings, include, Desirability of democracy, Restructuring of Nigeria and Resource control, Poverty Alleviation and Infrastructural development, Gender equality, Intra-Delta conflict management and Environmental protection. (IDEA; 2000:150). Since it is the basic duties of the state to facilitate the actualization of the above and thus provide for the security of its people, it must change its current perception of security to one that takes the local Niger Delta people and their participation into consideration.

This new thinking must be based on the argument that the participation of the Niger Delta people in the Nigeria project will be through dialogue, informed by the acknowledgement that they are equal partners in the project. This is apt especially as they lay the golden egg that feeds the nation. Such dialogue can only be initiated when the aggressor – in this case the federal government and the MNOCs-demonstrate good faith by demilitarizing the Niger Delta to give the air of freedom, and an enabling environment for reason to prevail. This gesture will justifiably provide the basis for the Niger Delta people to drop their guards, tone down their language of anger and meet the federal government at the negotiation table. At this short-term level, dialogue should revolve around rebuilding the confidence and trust of the people, firm reassurances and demonstrated commitment to redressing the past and present injustice as well as sound palliatives like the Niger-Delta Development Commission (NDDC) which will incorporate roles for local participation in enhancing nation-building. In the long term, the 1999 constitution must be reviewed to address the key issues of power devolution, self-determination and resource control, in line with the practice of true federation. These are the core issues in the neglect and marginalisation of the Niger Delta since 1958 till today.

Fortunately, the Niger Delta will not be the only beneficiary of a reviewed constitution to reflect true federalism. This is because other geopolitical zones of the country particularly the West and the East, have at different times in the last couple of years, called for the fundamental restructuring of the Nigerian state to reflect a true federal character. Devolution of power and self- determination are some of the critical issues they have raised. Development scholars like Claude Ake and Julius Ihonvbere have advanced a compelling argument for devolution of power and democratization of the development process in post-colonial African states like Nigeria, where centrifugal forces, propelled by competing demands of the elite of the various ethnic groups have made the task of nation-building very difficult (IDEA 2000). Subsumed under self-determination is the issue of Resource Control. Since people have a right to self-determination, they should also determine the control and usage of their resources. Thus resource control is a legitimate agitation.

Therefore in the light of the above issues of power devolution and self-determination, the review of the 1999 constitution to address them is in order. The first step of demilitarizing the Niger Delta to create a conducive atmosphere for dialogue, as a short term panacea, will prepare the grounds eventually for a Sovereign National Conference (SNC), from which a new federal constitution, commonly articulated, will emerge. This is the prerequisite for a stable, secured and developed Nigeria.

13.7. Concluding remarks

Development is the most contemporary issue in African studies and in the continent. Ake (1981) for instance, sees development not as a project, but a process by which people create and recreate themselves and their life circumstances to realize higher levels of civilization in accordance with their own choices and values

This definition of development falls well within the broad context of Michael Todaro's three core values of development, which include life sustenance (the ability to provide basic needs), self esteem and freedom from servitude which enables a people to choose from a wide range of choices, on how to live (Todaro; 1985:86). Is the Niger Deltan not entitled to this kind of life? Are the Niger Delta people different species of humans who have no right to aspire to and attain development? Perhaps these rhetorical but germane questions will help put this paper and its sympathy for the plights of the Niger Delta people in perspective, the question that should agitate our mind is; assuming the Niger Delta was in the North and its peoples Northerners, would there have still been a one Nigeria?

If the Niger Delta people are not paying a price for being a minority, then the Federal Government will need to sincerely answer this question. If it does, it will see the need for a paradigm shift from its current state-centric perspective of "securing the Niger Delta", to one based on human security. This is the first step, a necessary and sufficient pre-condition for peace, stability and development of the Niger Delta and *ipso facto* national security, nation building, democracy and development in Nigeria.

End notes

1. Emancipation as used here means the liberation struggle which according to Frantz Fanon "ought not to end with decolonization, but must manifest in the well being of the people, freed from all forms of alienation, domination and on the whole neo-colonialism".
2. Kwame Nkrumah in his work "Neo-colonialism: The last stage of Imperialism", claims that the Neo-colonial state is, in theory, independent and has all the outward trappings of international sovereignty. In reality its economic system and thus its political policy is directed from outside. We see the role of MNOCs in the Niger-Delta as imperial agents of Neo-colonialist tendencies.
3. Isike, while espousing on internal colonization, explains that following decolonization, African leaders have steadily recolonized their own people. This is portrayed by the sit-tight syndrome and massive corruption which have served to further widen the gap between socio-economic class formations inherited from colonialism. The imperial quests of Africa's new elites; its ruling class,

have watered their desire to continue the political domination of their own peoples in order to maintain the status-quo. The resultant political illegitimacy, and alienation of the people, which are antithetical to nation-building, and were visible during the colonial era, are a new form of colonization, only this time, from within.

4. On the shadow State see Christopher Clapham "Africa and the International system: The politics of survival" (London: Cambridge University Press 1996), pp 251-252.To him; "up to a point, an effective and functioning state might provide a useful mechanism through which rulers could help to assure their long-term survival by meeting the needs of their people and building up a sense of their legitimacy. Once the state ceased to be able to perform such functions, much of it became an encumbrance".

References

Ake, C. (1981) *A Political Economy of Africa,* London: Longmans Ltd.

Ayoob, M. (1995) *The Third World Security Predicament:* State making, Regional *Conflict and the International System.* Colorado: Lynne Rienner Inc, 1995.

Ayoob, M. (1997). "Defining Security: A Subaltern Realist Perspective" In Kath Krause and Michael Williams (eds) *Critical Security Studies,* Minnesota: Universit*Africa and the International system: The politics of survival,* London: Cambridge University Press

Fanon, F. (1968) Wretched of the Earth New York: Grove Press. 1968.

International Institute for Democracy and Electoral Assistance (IDEA) Democracy in Nigeria-*Continuing Dialogue(s) for Nation-building.* 2000.

Internet Document-www.google.com "The Lingering crisis in Nigeria's Niger Delta and suggestions for a peaceful resolution" March 2000.

Irele, D.(1993) The violated Universe: Fanon & Gandhi on violence, Critical Forum Monograph Series.

Isike, C(2002)."Beyond Imperialism and Colonialism: How Africa is underdeveloping Africa". Seminar paper, University of Benin, Benin Ciry.

King, M.C. (1988) *Localism and Nation-building,* Ibadan: Spectrum Books. 1988. Nabudere, D.(2000) *Globalisation and the Post-colonial African State* – Harare: AAPS Books.

Nef, J.(n.d)Human Security and Mutual Vulnerability: An exploration into Global Political Economy of Development and underdevelopment
www.idrc.ca/ubrary/document/103249/indeso.html

Ngubane and Solomon (2002) "Southern Africa's New Security Agenda" *Africa Insight.* Vol. 32. No.1.

Nkrumah, K. (1965) *Neo-colonialism: The Last Stage of Imperialism.* London: Panaf Books Ltd

Omari, A.H. (1995). *Regional Security: One view from the front line states.* Arusha Papers, Working Paper Series in Southern African Security,

Rowel, A. (1996) *Green Backlash*, London: Routledge.

Tarry, S. (1999) Deepening' *and "Widening' : An Analysis of security Definitions in the 1990s*. http://www.stratnet.ucalgary.ca/Journal/html

THISDAY (2002) Monday 30th December

Todaro, M. (1985) *Economic Development in The Third World*, Third edition New York: Longman Inc.

14. Implications of the niger delta conflict for national security in nigeria

Audu, GAMBO

14.1. Introduction

This paper analyzes the political, social and economic implications of the oil-induced conflicts in the Niger Delta region for national security in Nigeria. Broadly speaking, how have the articulation of discontent with the structure of the Nigerian State and particularly the demand by the minority ethnic nationalities in the oil-rich Niger Delta region that their expectations, preferences and aspirations be factored into the Nigerian project, stressed the resilience of the Nigerian federation? It should be emphasized here that the oil-induced conflicts in the Niger Delta region are deeper expressions of the "National question" to which the Nigerian State at the moment, has not articulated a mutually agreeable solution. It is this failure of the state to articulate appropriate responses to the "National question" that has spawned a groundswell support for sub-nationalist agitations for self-determination, resource control, equity and a carefully negotiated basis for cooperation. This paper seeks to bring into bold relief, the interface between oil-induced conflicts in the Niger Delta region and national security in Nigeria. It seeks to explicate the ways in which the oil-induced conflicts in oil producing multi-ethnic developing society like Nigeria could threaten national security broadly conceived as the stable, dynamic and corporate survival of the Nigerian federation.

14.2. Theoretical perspectives on security:

Two major theoretical perspectives on security could be gleaned from available literature on the subject. These are: classical or conventional and revisionist or new thinking perspectives (Ate, 1992; Buzan, 1992; Imobighe, 1998; Don 1999). These two contending perspectives only support Buzan's (1992) argument on the need to "habi litate" the concept of security because of its inadequate conceptualization and articulation. While the classical or conventional perspective equates security with massive military build-up, the revisionist or new thinking perspective is socially based with strong emphasis on the desire to utilize the available resources of any nation to meet enduring welfare ends of the citizens. Let us examine in detail, the arguments of these two theoretical perspectives on security in order to appreciate the extent of their divergence or convergence.

14.3. Classical or conventional perspective:

This perspective on security is rooted in the state-centric view of International Relations. States as the key actors in International Relations are believed by proponents of this perspective, to operate in a system that is devoid of a well structured and centralized political authority. Kenneth (1979) stresses the point that the anarchical nature of the international system makes it a matter of the practical necessity for states to look out for their protection and survival in the system. He describes the system as a 'self-help' system in which each state must take adequate and effective measures to provide for its own security largely conceived in terms of territorial integrity, independence and sovereignty. Proponents of this perspective argue that nation-states compete for many of the same things including self-preservation, national identity, sovereignty, status and wealth, which are acutely scarce.

It is therefore, the acute scarcity of these values that provides strong motivation for sovereign States to amass as much power as they can. Security in this context is largely perceived as a function of power and that power is a demonstrable function of military capability. Quite implicit in the orientation of this perspective is the lack of mutual trust among sovereign States that accounts for their unyielding disposition towards the idea of entrusting their protection to international organizations and law. States are believed to have unfaltering faith in the efficacy of "build up of strong military defence as a means of protecting the territorial or national integrity of the State from both internal and external violations." (Imobighe, 1998:14).

This militaristic thinking on the concept of security could be traced to the post World War II when the United States of America and the then Soviet Union were almost inextricably locked up in ideological conflict which came to be known as East-West cold war. While the United States of America was determinedly committed to not only shielding the capitalist bloc but also expanding its frontiers, the Soviet Union sought to construct a vast sphere of influence for the socialist bloc. That ideological schism ignited an arms-race between the two blocs.

It is important to stress the point that the "classical' or "conventional' perspective of security focuses exclusively on the state and security is measured by the military capabilities of nation-states. This notion of security gives the statewide latitude to clamp down on its citizens even when objective conditions call for mass protests against certain unprogressive policies of the state such as the Structural Adjustment Programmes being implemented in most third world countries. This is because, the realist or militaristic view of security equates it with regime survival and therefore anything that threatens the internal coherence and strength of the state must be hurriedly tackled using the most effective instruments of coercion. This however, does not promote mutually progressive relationship between the state and the citizenry.

Exclusive focus on the security of the state while neglecting human security could indeed be counter-productive in the long run. Widespread social unrest and violence that often accompany economic crises in most developing societies is a potent indicator that security cannot be defined exclusively in military terms. This kind of manifest expression of people's discontent with their social and economic conditions could weaken the legitimacy, stability and security of the state.

14.4. Revisionist or new thinking perspective:

The revisionist or new thinking perspective stresses the non-military aspects of security. It seeks to reconstruct the concept of security in a more inclusive sense. The new thinking perspective on security emerged with the collapse of the Soviet Union in the late 1980s. Since the cold war itself was the basis for the militaristic conception of security, its demise has consequently cast a dark shadow of doubt upon the strategic relevance of security through build-up of a formidable military defence. This perspective is evidently, a post-cold war peace dividend, which seeks to redirect efforts towards expanding the frontiers of human-centred development. It argues that the security of any nation is best guaranteed by utilizing disproportionate amount of its available resources to attain the enduring welfare ends of the citizens (Ate, 1992; 1999; Ibeanu, 1997; Imobighe, 1998).

Proponents of the revisionist or new thinking perspective on security (Michael, 1998; Imobighe, 1998 James, 1995; Luckham, 1983; Don, 1999; Obi, 1997; Buzan, 1995;) contend that the militaristic notion of security largely dictated by the cold war phenomenon, has lost its soul with the demise of the cold war. Emphasis in articulating a more inclusive notion of security is placed on building a well contented, prosperous, freer and fulfilled citizenry that is radiating with happiness. The strength of this view point is found in the fact that since the 1980s., most conflicts across the globe are largely intra-state in nature and more frequently ignited by the repressive actions of the privatized state especially in the developing societies that are notably characterized by social and economic under-development, widespread deprivation, neglect and marginlization of minority ethnic nationalities (Ibeanu, 1997). Once the state has been captured and privatized by a tiny fraction of the elite that use public institutions and resources to terrorize non bourgeois communities, abuse human rights, loot public funds, and mortgage the future of the vast majority of citizenry, it became bereaved of the loyalty and support of the citizenry (Ihonvbere, 2000). Consequently, the state is pitched against the citizenry with growing reliance on the use of brute force to extinguish the flaming anger of the people.

The notion of security as articulated by the proponents of new thinking" or "revisionist' perspective includes the ability of a social entity to provide self-protection and development and the enhancement of the general well being of the people (Imobighe, 1999). It is this ability to attain the enduring welfare ends of the citizens by the state that cultivates in them the feeling of patriotism and deeper commitment to the security of the state. Security in this sense means the capacity of groups to provide for their physical and psycho-social needs and livelihoods. This connotes a progressive elimination of objective conditions that limit or threaten to erode this capacity as well as reduction of fears and anxieties about their abilities to meet these needs (Ibeanu, 1997).

Revisionist' or new thinking' perspective of security is therefore, an attempt to reconstruct the notion of security in a broader and more inclusive term. It would be misleading however to assert that proponents of this perspective have stripped the concept of security of its conventional meaning, as the military is still an essential component of security of states in the post cold war period. The point of divergence between the "classical or conventional and "revisionist or new thinking' perspectives on security is that the latter has shifted focus from the state-centric and militaristic conception of security as upheld by the former, to human-centred notion of security. The human security perspective asserts that the security of the state is not an end in itself but rather, it is a means of ensuring security for its people. In this context, state security and human security are mutually supportive. Perhaps, we should note here that building an effective democratic state that values its own people and protects minorities is a critical factor for promoting human security.

14.5. The political implications of the niger delta conflict for national security

The articulation of dissatisfaction with the structure of the Nigerian federation by minority ethnic nationalities in general and the oil producing minorities in particular has largely remained an enduring source of sectional tensions in Nigeria. National unity and integration are significantly threatened as the minority ethnic groups become deeply socialized into the culture of violence in their struggle to rise above the slave culture of silence. Yet the Nigerian state finds it compelling to employ any available means to preserve the fledgling federal compact foisted on the country at the twilight of colonial rule. This inevitably pitches sub-ethnic groups agitating for self-determination against the state intent on preserving the corporate survival of the country thereby deepening the rift between citizens and the state. Where minorities in a multi-ethnic developing society like Nigeria are structurally marginalized in a federal system with no adequate channels for expressing their legitimate grievances, they often, as Osaghae (1994:25) has asserted, "pursue violent underground means of seeking justice and in some cases, they become separatist."

The minority ethnic nationalities in the Niger Delta have repeatedly pointed to the fact that the structure of the Nigerian federation as it exists today, is defective and consequently, as minorities, they have been emasculated politically. This widespread perception of the people has fueled an un-yielding agitation for the convocation of a "Sovereign National Conference' as an inclusive platform for re-inventing a mutually acceptable federal framework in Nigeria (Ihonvbere, 2000; Akhakpe, 2002; Okonta and Douglas, 2001; and Odia, 2000). Ihonvbere (2000:75) while providing a rationale for the convocation of "Sovereign National Conference", opines that:

> Nigerians have never had the opportunity to discuss and reach some consensus on how the nation should be structured, power defined, contestations for power organized, resources generated and allocated, rights protected and the larger democratic project articulated and compacted in a truly democratic constitution. It is embarrassing that almost four decades after political independence, Nigerians still find data, political arrangements and institutions designed by the brutal and totally undemocratic and illegitimate Colonial State as the point of reference.

The Niger Delta conflict is therefore waged in part to draw attention to what the oil bearing minority ethnic groups perceive as some critical defects in the subsisting federal structure of the Nigerian State. The Movement for the Survival of Ogoni People (MOSOP) in August 1990, adopted the Ogoni Bill of Rights (OBR). In October 1992, the Movement for the Survival of Ijaw Ethnic Nationality in the Niger Delta (MOSIEND) presented an Ijaw people's charter to the Government and people of Nigeria. In November 1992, the Movement for Reparation to Oloibiri (MORETO) adopted the charter of Demands of the Oloibiri people while in August 1997, a more inclusive platform christened Chikoko Movement called for the abrogation of obnoxious laws like the land use Decree and the Petroleum Decree that rob their people of the right to control land and mineral resources for sustainable development in the Niger Delta. In December 1998, Ijaw youths numbering over five hundred congregated at Kaiama and adopted the "Kaiama Declaration' (Human Rights Watch, 1999:124-130). In all these carefully articulated declarations, the key demands that are emphasized include self-determination and political autonomy, resource control, abrogation of all oil-related mineral laws which dispossess the minority ethnic groups in the Niger Delta of oil resources and a halt to further environmental degradation. Further, they call for restructuring as a platform for the remediation of what in their own estimation, constitute the pitfalls of the Nigerian federal system.

In a federation such as Nigeria where resource generating units harbour strong ill-feelings towards the state such as the Niger Delta experience, and where the state is starkly ill-predisposed to tackling the grievances that have induced such ill-feelings,there is spawned a groundswell of support for sustained opposition to the authority of the state. This kind of organised sectional opposition could easily undermine political order and state cohesion in a multi-ethnic developing society such as Nigeria. Though the Nigerian State may secure temporary respite from the use of its coercive apparatus to harass and liquidate the leading and articulate elements of organized opposition such as the experience under the late General Sani Abacha indicated agitations continue to swell and threaten the very foundation of the Nigerian federal system with grave implications for national security (Maier, 2000).

The inflexible disposition of the oil producing minority ethnic nationalities in the Niger Delta towards the key agenda of restructuring of the federation as the only effective response to the present Niger Delta conflicts, the specter of disintegration all over the country and the obvious reluctance of the government to meet this aspiration of the people could cause severe rifts between this component of the federation and the Nigerian state. Already, one of the leading proponents of the restructuring and resource control agenda, Major-General David Ejoor, (Rtd) a former military governor and Chief of Army Staff, has, in the wake of the Supreme Court ruling over the off-shore/ on shore dichotomy between the Federal Government and the oil producing states, questioned the basis of his continued loyalty to the Nigerian State. According to him,

> Why should I be loyal to Nigeria again? My loyalty in keeping Nigeria together was not reciprocated. The area (Niger Delta) I came from that is being treated as not part of the Nigeria. They took all the wealth and money away and the place is neglected (Thisday, 2002:72).

This is quite a serious issue because, for someone who fought a 30-month civil war to keep Nigeria one to turn round to express this level of discontent with the operation of the federation, shows the extent of frustration he has been driven to. The fact is that whether the issue of political marginalization of the Niger Delta people is real or imaginary, the fiery emotions of the people have been awakened through intense and wide mobilization by the various associational lives in the socio-political landscape of the region. It is exceedingly difficult for government to employ the instrument of force as it is habituated to, to destabilize a carefully mobilized and organized minority ethnic group agitation for self-determination and socio-economic and political autonomy. If the former Soviet Union and Yugoslavia in spite of their awesome military might could disintegrate because they could no longer use force to keep people together, the existing conflicts in the Niger Delta if not carefully managed and resolved, could lead to the deconstruction of the fragile and intensely stressed federal compact.

The apparently unabating nature of the conflicts in the oil producing region of the federation shows in a fundamental sense, a failure of the Nigerian state to be responsive, fair and effective in the management of legitimate group grievances. The agitation of the oil producing minority ethnic nationalities for resource control is heavily predicated on the general perception of the federal system as too centralized and therefore, only those who control political power at the centre also control the vast oil resources, including their extraction and distribution. Even though the extant 1999 Constitution of the Federal Republic of Nigeria has placed mining under exclusive legislative list, the agitation for resource control by the oil-producing minority ethnic nationalities is unyielding. One of the leading proponents of the resource control agenda, Lemmy Owugah (2000: 114) has attributed the insistence of the oil producing minority ethnic nationalities on resource control and autonomy "to the loss of confidence in the Nigerian State to meet up to the expectations and aspirations of the people". This brings into bold relief, the serious security implications of the peoples agitation for autonomy and resource control in a federation whose life-line is oil and which is held together by this same strategic commodity. If there is a feeling of loss of confidence in the Nigerian State by the Niger Delta oil minority ethnic nationalities, the prospect of arriving at a mutually agreeable solution to the legitimate grievances of the people may be pretty remote.

This may be so because, the loss of confidence as expressed above by Lemmy (2000) could mean that they have equally lost the feeling of stake hood in the corporate survival of Nigeria as a federation strongly supported by the oil resources derived from their land. Since there is loss of faith in the Nigerian State, the most effective response is to agitate for the control of oil resources to be used to effect the desired development in the area. National security is significantly threatened where a resource generating group feels itself excluded from the benefit of the resources and is "discriminated against in the existing socio-economic and political order, and cannot tap credible, representative and responsive channels for the resolution of its grievances" (Olukoshi and Agbu, 1996:97). The growing militancy and minority nationalism and the struggle to wrest control of oil resources from the Nigerian State could inevitably weaken the welfare basis of state legitimacy and consequently spawn mass discontent with its responsive capacity. Since the foundation of national security is essentially predicated on state's policies designed to abate "sources of deprivation, social discontent, inequalities and neglect and since no government in the face of widespread or mass discontent can retain legitimate power in the society, what could happen is a drifting of the society into a state of anarchy and consequent self-destruction" (Luard, 1988:83). The Nigerian State as it is presently configured, owes its survival to the oil resources embedded in the Niger Delta environment and should the struggle of the oil minorities result in their total control of the resources, it is likely to torpedo the foundation of the state.

The vulnerability of the Nigerian state to the threat generated by the agitation for resource control and political restructuring by the oil producing minority ethnic nationalities could be better appreciated against the background of the fact that as it exists today, it is buffeted by some strong ethno-regional strains and stresses clamouring for a federally desirable autonomy. While the Yoruba have been the leading proponents of political restructuring to guarantee them federally desirable autonomy, the Hausa/Fulanis have found the adoption of the Sharia legal system an effective response to the Yoruba¶s agitation for autonomy. In the same manner, the Igbos have been leading a campaign for the actualization of sovereign state of Biafra. The oil producing ethnic minority nationalities have responded to all these ethno-regional demands by stepping up efforts in their resource control and political autonomy agitation. The Middle Belt minorities have also carefully articulated their political preference for a federation with a strong centre in order to be protected (Elaigwu, 2002).

The implication of these discordant preferences of ethno-regional groups is that there is widespread dissatisfaction with the structure of the Nigerian federation and as each group strives to give expression to its feelings about the system, it faces intense, and sometimes, violent opposition from others. The Nigerian federation is driven by these conflicting ethno-regional political preferences. There has been such a massive a loss in faith in the Nigerian federation since the 1990s as a result of the repeated failure of the system to promote the developmental aspirations of the diverse constituent units (Olukoshi & Agbu 1996; Obi 1998). The ethnic minorities, particularly the oil producing minority ethnic nationalities feel more embittered about the inadequacy of the Nigerian federation to guarantee them a sense of inclusion. Consequently, there has been a steady build up of pressures for the reconfiguration of the federation to reflect their articulated wishes and aspirations.

The dominant feeling among the ethnic nationalities in the Niger Delta is that the demands of other ethno-regional groups are carefully articulated to blunt their determined struggle for resource control and political autonomy since the time of military dictatorship. According to Glory (2002), the National President of the Movement for the Survival of Ijaw Ethnic Nationality of the Niger Delta (MOSIEND),

> "I cannot see how the agitation of these other ethnic groups deliberately sponsored to confuse our struggle for equity and justice can undermine our collective resolve to agitate for resource control and political autonomy. This is a struggle we have invested in by way of the execution of Ken Saro-Wiwa and other compatriots. If it takes only oil to keep the Nigerian federation afloat, let it sink in the process".

This sentiment brings to bold relief, the unyielding determination of the people of the Niger Delta oil producing region to ensure the restructuring of the Nigerian federation to reflect their wishes and aspirations. The implication of this for the Nigerian state is that whether or not restructuring of the federation is agreeable to the interests of other component units, it is an agenda that must be executed even if it means risking disintegration.

The view of Nana Kumo (2001), an Ijaw youth activist, on the restructuring agenda agrees with the preceding position. He argues that the;

> "key demands of the Niger Delta people for resource control, political autonomy, self-determination and environmental restitution have been totally neglected by the 1999 Constitution of the Federal Republic of Nigeria which he describes as insensitive, a recipe for anarchy and therefore incapable of resolving the conflict in the region".

His perception of the Constitution is perhaps, informed by the fact that the document still retains the land use Decree of 1978, the Petroleum Act of 1969 and a couple of other "unsympathetic" legislations which deny the oil producing communities in the Niger Delta region access to the natural resources. If the oil producing communities in the Niger Delta region are seeking to have these legislations repudiated, it is but a recipe for the dismemberment of the Nigerian Federation which thrives on the oil-driven economy. This therefore, implies that the future of the Nigerian federation is gravely at stake as the oil producing communities struggle to claim total control of the vast oil resources embedded in their environment.

What could be gleaned from the preceding analysis of the political implications of the Niger Delta conflicts for national security is that there is strong feeling of discontent with the structure of the Nigerian federation among the oil producing minority ethnic nationalities which has led to the articulation of some demands that are obviously beyond the carrying capacity of the federation. This could easily stress the resilience of the federation to the brink. Reference is often made to the experiences of older and more established federations such as the United States of America and Canada where the ownership and control of resources are vested in the states or provinces, but these may not be quite helpful in the enterprise of nation-building. Our peculiarities as an emerging federation that is sustained by the oil-driven economy should be appreciated and balanced against the "ideal" experiences of these countries. While these countries have attained significant level of industrialization with corresponding high profile of political stability, any attempt by the Nigerian federation to pattern itself after these two countries could precipitate either peaceful, as advocated by Adedapo (2002) or violent dissolution. However, as it is now, the issues of self-determination, environmental and social justice for the Niger Delta people remains central and critical to achieving national cohesion and stability in the Nigerian nation-state.

14.6. The social implications of the niger delta conflict for national security

The social implications of oil-induced conflicts in the Niger Delta for national security are as grave as the preceding political implications we have analysed. Our analysis of the social implications of the Niger Delta conflicts for national security is situated in the context of the three identifiable structures of the conflict. These are: the state versus oil producing communities, oil companies versus oil producing communities and inter/intra-community (ies) conflicts. The aggregation of these three patterns or structures of conflicts in the Niger Delta have fundamentally undermined the peace, harmony and communal cohesion of the oil producing communities. The spate of inter-community and intra-community conflicts in particular has resulted not only in the destruction of lives and property but also other forms of social dislocation as well as the militarization of the youths and consequent state repression.

The relations between the various communities in the Niger Delta is characterized by tension, violence, mistrust and attempts by each community to demonstrate that it is first among the exploited and despoiled victims of the oil exploration and production in the region (Iyayi, 2002). Consequently, the various ethnic minority nationalities in the Niger Delta have carefully constructed resilient platforms through which grievances could be assiduously articulated and responses planned and executed. The development of competitive relations among the minority ethnic nationalities in the Niger Delta, who though share similar problems and harbour the same grievances, have spawned the culture of restiveness among the youths of each of the groups. Youth restiveness in the region is cultivated and sustained by the psychological feeling that it is more rewarding to attack and hold expatriate staff of oil companies' hostage for ransom than attending school. Possession of sophisticated weapons for offensive warfare is a common thing among the youths of the various ethnic minority nationalities of the region (Benjamin,

2002). The implication of this is that the youth have lost faith in education as a liberating social tool in preference for a trade (warfare) that yields them immediate tangible benefits. It means then that even if the existing grievances of the region are adequately resolved in the future, disarming the youth who have found cheap means of livelihood in warfare may create a new context for a war and yet, leaving them to continue to bear arms in post peace Niger Delta region will heighten the state of insecurity both at the micro and macro levels.

Illiteracy, which is the inevitable consequence of youth self-withdrawal from schools, is one of the most potent threats to the security of Nigeria. This is so against the background of the fact that illiteracy renders them vulnerable to the manipulation of social forces that may have axes to grind with the government of the Nigerian state (Vogt, 1990). This is more so that they are exceedingly habituated to cheap subsistence. Indeed, foreign adversaries who may want to destabilize or undermine national security could, painstakingly recruit these illiterate elements with weak sense of patriotism to the Nigerian state for this purpose. By and large, the bearing of arms by illiterate youths in a multi-ethnic developing society like Nigeria poses grave and credible threat to national security. The killing of policemen and soldiers by youths in Odi Community in Bayelsa in 1999 is a concrete case in point to stress the seriousness of the point being articulated here. As a reaction to this orgy of violence inflicted on the men of the Nigerian army and the police force, government ordered military action, which rendered the Odi community desolate. These responses by government generated intense furore in the country as it was widely considered to amount to an over kill in a democratic setting. This brings to bold relief, the acute limitation of illiterate youths in seeking the resolution of the grievances through dialogue, compromise and persuasion.

Another obvious social implication of the Niger Delta conflicts for national security is the near total breakdown of social order in the region. Preceding the advent of violent agitational politics in the oil-rich Niger Delta region in the 1990s, the hierarchy of authority rested with the elders whose views and opinions on issues of common relevance were scarcely contested by the youths. This established traditional social order according to Benjamin (2002) has suffered grave setback as youths have now come to view elders as collaborators in a system that has deeply traumatized them. Elders are often accused by the youths of colluding with the state and oil multinational companies to deny them appropriate volume of compensation for oil spillage and other forms of environmental pollution (Okoko and Nna, 1997). Consequently, the voice of elders is easily overwhelmed and submerged by the voice of the youths who are more readily predisposed to using violence to squeeze out the appropriate quantum of monetary compensation from the oil companies operating in their communities. A breakdown in established traditional social order of a society may drift it into a state of anomie thereby casting it in the image of a Hobbessian "state of nature" generally described as "nasty, brutish, short, solitary and poor". It constitutes a threat to national security in the sense that this tendency may grow beyond the frontiers of the oil producing communities to manifest in non-recognition of any constituted civil authority as the killing of police and soldiers deployed on peace mission in Odi community in 1999 would show.

Further, as different ethnic nationalities in the Niger Delta strive to maximize oil-related benefits in terms of Community Assistance Projects (CAP) being sponsored by oil companies operating in the industry, they are pitched against one another thereby fracturing communal cohesion as a critical factor in the administration of national security. Contestation of ownership of land known to be endowed with crude oil has riven the institution of kinship in the Niger Delta oil producing minority communities. People belonging to the same family have violently contested ownership of oil well and land. This violence is often spawned by the belief that one stands to derive immense benefits from ownership of oil well and land through payment for pipeline and other sundry uses should such land be acquired by the MNCs. In 2002 in Nembe community for instance, Bassambiri and Ogbologbo mabri who were members of the same Ijaw clan in Bayelsa State inflicted large-scale violence on each other leading to the death of 200 persons (Benjamin, 2002). The Ijaw-Ilaje conflict, Ogoni-Andoni conflict, Ogoni-Okrika conflict and the Urhobo-Itsekiri conflict are some of the numerous examples of inter and intra-ethnic/communal conflicts in the Niger Delta (Ibeanu n.d). The basis of communal cohesion, which is mutual trust, is torpedoed as people painstakingly struggle to appropriate economic resources without any consideration for the morality of doing so.

From the below tables which represents the nature, actors and location of inter and intra-communal conflicts in the Niger Delta oil producing region of the Nigerian federation, it could be gleaned that most of them were resource-motivated. The Ogonis in particular, is understandably so because of the resilient platform of resistance which this minority ethnic nationality had carefully constructed. The Movement for the Survival of Ogoni People (MOSOP) had effectively mobilized the entire Ogoni people to halt further exploration and production of oil in their land since 1993. The Guardian newspaper captures vividly the implication of this development for the Nigerian governments.

"[3] The Ogoni's defiance posed one of the most serious challenges to Nigeria's power structure since the Biafran war because it was an example of effectively organized resistance that could not be quelled with money or threats. It also raised the spectre of separation in a country of over 250 ethnic groups. And it touched one of the army,s raw rest nerves, its source of cash. When Shell was forced out of Ogoniland in 1993, the army lost a substantial slice of its pie (Guardian Newspaper, 1995:7).

Table 1: Showing the Frequency, Nature and Actors in Communal Conflicts in the Niger Delta from 1993 – 2001.

S/N	*Date*	*Nature of Communal Conflict*	*Actors*	*Location*
1.	August 5, 1993	Conflict between the Ogonis and Andonis over land and fishing rights believed to have been carefully fueled by government	Ogoni and Andoni Neighbouring groups	River State
2.	April 3, 1994	Conflict between the Ogonis and Ndokis over land alleged to have been fueled by government in which several lives were lost on both sides	Ogoni and Ndoki	Rivers States
3.	December 10, 1993	Conflict between Ogoni and Okrika believed to have been precipitated by some Ogonis resident in Port-Harcourt, celebrating Abacha Coup which ousted Governor Ada George, an Okrika man.	Ogoni and Okirika	Rivers State
4.	October 1999	Violent conflict between Eleme and Okrika communities over traditional and legal titles to a piece of land on which the Port Harcourt refinery is located	Eleme and Okrika	Rivers State
5.	November 16, 1999	Bloody communal conflict between Oleh and Olomoro over sharing of used pipes from Oleh-Olomoro flow station.	Ole and Olomoro	Delta State
6.	November 26, 1999	Renewed hostility between the two neighbouring communities of Eleme and Okrika over land.	Eleme and Okrika	Rivers State
7.	March 18, 2000	Communal hostility between the Eleme and Okrika over land which degenerated into violent conflict claiming several lives.	Eleme and Okrika	Rivers State
8.	April 12, 2000	Bloody conflict involving six communities in Gokana LGA of Rivers State over land.	Intra-Communal Ogoni	Rivers State
9.	May 16, 2000	Violent conflict between the Akaasa and Ignama communities over land.	Akaasa and Igwama	Bayelsa State
10.	October 12, 2001	chieftaincy tussle which later snowballed into violent clannish conflict which claimed considerable lives and large scale destruction of property	Intra-Communal Okrika	Rivers State.

Source: Fieldwork, 2001.

Given the grave implication of the Ogoni action for steady revenue flow from oil, the Nigerian state carefully sought to deconstruct the MOSOP platform by using the notorious Rivers State Internal Security Task Force (RISTF) headed by the much dreaded Major Paul Okuntimo, by pitching them against their neighbours who were largely assisted by the aforementioned security task force to rout the Ogoni (Human Rights Watch, 1999). In a sense, the so called conflicts between the Ogonis and their neighbours whom they have maintained harmonious relationship "contrived by the government and Shell Petroleum Development Company (SPDC) was to provide an opportunity for the military crackdown to enable Shell resume operation in Ogoni as well as act as a deterrent to other oil-producing communities" (Olorode et al, (eds), 1998:398). This tendency by government has obvious security implications because it shared that communities that hitherto maintained peaceful coexistence, could be instigated and armed to turn against one another thereby providing a fertile ground for mutual suspicion, distrust, violence and insecurity. This divide and rule strategy adopted by government has achieved some results in the sense that it worked against the emergence of a Pan-Niger Delta Platform capable of replicating the Ogoni experience across the region which would obviously have spelt doom for the Nigerian State. It is however, genuinely feared that with increased mobilization and awareness campaigns in the region, the people may soon discover this trick with possible boomerang effect against the state.

This same tendency is played out at the macro-level (national) where minority agitations for resource control have fueled yet another contradiction in the Nigerian federation. As Kirk Greene (1971) has observed, the Nigerian project is bedevilled by a set of oppositions each widening and reducing the hope of constructing a united and indivisible Nigeria; North versus South, Islam versus Christianity, alleged feudalism versus assumed socialism, federal versus unitary preferences, haves versus have-nots, and recently oil producing versus non-oil producing states, each with sinister undertones of tension, irreconcilability and threatened withdrawal from the Nigerian project. This contradiction between oil producing states versus non-oil producing states as Obi (1998) has noted is deeply embedded in intra-class squabbles over oil rents. The battlefield of this struggle has been the revenue allocation system which is being challenged by the oil producing states as devoid of equity and fairness. Oil producing states have always expressed preference for a sharing formula founded on derivation against the preference of the non-oil producing states for equality of states and the size of population. To justify their preference for oil revenue sharing formula on the basis of principle of equality and size of population, the non-oil producing states have always "accused their oil producing counterparts of greed, insisting that they do not have a sole right to the oil within their territory since they did not put it there, nor contribute anything to it" (Adebayo, 1993:172).

This kind of inflammatory statement can harden the resolve of a people with an acute historical sense of marginalization, underdevelopment and environmental degradation in a multi-ethnic competitive federation. The increase from 3% to 13% of statutory allocation to mineral producing areas have neither appeased the oil minorities nor allayed the deep seated fears of the non-oil producing states, thus worsening and exaggerating the tension in inter-state relations. National cohesion as a vital national security input could be threatened when the oil producing minority ethnic nationalities are made to feel that they do not have a stake in the survival of the Nigerian federation. It is in this context that the oil producing minority ethnic nationalities in the Niger Delta have viewed the agitation by the people of Niger State for the establishment of Hydro-electricity Power Producing Area Development Commission (HYPPADEC) and the Solid Mineral Producing Areas Development Commission (SOMPADEC) by some states in the Middle Belt geo-political zone of the Nigerian federation, as carefully articulated demands to counter their demand for a greater share of the oil proceeds in the custody of the federal government (Thisday, 2001). Although these could be legitimate demands of the people of these states in a multi-ethnic competitive Nigerian federation, the earlier accusation of the oil producing minority ethnic groups of greed creates in them, the belief that these demands of other groups are, but an attempt to weaken their resolve to effectively push their case through.

The social phenomenon of unemployment in the Niger Delta region as in any other part of the Nigerian federation constitutes a fundamental threat to national security. The youths who are the worst victims of this social problem and who are endowed with vast amount of energy, have taken recourse to violence as an alternative means of livelihood. Oil exploration and production have diverted the attention of the youths of the Niger Delta region from the traditional farming and fishing occupations. Even for the rest of the people who subsist on these enterprises, frequent oil spillage and exploratory activities have polluted and degraded lands in the region leading to acute contraction of the range of fishing and farming activities (Okonta and Douglas, 2002). Oil companies however, are reluctant to engage the people, particularly the youths who are said to lack the basic qualification to work in the oil industry. Meanwhile pollution of arable land by oil spillage as noted earlier, has caused an acute shrink in the size of cultivable land. A direct fall out from this is that most of the people whose main occupations are farming and fishing have thus been deprived of their source of livelihood (Aluko, 2000).

Consequently, there has been a diminution of opportunities for the people to realize their developmental aspirations. This has heightened the sense of exclusion and marginalization. Years of unfulfilled government promises and growing inability to find gainful employment opportunities have left the people deeply frustrated and restive. There is a steady build-up of an army of unemployed youths who have, for want of any productive engagement, taken to violence and armed struggles in the Niger Delta region. Inevitably, the youths driven by the problem of unemployment are being socialized into the culture of violence with grave implications for national security.

Innocent (2002) observes that the sophisticated nature of arms used by the unemployed youths to perpetuate violence in the Niger Delta region poses critical threat to not only the security of the region but that of the nation as a whole. This is so because of the belief already cultivated in the minds of the unemployed youths that violence guarantees them survival in the absence of gainful employment opportunities. In a region that has remained historically restive and agitated, the vast army of unemployed youth could easily be transformed into an alternative armed force in opposition to the armed forces of the federation which they may perceive as a tool used by both the state and the oil companies to terrorize them. If there should emerge a pan-Niger Delta revolutionary front, which is quite pretty remote but not totally impossible, with enhanced mobilization in the region, the scale of threat to national security would be more pronounced in view of the fact that youths in the region are already habituated to the swampy terrain and may find it quite cheap to engage conventional armed forces of the Nigerian state in a war of sub-national liberation.

As Buzan (1992) has argued, where state and the citizens are severely at odds, domestic disarray could sufficiently threaten the coherence of the state in a manner that makes the value of national security exceedingly difficult to achieve. This fits into the experience of the Nigerian state as the youths who engage the oil companies in violence, are pitched against government and disrupt the traditional social order. There is in effect, a fundamental threat to national unity and integration in addition to the growing socialization of the youths into the culture of violence thereby rendering them anomie.

14.7. Economic implications

The economic implications of the Niger Delta conflict for national security in Nigeria could be better analyzed in the context of oil as the soul of the Nigerian nation-state. The conflict in the Niger Delta region is invariably conflict over the soul of Nigeria and its future. Oil, arguably, is the critical stuff of contemporary Nigerian politics and the Niger Delta which harbours the oil industry is the field on which the vicious battle to control this money spinner is waged" (Okonto & Douglas, 2002). The economic implications of the conflict in the Niger Delta for national security are predicated on the fact that oil is at the heart of the conflict involving the Nigerian state, oil multinational corporations and the ethnic minority oil producing communities. If the oil commodity is viewed as the foundation of the Nigerian federation and that it is deep at the core of the enduring conflict in the oil producing Niger Delta region, the obvious and predictable options are between collapse and redemption depending on how it is managed and resolved. The ability of the tenuously compacted federation to survive the strains and stresses generated by the push and pull forces and to promote the developmental aspirations of the diverse ethnic nationalities is inseparably linked to the state of health of the oil industry.

For a federation that depends almost exclusively on oil revenue, the consequences of conflict in the oil industry could be grave. Since its manifestation in the early 1990s, the unending conflict in the Niger Delta is taking great toll on the nation's financial well-being. Oil multinational corporations responsible for the exploration and production of the critical oil resources are facing a distressful political and economic environment both nationally and at the level of the oil producing communities where their facilities are located. At the community level, the companies are faced with growing protests directed at their activities over lack of development in spite of their harbouring the providential wealth.

In a carefully researched account of cases of human rights violations in Nigeria's oil producing communities, Human Rights Watch, a non-governmental organization based in the United States of America, has observed that "successive governments have misspent the oil wealth which the oil companies have helped to unlock, salting it away in foreign bank accounts rather than investing in education, health and other social investment, and mismanaging the national economy to the point of collapse" (Human Rights Watch, 1999: 2-3). There is in effect a causal relationship between the spend thrift of successive Nigerian governments on the region and the conflict. The anger of the oil producing communities is fueled by their perception that they have remained inadequately touched by the stupendous wealth derived from their backyard.

It is the anger of the oil producing communities expressed in armed attacks against the oil companies in the oil industry and disruptions of the production of oil on sustainable basis that has provoked the armed response by the Nigerian State. Armed troops were quickly deployed to provide adequate security for oil installations and to crush the protests by oil producing communities in the Niger Delta region. The armed response by the state to the armed violence perpetrated against oil companies by the oil producing communities has only fueled tensions in the region and consequently, made it exceedingly difficult for the oil companies to sustain optimum production of oil (Pegg, 1999). In the wake of violent protest by oil producing communities in 1993, Shell claimed that it lost N9.9 million daily in Ogoniland when it was forced to stop production by the angry villagers (Obi, 1999). Similarly, Shell, Elf and AGIP collectively lost about $200,760,000 in 1993 to violent protests and blockages (Lukula, 1994). While these tensions seethe and the conflicts rage, both oil companies and oil producing communities have sought to outdo each other in terms of arms acquisition. Though the oil companies maintain a lead in this arms race, the irrepressible resolve of the oil producing communities to destabilize the oil industry poses credible and tremendous threat to the monocultural oil economy (Human Rights Watch, 1999).

The unspeakable scale of armed violence and its negative impact on the oil dependent economy became so pronounced in 1998 that Gen. Abdulsalami Abubakar, the then Head of State, in his January 1, 1999 budget speech, stated inter-alia:

> This administration is also aware of the dissatisfaction among certain segments of our population arising from certain government actions or inaction in the past. Genuine as such grievances may be, we cannot allow the continued reckless expression of such feelings. The development in the oil producing areas of the Niger Delta region is a case in point. While we appreciate the feelings of the people in the area over their sad condition, this administration notes with great displeasure the disruptions of the activities of oil companies and private enterprises by rampaging youths. Seizure of oil wells, rigs and platforms as well as hostage-taking and vehicular hijacking, all in the name of expressing grievances are totally unacceptable to this administration. ... Government will not accept brazen challenge to the state authority under threat of violence as recently happened in the Niger Delta region. Government has a responsibility to safeguard the state and the security of life and property of all its citizens and those of foreign nationals on our soil carrying out their legitimate preoccupations'.

The preceding statement is an explicit acknowledgement by government of the seriousness of threat generated by the violent protests of oil producing communities to the viability of the economy and impliedly, to national security. It is obvious why General Abdusalami had to reel out the statement in view of the Kaiama Declaration of December 11, 1998 which urged all multinational oil companies to withdraw from Ijaw territories pending the resolution of issue of resource ownership and control" (Kaiama Declaration, 1998). Shortly after this historic declaration, thousands of troops and navy personnel were brought into the region in response to this threat to the operations of oil companies in the Niger Delta region. The Nigerian economy is therefore, inevitably caught up in her own contradiction. Its failure to empower the communities from which it derives its strength has spawned widespread poverty, which in turn, has provoked and fueled the revolutionary anger of the people seeking to delink it from its lifeline. The violent protests in the region do not only threaten oil production but also the security of the nation at large. It undermines the responsive capability of government to the competing demands of the citizenry.

Apart from threatening the viability of the economic foundation of the Nigerian State, the violent protests in the Niger Delta region do not project a favourable image of the country to foreign investors who may want to come and do business.

The country is considered by the outside world as one perpetually drawn into the vortex of a seething cauldron with scant promise of achieving national stability and progress. This negative perception of the country by the outside world robs it of its much-desired foreign investment to lessen intense pressures on the oil commodity as the sole foreign exchange earner.

In its editorial of August 14, 1999, the Post Express, one of Nigeria's dailies, estimated that the federal government and the eight joint venture partners in the nation's oil industry may have lost well over one billion dollars (about N100 billion) in the six months between January and June of 1999 (Post Express, 1999:1). This is no doubt, a whooping sum of money lost to the spate of violence in the Niger Delta region which could have been prudently utilized to induce the much desired socio-economic transformation that is being sought in the oil producing communities. The violent protests in the oil industry heightened in 1997, as the late General Abacha, the then Head of State and Commander-in-Chief of the armed forces sought to tighten federal government¶s grip on the oil resources. The steady rise in violent protest during the Abacha¶s era is graphically presented in the tables that follow;

Table 2: Showing the Frequency, Actors and Location in Community/Oil Companies Conflicts from 1992-199 7.

	Date	*Nature of Conflict*	*Location*	*Oil Company*
1.	October 10, 1992	People of Umuechem Community protested against Shell Petroleum Development Company on the ground of discrimination in the employment of even non-skilled workers and lack of basic amenities in the community.	Rivers State	SPDC
2.	August 17, 1993	Ogoni People forced the Shell Petroleum Development Company out of Ogoniland thereby halting the production of 108,000 bpd. Since 1993, Shell has remained fenced out of Ogoniland.	Rivers State	SPDC
3.	October 9, 1993	Obagi Community stormed the Elf Petroleum Nigeria Limited flow station and halted the company¶s operations for several weeks, protesting non compensation of oil spillage.	Rivers State	Elf
4.	April 30, 1993	Demonstrations and protests by Ogoni farmers against laying of an oil pipeline by Shell and its agent, Wilbros Construction Company.	Rivers State	SPDC

5.	March 5, 1997	Youths captured a barge delivering goods to a Chevron Installation and demanding jobs on the vessel	Delta State	Chevron
6.	August 12, 1997	Iyokiri community blocked access to Shell Petroleum Development Company employees and demanded compensation leading to the closure of 3 flow stations for several days.	Rivers State	SPDC
7.	September 3, 1997	Peremabiri Community demanded compensation for fishing nets damagedby an oil spill in June leading to the closure of 10,000 barrel per day Dieubu flow station for several weeks.	Bayelsa State	SPDC
8.	October 6, 1997	Youths demanded that 50 of them be employed, this led to the closure of Odeama flow station for several days	Bayelsa State	SPDC
9.	October 14, 1997	Youths protesting the effect of the gas flaring on their village in Gelegele village which halted production for several days at a well yielding up to 2,000 bpd.	Delta State	Dubri Oil Company an indigenous Nigerian operator.
10.	November 14, 1997	Ekakpamre community forced the closure of ughelli West flow station for several days, demanding N20 million (U.S. $2222,000) compensation for encroachment on their land, a new access road and other projects. About 6,500 bpd has been shut in by the protest.	Delta State	Oil Company
11.	November 25 – December 23 1997	Closure of Tunu and Opukoshi flow stations together pumping 80,000 bpd by villagers.	Bayelsa State	SPDC
12.	December 13-17, 1997	13 employees of Western Geophysical were held hostage by youths in a barge off the coast of Ondo State	Ondo State	SPDC

Source: Fieldwork 2001 and Human Rights Watch, 1999

Table 3: Showing the Frequency, Actors and Location in Community/Oi Companies Conflicts from January 1998- December 1998.

1.	January 20, 1998	Odeama Creek flow station, pumping 18,000 bpd, was closed for several days by youths demanding environmental tests, a reduction of gas flaring, clean water supply and other projects.	Rivers State	SPDC
2.	March 10-20, 1998	Youths from Koluama occupied for over a week, Texaco's Funiwa Platform causing an eleven day spillage in the export loading schedule from the plumingto terminal	River State	Texaco
3.	April 28, 1998	Shel's Flow stations at Odidi and Egwa were closed by protesting youths	Delta State	SPDC
4.	May 25, 1998	Youths occupied Chevrons Parabe Platform Onshore from Ondo State, and held workers hostage.	Ondo State	Chevron
5.	June 11, 1998	Villagers protested a spillage leading to closure of four wells and distrupted exports from AGIP's terminal.	Bayelsa State	AGIP
6.	July 15, 1998	400,000 bpd had been shut in at SPDC's Nun river flowstation by protesting youths who also hijacked a helicopter and forced the evacuation of staff.	Rivers State	SPDC
7.	August 11, 1998	60,000 bpd production from Nembe was closed for several days by youths protesting non-fulfillment of employment and compensation promises.	Rivers State	SPDC
8.	August 29, 1998	Youths disturbances at Nembe shut in at least 440,000 bpd of output and forced both Shell and AGIP to stop exports. Also 10 Shell facilities, producing about 200,000 bpd were shut for some days.	Rivers State	SPDC and AGIP.

9.	October 9, 1998	400 youths occupied Shell's Forcados terminal for several hours, protesting non-payment of compensation for Mobil's January Oil Spill.	Bayelsa State	SPDC
10.	October 14, 1998	Youths seized control of two Chevron flow stations at Abiteye and Olero Creek, near Escravos taking some thirty workers hostage for two days	Delta State	Chevron
11.	November 12-18, 1998	Eight oil workers employed by Texaco were held hostage by youths from the community of Eorupah, near Warri, demanding social investment in their village and compensation for a recent oil spill	Delta State	Texaco
12.	December 9, 1998	Youths occupied Shell Forcados terminal protesting the Mobil Spill	Bayelsa State	SPDC

Source: Fieldwork 2001 and Human Rights Watch, 1999

As could be gleaned from the foregoing tables, the various attacks on oil company installations were motivated by one grievance or the other with the youths as the leading protesters in the region. Youths of the various communities in the Niger Delta have always engaged in different forms of protests against unemployment, environmental degradation and neglect. The implication here is that so much has been lost in terms of oil proceeds, which could have been invested socially to improve the lot of the citizenry. Thus, all these attacks inflicted on the oil companies are, in a fundamental sense, attacks against the Nigerian state, which has entered into partnership with the oil companies as majority shareholder in order to ensure the good health of the oil driven economy. Men of the armed forces and mobile police have always been drafted to conflict scene to secure the embattled oil companies.

If national cohesion is viewed as the soul of national security (Pegg, 1999), the former can hardly be achieved in a multi-ethnic federation characterized by weak economic foundation. This is so because the weak economy may not adequately be responsive to the competing demands of the various social forces thereby pushing them, as Ake (1994:7) has noted, to "seek safety and fulfillment in their community, ethnic or nation. The demands which they make on these social formations have turned them into informal polities, in competition with the state".

14.8. Concluding remarks

From the preceding analysis of the political, social and economic implications of the Niger Delta conflict for national security in Nigeria, some objective threats spawned by the conflict could be identified. Politically, the conflict, if not carefully managed and resolved could lead to a deconstruction of the federal compact because as governments responsive and distributive capabilities are overstretched, repression, intimidation and wanton violation of human rights become pronounced thereby deepening the rift between government and its citizens. Socially, national cohesion is undermined as sub-national groups take up arms against one another in their agitation for an enhanced profile of inclusion and belonging in the mainstream affairs of the nation-state.

Economically, the oil-induced conflict in the region could pose credible threat to the future economic viability and prosperity of the Nigerian state. Frequent disruption of oil exploration and production by youths in the Niger Delta could spawn social ills such as poverty, illiteracy, unemployment, environmental degradation, cleavages and acute inequality. All of these could easily combine to make it near impossible, the task of general social mobilization in Nigeria for the enterprise of building a united, strong and self-reliant nation; a great and dynamic economy; a just and egalitarian society; a land of high and full opportunities for all citizens; and a free and democratic Nigerian society (Federal Republic of Nigeria, 1971:141).

National Security in Nigeria would remain elusive if the state cannot aggregate the social elements to construct a polity that would concretize the aforementioned national goals and aspirations. Failure to achieve these goals and aspirations could widen the frontiers of Nigeria's vulnerability and yet if we must achieve these enduring national cohesion, shared aspirations preferences and perceptions are critical psychological inputs that cannot be discounted. Heavy reliance on the use of force to destabilize a carefully mobilized people not only presents the image of a state which is low in legitimacy and desperately struggling to survive, but also in the long run can do more to threaten state coherence than to aid it (Edmond, 1983)

References

Adebayo, A. G. (1993) *Embattled Federalism: History of Revenue Allocation in Nigeria. 1946-1960:* New York: Peter Lang

Adedapo, A. (2002) *Nigeria: The Case for a Peaceful and Friendly Dissolution,* Lagos: Unitype Enterprises.

Ake, C. (1994) *Democratization of Disempowerment in Africa,* CASS Occasional Monograph No. 1, Lagos: Malthouse Press Limited.

Akhakpe, I. (2002) "Federal and Resource Control Agitations in Nigeria: Towards an Alternative Framework for Peace" in The *Constitution, A Journal of Constitutional Development,* Vol. 2, No. 3.

Aluko, M.A.O. (2000) "Social Dimensions and Consequences of Environmental Degradation in the Niger Delta of Nigeria: Suggestions for the next Millennium" in Oshuntokun (ed) *Environmental Problems of the Niger Delta,* Lagos: Friedrick Ebert Foundation.

Ate, B. (1992) "Towards An Integrative Approach to Nigeria's Security Relations with is immediate neighbours" in Ate and Akinterinwa (eds) *Nigeria and its Immediate Neighbours: Constraints and Prospects of Sub-Regional Security in the 1990s,* Lagos: NIIA

Benjamin O., Interview on 16th October 2002 at Abuja, FCT.

Buzan, B. (1992) *People, State And Fear,* New York: Harvest Wheatsheaf Press.

Don, H. (1999) "Human Security: Safety for people in a Changing World" Unpublished paper Presented at a Two-day *Regional Conference on*

Management of Africa Security and Challenges of 21st Century, Organised by NIIA, Lagos, 23rd – 24th June, 1999.

Edmond, K. I. (1983) "The State, Public Policy and The Mediation of Ethnic Conflict" Donald and Victor (eds) *State versus Ethnic Claims: African Policy Dilemmas,* Voulder; West View Press.

Elaigwu, J. I. (2002) "Federalism in Nigeria's New Democracy Polity" in *Publius; The Journal of Federalism,* 32:2 (Spring, 2002).

Federal Republic of Nigeria (1971) *National Guidelines for the Projection of Nigeria's Philosophy and Action,* Lagos: Federal Ministry of Information.

Glory, F. M., Interview on 18th January 2001 at Yenagoa, Bayelsa State.

Human Rights Watch (1999) *The Price of Oil: Corporate Responsibility and Human*

Rights Violations in Nigeria's Oil Producing Communities, printed in USA. Ibeanu, O. (Not dated) "Janus Unbound: Antimonies of Petro business and Petropolitics in the Niger D elta" unpublished commentary.

Ibeanu, O. (1997) *"Oil Conflict and Security in Nigeria: Issues in the Ogoni Crisis,* AAPS Occasional Paper, Vol. No. 2.

Ihonvbere, J. (2000) "A Recipe for Perpetual Crises: The Nigerian State and the Niger Delta Question" in Wumi (eds) *Boiling Point: The crises in the Oil Producing Communities in Nigeria,* Lagos: Committee for the Defence of Human Rights (CDHR).

Imobighe, T. (1998) *The Management Of National Security,* Ekpoma: Edo State University Publishing House.

Imobighe (1999) "An Overview of the Theoretical issues in African Security" Unpublished paper Presented at the Two-day *Regional Conference on the Management of African Security and the Challenges of the 21st Century,* NIIA, Lagos 23-24the June.

Innocent, A. O. (2000) "The Nigerian State and the Environmental Question" in Osuntokun (ed) *Environmental Problems of the Niger Delta,* Lagos: Fredrich Ebert Foundation.

Iyayi, F. (2000) "Oil Companies, Community Relations and International Interests: Oil Corporations and the Politics of Community Relations in Oil Producing Communities" in Wumi et al (eds) *Boiling Point: The Crises in the Oil Producing Communities in Nigeria,* Lagos: CAHR.

Kenneth, W. (1979) *The Theory of International Politics,* Addison: Wesley Reading Mass.

Kirk-Greene, A.H.M. (1971) *Crisis and Conflict in Nigeria: A Documentary Source Book 1966-1970,* Vol. 1, London: Oxford University press.

Lemmy, O. (2000) "Political Economy of Resistance in the Niger Delta" in Ola (ed) *The Emperor has no clothes: Report of the conference on the Niger Delta and The 1999 Constitution,* Benin City: Environmental Rights Action/Friends of the Earth Nigeria.

Luard, E. (1988) *The Blunted Sword: The Erosion of Military Power in Modern World Politics,* London: Turis and Co. Ltd.

Maier, K. (2000) *This House has Fallen: Nigeria in Crisis,* London: Penguin Books. Nanakumo, S., Interview at Port-Harcourt on December 10, 2002.

Michael, T. K. (1998) "Redefining Security: The New Global Schisms" in Bernard Martin (eds) *Violence and Terrorism, 1998/99 Connecticut:* Dushin Publishers. Obi, C. (1997) *Oil, Environmental Conflict And National Security In Nigeria: Ramifications of The Ecology-Security Nexus For Sub-Regional Peace.* Arms Control, Disarmament And International Security Occasional Paper, University Illinois At Urbana-campaign.

Obi, C. (1998) "The Impact of Oil on Nigeria's Revenue Allocation System: Problems and Prospects for National Reconstruction", in Amuwo et al (eds) *Federalism and Political Restructuring in Nigeria,* Ibadan: Spectrum Books Limited.

Okoko, K. & Nna, J. (1997) "Emerging Trends and Community Perception in the Nigerian Oil I ndustry" in *Nigerian Journal of Oil and Politics,* Vol. 1, Number 1.

Okonta, I. & Douglas, O. (2001) *Where Vultures Feast: 40 Years of Shell in the Niger Delta,* Benin City: Environmental Rights Action/Friends of the Earth.

Olukoshi, A. and Agbu, O. (1996) "The Deepening Crisis of Nigerian Federation and the Future of Nigerian State" in Olukoshi and Laakso (eds) *Challenges to the Nation-State in Africa,* Uppsala: Nordiska African Institute.

Pegg, S. (1999) "The Cost of Doing Business: Transnational Corporations and Violence in Nigeria: *Security Dialogue,* Vol. 30, Number 4.

Thisday Newspaper (Lagos), 2002.

Thisday Newspaper, (Lagos) 2001.

Vogt, M. A. (1990) "Nigeria's Defence Policy: An Overview" in Ekoko and Vogt (Eds) *Nigerian Defence Policy: Issues and problems,* Lagos, Mathouse Press Limited.

Section V
the challenges: charting the way forward

15. Environmental diseconomies and management strategies in the nigerian petroleum industry: the need to evolve sustainable partnerships with host communities

Christopher O. ORUBU, A Yodele ODUSOLA, and William EHWARIEME

15.1. Introduction

Petroleum industry activities have been known to be associated with substantial environmental degradation and social crises that pose a potential threat to sustainable development, in the Niger Delta, where the bulk of Nigeria¶s petroleum resources are found. Although most of the negative environmental consequences of oil industry activities are localised and more intense in the areas of primary activities, some of the effects have trans-boundary implications. For example, gas flaring which is a common feature of the Nigerian petroleum industry has been known to be a factor in the problem of global warming (World Bank, 1995). In the same vein, mangrove swamp and rain forest destruction as a result of oil industry activities can have long-term consequences for both ecological and climatic balances.

Although there have been several environmental policy and management strategies in the Nigerian petroleum industry to ameliorate degradation, the allegation of involvement of communities and other local interests in the destruction of oil facilities resulting in large-scale environmental hazards in the Niger Delta, particularly in the last five years or so, adds a new dimension to the resolution of the environmental question. While the local people may suffer from the constraint of incomplete scientific knowledge about the long- term welfare implications of environmental degradation, events particularly since the 1990s indicate that environmental awareness among the people in the oil-producing communities in the Niger Delta has been on the increase (Orubu *et al,* 2002). For example in 1991, the Ogoni community had sent a representation to the United Nations, demanding for the payment of US $20 million from the Federal Government of Nigeria, as compensation for environmental degradation and for over 30 years of oil exploitation in the area. Between 1995 and 2000, the Niger Delta had been turned into a virtual battle field in a large-scale crisis which saw the youths of the area pitched against the *Industry: The need to Evolve Sustainable Partnerships with Host Communities,* oil companies and government – shutting down oil installations, destroying oil facilities, and kidnapping employees (sometimes, expatriates) of oil companies and came a little short of declaring a *Republic of the Niger Delta*, through the *Kaiama Declaration* of 1998.

The developments highlighted above, no doubt represent dangerous currents in the political economy and history of petroleum resources in Nigeria, and highly underscore the need for sustainable partnerships between the host-communities and oil companies in order to address the environmental, and other related problems. This is necessary in order to establish peace in the Niger Delta, which for now is the treasure base of the Nigerian economy. This paper is therefore concerned specifically with the identification of the major environmental problems and issues arising from oil industry activities in the Niger Delta, with a view to highlighting those areas in which co-operation between oil companies and their host-communities could be essential to the significant reduction of negative environmental incidents and restoration of peace in the region.

This paper is structured into seven sections. The first is the introduction. The next section (2) is concerned with the identification of the environmental consequences of oil industry operations. This is followed by section 3, which examines the interaction between environmental degradation and economic environment, while section 4 reviews the essential features of environmental policy and management strategies in the petroleum industry. In section 5, we discuss the issue of community awareness of petroleum-induced environmental problems. In section 6, we identify possible areas in which sustainable partnerships could be forged between oil companies and their host-communities in order to protect the environment and reduce to the minimum, the social crises usually associated with environmental incidents of catastrophic dimension. Finally, section 7 summarises and concludes the paper.

15.2. Identifying the environmental consequences of petroleum operations

Several researchers such as Awobanjo, 1981; Famuyiwa, 1998; Eromosele, 1998; Chukwu *et al*, 1998; World Bank, 1995; Moffat and Linden, 1995; Grevy, 1995; Olomo and Omene, and NDES, 1997) have investigated the environmental consequences of oil industry activities. It should be noted that the petroleum industry shares in the essential characteristics of modern industry, although at the base, it is an extractive activity. The combination of modern industry and extractive features therefore makes the environmental consequences of oil industry more pervasive in terms of immediate externality effects and their implications for sustainable development. Table 1 summarises the main sources of environmental diseconomies arising from oil industry activities as well as their actual/potential impacts.

A careful analysis of Table 1 gives rise to some definite and useful observations. The first is that, every aspect of oil operations, in varying degrees has significant negative implications for the environment. The second, and closely related to the first, is that in most cases all the facets of what constitute the environment are affected in one single operational line. Third, the effects of the various aspects of oil operations on the environment are not mutually exclusive, but rather reinforcing. Fourth, is that the environmental consequences impose economic effects on the people. And finally, social tension tends to result from compensation disagreements arising from environmental damage claims by host communities.

Let us examine briefly, some of the sources of environmental diseconomies listed in Table 1. Gas flaring for instance, has negative effects on the immediate environment, particularly on plant growth and wildlife (Orubu, 1 999a) as well as on human life. Some of the green house gases such as methane and carbon dioxide emitted from gas flares contribute to global warming, which could accelerate the problem of climatic change and harsh living conditions on earth, if not checked. Apparently, it is this which explains the interest of the World Bank and the Global Environmental Facility (GEF) in proposing a gas flare reduction project for the oil fields of the Niger Delta in the 1990s (Moffat and Linden, 1995). It has been estimated that the total emission of carbon dioxide (CO_2) from gas flaring in Nigeria amounts to about 35 million tons per year. It should be noted that methane and carbon dioxide are the main green house gases responsible for global warming that has probably raised the average global temperature by around 0.5^{o}C within the last century, and it is on record that Nigeria is the highest gas flarer in the world (World Bank, 1995, 2000/2001).

Massive oil spills occurring in riverine areas in the Niger Delta have also done untold damage to the aquatic ecosystem, particularly in the mangrove swamp forest zone. Table 2 summarises the legend of oil spills in the oil industry in the period 1976-1996. As shown in the table, the total number of reported spills is put at 4,835, resulting in a cumulative spill volume of 2,382,373.7 barrels of crude oil over the period. Of this amount only about 15.91 percent was recovered, on the average, implying that about 84.09 percent of the cumulative spill was lost to the environment!

Table 1. Oil Operations and Their Impacts on the Environment

Activity/Event	*Actual and Potential Impact on the Environment*
1 Exploration – including geological surveys and, geophysical investigations.	Destruction of forest land, vegetation and farm land/human settlement. Noise pollution and vibration from seismic shooting. Effects on animals and nearby settlers (on shore) and on fisheries (near/offshore) Disturbances of flora and fauna habitats. Dislocation of economic activity. Tension on the social environment due to compensation disagreements.

2 Drilling	Accumulation of toxic materials from drilling materials, oil pollution of the sea, beaches or land. Destruction of fisheries production. Destruction of breeding ground for some marine fisheries. Alteration of the taste of fish. Killing of bottom dwellers. Pollution of underground water (waste pots). Adverse health effect on human, social tension arising from compensation disagreements from accidental spills from locations.
3 Production/Process i Platforms and tank farms ii Gas flaring	Water pollution from long term cumulative effects of produced water (with high salinity). Water pollution from salinity waste, used lubricating oil and solid waste. Main effects on marine life
	Air pollution from gas and processing evaporation and flaring. Production of heat kills vegetation around the heat area. Suppresses the growth and flowering of some plants. Reduces agricultural productivity and wild life concentration in area
4 Refining Petroleum	Air pollution and waste water impacting negatively on human health and ecosystem.
5 Oil spillage	Destruction of farmland, fishery and aquatic resources and mangrove ecosystem. Water pollution. Creates social tension due to compensation disagreements.
6 Tanker loading, location (on shore and off shore)	Water pollution from ballast and tank washing. Deck drainage, spillage during loading operation with accompanying effects on the fauna and flora. Disruption of seabed by dredging (i.e. canalisation).
7 Storage Depot	Land pollution from effluent water and solid waste of chemical cans and drums. Destruction of farmland for the establishment of storage depots, water pollution from effluent water. Air pollution from gaseous fumes during loading.
8 Transportation	Disruption of the sea-bed by dredging for pipeline installation. Sedimentation along pipeline routes. Water pollution from consequences of leaks from fracturing or breaking of pipe caused by metal fatigue, trawlers and dredges or sea floor failures, and sabotage. Air pollution by transport tankers. Destruction of environmentally sensitive area e.g. lowland, where estuaries, wet lands and sand dune fields exist. Erosion and flooding of the area drastically affected.
9. Marketing	Pollution of immediate environments from retail outlets. High hazard potential where located near residential buildings.

Source: Orubu et al (2002). See also Onwuoduokit (1998).

Transportation of petroleum products through the system of pipelines has been one of the most visible sources of environmental hazards in the industry. As a result of oil industry activities, more than 3,000 kilometres network of pipeline links different parts of the country. Without effective monitoring by the appropriate authorities, this system of transportation has continued to be a source of environmental hazard, when pipelines are either accidentally damaged, or give way due to old age, or through sheer sabotage. The Jesse fire incident (in Delta State) of October 1998 is still very fresh in our national memory. Usually, compensation is paid to victims when oil spills from pipelines and production facilities impact negatively on surface rights. However, if oil spillage is due to sabotage, oil companies do not pay compensation. Over the years, many oil companies have been accused of always invoking the sabotage clause in order to avoid the payment of compensation, and this has always been a source of conflict between oil companies and host-communities.

In the most recent incidents of massive oil spills and leakages, it has been very difficult to differentiate cases of sabotage from accidental cases. Between January and June 2000 for example, the Pipelines and Products Marketing Company (PPMC) recorded about 800 cases of pipelines rupture in different parts of the country. Out of this number, 764 were attributed to sabotage (FGN, 2000), indicating that sabotage accounted for about 96 percent of cases of pipeline ruptures in the period of six months And yet, as noted by Nwankwo *et al* (1998), many of these pipelines have been overdue for replacement for a long time. Many oil companies did not start replacing some of these pipelines, some of which are as old as thirty years, until the mid-1990s. The environmental consequences of oil industry activities have definite economic impact on the people in the areas of primary activities.

15.3. Interaction between environmental degradation and the local economy

The ultimate economic effect of environmental impacts of oil production activities is on the standard of living, that is, it catalyses a reduction in the standard of living of people in the area of primary activities. The economic effects are indeed very extensive and in the Niger Delta, these include dislocation of traditional economic activities and associated livelihood pursuits as well as danger to human health. Expectedly, all of the economic effects translate to pecuniary economic effects, which can be measured in terms of reduced real incomes and the loss of alternative uses of resources consumed by oil companies (Ekuerhare and Orubu, 1996).

We can refer to the effects described above as *first round Effects*. The *second round Effects* have longer-term welfare implications at the microeconomic level as well as for sustainable livelihood and development of the Niger Delta region. Each round of environment–degrading activity tends not only to increase the incidence of poverty among vulnerable groups such as farmers and fishermen and

Table 2: Oil Spills in the Petroleum Industry (1976-1996) in Barrels

Year	*Number of Spills*	*Quantity Spilled*	*Quantity Recovered*	*Net Quantity Lost to Environment*	*Percentage of Quantity Lost to Environment*
1976	128	26157.00	7135.05	19021.50	72.72
1977	104	32879.25	1703.01	31176.24	94.82
1978	154	489294.75	391445.00	97849.75	20.00
1979	157	694117.13	63481.20	630635.93	90.85
1980	241	600511.02	42416.83	558094.19	92.94
1981	238	42722.50	5470.20	37252.30	87.72
1982	257	42841.00	2171.40	40669.60	94.03
1983	173	48351.30	6355.90	41995.40	86.85
1984	151	40209.00	1644.80	38564.20	95.91
1985	187	11876.60	1719.30	10157.30	85.52
1986	155	12905.00	552.00	12353.00	95.72
1987	129	31866.00	6109.00	25757.00	80.83
1988	208	9172.00	1955.00	7217.00	78.69
1989	228	5956.00	2153.00	3803.00	63.85
1990	166	14150.35	2092.55	12057.80	85.21
1991	258	108367.01	2785.96	105581.05	97.43
1992	378	51187.90	1476.70	49711.20	97.12
1993	453	8105.32	2937.08	5168.24	63.76
1994	495	35123.71	2335.93	32787.78	93.35
1995	417	36677.17	3110.02	33567.15	91.52
1996	158	39903.67	1183.81	38719.86	97.03
Total	*4835*	*2382373.7*	*550234.19*	*1832189.49*	

Source: Niger Delta Environmental Survey, Phase 1 Report (1997), Vol. I. NNPC (1997) – Annual Statistical Bulletin

their dependants, but also involves intensified exploitation of existing natural resources, such as timber and non-timber forest resources. For example, pollution of major fishing waters leads to massive exploitation of marginal fishing waters. The pressure on land as a result of pollutive oil industry activities also leads to the exploitation of marginal farmlands, over-farming and deforestation, all of which result in a new wave of environmental degradation. In this way, a kind of vicious circle relationship between environmental degradation and poverty incidence is created,particularly in the face of inappropriate compensation programmes of oil companies.

In many rural oil-producing communities, individuals displaced from their land and water resources, including their dependants often migrate to the urban centres in the hope of securing paid employment in the oil industry or other sectors of the urban economy. Unfortunately production in the petroleum industry is largely skill – and technology–intensive, thereby reducing the ability of oil companies to absorb the largely unskilled migrant-labour from the rural oil-producing communities. Two definite effects are observed here. First is the depletion of the active labour force in rural oil-producing communities. Then there is the second effect of population pressure on the "oil city", swelling the pool of the unemployed.

In the most recent years, the major oil companies operating in the Niger Delta have stepped up their corporate social responsibility based development efforts as evidenced in the implementation of several community development projects such as the building of classrooms, health centres, introduction of youth empowerment schemes, and construction of access roads. Such efforts may be seen as one way of compensating the people for environmental damage resulting from oil industry activities, apart from the desire of the oil companies to improve the living conditions of their host-communities and create good will. However, oil companies and host-communities continue to live in mutual suspicion, and the situation seems to have heightened since the advent of democratic administration in 1999. To the extent that oil operations require a peaceful and secure social environment, the need for oil operators and communities as important co-stakeholders to foster good relationships through sustainable partnerships or memoranda of understanding cannot be overemphasised. However, evolving such partnerships must be based on an understanding of the current framework for environmental management in the petroleum industry. This is therefore the subject of brief review in the next section.

15.4. Environmental policy and management strategies in the petroleum industry

The evolution of environmental policy in the Nigerian Petroleum industry dates back to 1914 when the *Minerals Ordinance* was enacted by the colonial administration. The main objective of the ordinance was to prohibit the pollution of watercourses in the process of mining and prospecting for any mineral, including petroleum. The *Mineral Oils (Safety) Regulations,* 1963, *Petroleum Regulations,* 1967, as well as the *Oil in Navigable Waters Act* 1968 (among others) are also examples of post-independence statutory efforts directed at environmental protection in the petroleum industry. Environmental policy in the industry did not however enter into an active phase until the enactment of the *Petroleum Act,* 1969 which gave the Minister in charge of petroleum matters, vast powers to make regulations relating to all aspects of petroleum operations, including protection of the environment. The statutory framework for environmental policy in the Nigerian petroleum industry is depicted in Figure 1, while some of the relevant legislations on environmental policy relevant to the petroleum industry are summarised in Table 3.

More specifically as from 1988 after the establishment of Federal Environmental Protection Agency (FEPA), the Department of Petroleum Resources (DPR) became more aggressive in its role as environmental watch-dog of the petroleum industry, with its activities closely guided within the framework of the National Policy on the Environment (NPE), which was launched in 1989. Drawing its authority from the numerous statutes and regulations, which conferred on the Director of Petroleum Resources the power to set up strict environmental standards for the petroleum industry, in 1991, DPR came up with the *Environmental Guidelines and Standards for the Petroleum Industry* (EGSPI). The EGSPI were reviewed late in 1998, and a draft of the review was ready by the end of 1999.

A critical examination of the various legislations and regulations however reveals one flaw, namely that no explicit provision is made to incorporate the host communities of oil companies in the process of implementing environmental protection and management strategies – except for the institution of a Joint Investigation Team (JIT) in the management of spills. Even at that, experience has shown that most oil companies have not generally complied with this statutory requirement in the EGSPI to report all cases of spills within 24 hours, with the JIT being responsible at all stages for abatement efforts. There is no doubt that if the oil companies have confidence in their host communities, and carry them along in JITs for the purpose of clean – ups and educating the local people, a number of community level crises arising from oil industry – related environmental incidents in the Niger Delta could have been avoided. The need to involve the local people in environmental matters in oil industry, particularly in the Niger Delta where the largest proportion of Nigeria¶s oil is produced is further justified by the fact that the people have become very sensitive to issues of environmental degradation in the recent years.

Table 3: Some Relevant Statutory Instruments of Environmental Policy and their Objectives

S/N.	*STATUTORY INSTRUMENT*	*OBJECTIVE / REMARK*
1	Minerals Ordinance (1914), amended 1925, 1950, 1958	Prohibits the pollution of water courses in the process of mining and prospecting for any mineral, including petroleum.
2	Oil Pipeline Act (1956), amended 1965.	Provides among others for the prevention of pollution of land and water resources as a result of petroleum and production activities.
3	Public Health Act (1958)	Provides legal framework for the preservation and management of public health.
4	Criminal Code (1958)	Provides legal framework for seeking redress From environmental diseconomies, among others.

5	Mineral Oils (Safety) Regulations (1963)	Provides framework for health, safety and environmental–friendly exploration and production activities.
6	Petroleum Regulations (1967)	Provides framework for safe petroleum operations, including environmental protection
7	Oil in Navigable Waters Act (1968)	Prohibits discharge of oil into navigable water courses and other areas.
8	Petroleum Act (1969) and Related Regulations	Major legislation on petroleum industry to date. Provides encompassing framework for the Regulation of upstream and downstream Petroleum activities so as to protect the environment.
9	Land Use Act (1978)	To reform existing land ownership rights Through nationalisation. Adequate and fair compensation to be paid for loss of surface rights.
10	Associated Gas Re-injection Act (1979), amended 1984, 1985.	Provides statutory basis for the regulation of gas flaring in Nigeria.
11	Harmful and Toxic Wastes (Criminal Provisions) Decree No. 42 (1988)	Provides legal anchor for redressing the dumping of toxic and hazardous wastes.
12	Federal Environmental Protection Agency (Decree No. 58, 1988), and related legislations.	Provide a quasi legal framework for checking environmental crimes, and to set environmental standards for different pollutants
13	Industrial Pollution Abatement Regulations (1991)	To regulate the generation and disposal of Industrial waste through the principle of environmental permits.
14	Effluent Limitations Regulations (1991)	Provision of standards for industrial effluent discharge and emissions into the atmosphere.
15	Environmental Impact Assessment Act (1992)	Provides statutory basis for EIAs, as part of project development authorisation process.
16	Environmental Guidelines and Standards for the Petroleum Industry (DPR), 1991, 1999.	Most comprehensive framework for environmental policy and management in the petroleum industry.

Source: Lagos Chamber of Commerce and Industry, (OPTS), SPDC and DPR records. See also L.F.N. (1990)

15.5. The level of community awareness of environmental issues

As already noted, there is some evidence that since the 1990s, oil-producing communities particularly in the Niger Delta have been very sensitive to oil-related pollution problems, and have tended to develop a common framework of response to the problem through a well-established system of traditional flow of information/authority in which traditional rulers and chiefs, councils of elders and community development committees (CDCs) play a significant role (see Figure2). Within this context of the traditional society, actions taken by the youths

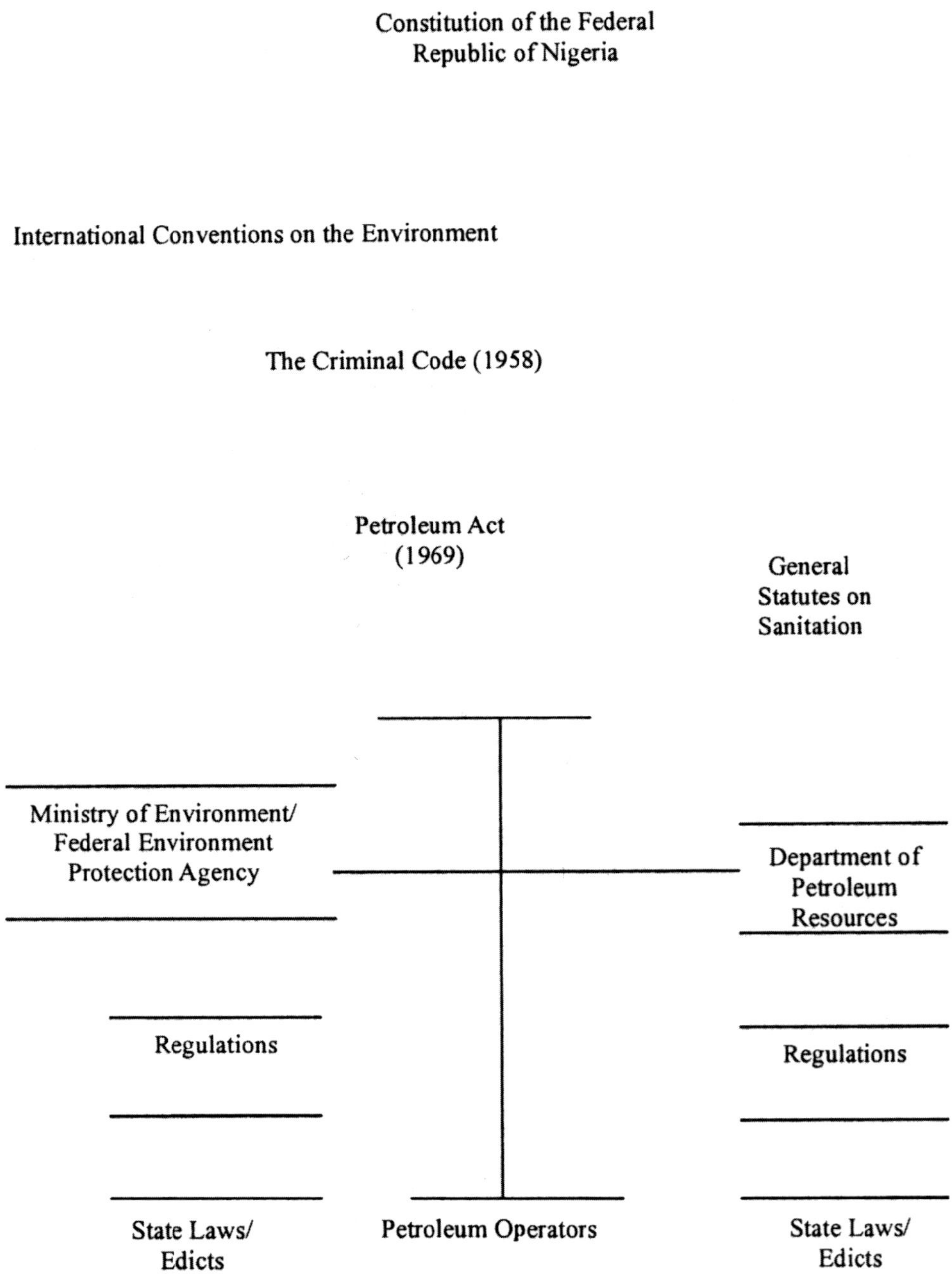

Fig. 1: Statutory Framework for Environmental Policy in the Petroleum Industry

(and sometimes women associations) are not to be considered as isolated actions. Rather, the youths are to be seen as expressing views that receive the blessing and support of the whole community. Consequently, the traditional organisational structure is one in which consultation is mutually reinforcing, with explicit support from all subsystems, including the religio-cultural. It turns out that the recognition of this intricate system of authority flow would be fundamental to the resolution of any crisis arising from the exploitation of the country's petroleum resources in general, and particularly from environmental diseconomies, which specifically affect the people living in the oil-bearing communities.

In the past, a number of pro-community NGOs have also been active in Nigeria's oil – bearing communities, publicising their experiences and encouraging the people to demand for their rights. Some of these NGOs, whose activities are focused primarily on the Niger Delta area, include the Environmental Rights Action (ERA), Niger Delta Human and Environmental Rights Organisation (NDHERO), the Rivers Chiefs and Peoples Conference and Wetlands Environmental Protection Association. Some of these non-governmental efforts, working with the media, have over the years been able to sensitise the people to the negative environmental consequences of oil industry activities. For example, the Rivers Chief and Peoples Conference had discussed the environmental and social issues of the oil-bearing Niger Delta at the Indigenous Peoples Conference under the auspices of the United Nations Conference on Environment and Development (UNCED) in Rio de Janeiro in 1992. This NGO also prepared the first regional assessment of the major environmental and social problems of the Niger Delta (World Bank, 1995).

In a recent study, Orubu *et al* (2002) also found that communities in the Niger Delta are well informed about the environmental consequences of oil industry activities. Table 4 summarises communities' knowledge of environmental problems associated with the oil industry.

It can be strongly argued that the crisis that has bedevilled the Niger Delta area since the 1 990s has been closely associated with the environmental question. There are also the twin issues of neglect and inadequate compensation programmes (Orubu, 1999b, 2001). As noted by the World Bank (1995);

> The current compensation programs aggravate community relations, and reinforce the perception that oil activities cause most of the problems of the Delta. Riverine people feel that the oil companies do not consider themselves accountable to the local people. Resentment of their marginalisation in contrast to the value of the oil reserves has resulted in clashes with oil company personnel and federal police/military forces...

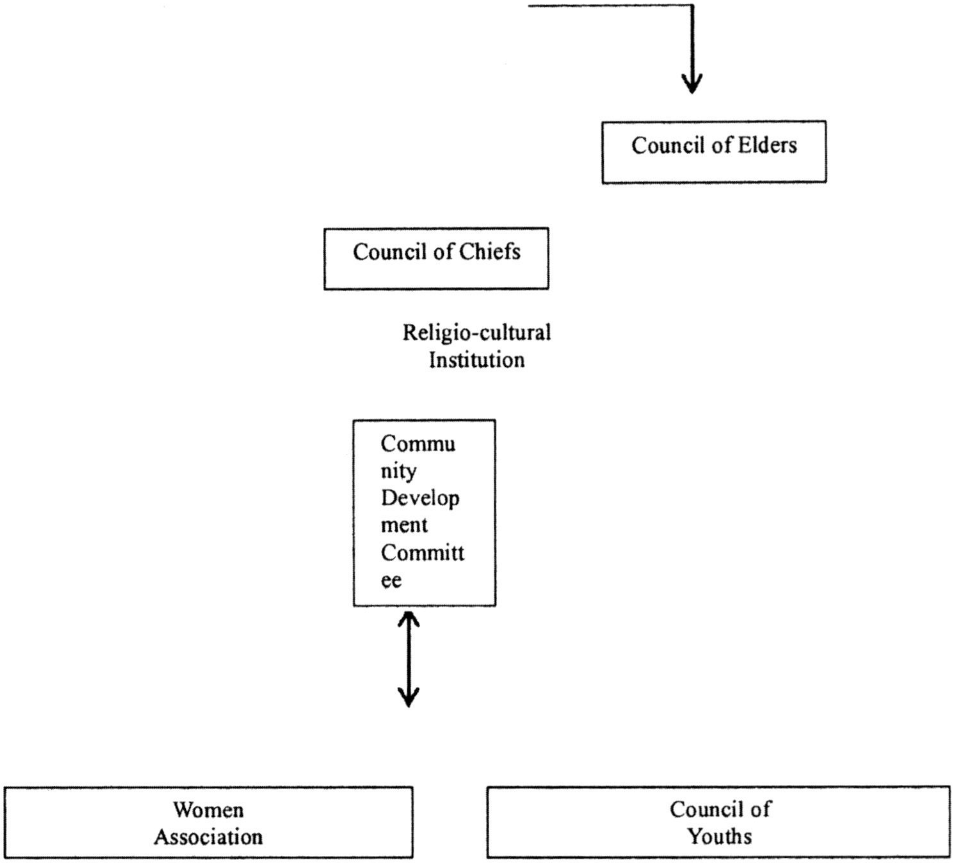

Fig. 2: Structure of Traditional Authority and Information Flows in Oil-Producing Communities

Table 4: Community Awareness of Environmental Problems in the Niger Delta

S/N	*Environmental Problems Identified*	*Perceived Effects of Environmental Problems*
1	Oil spillage	(a) loss of soil fertility (b) pollution of fishing waters, leading to loss of fish population and other aquatic organisms (c) pollution of drinking water affect health of the people (d) associated with social strife (e) negative effect on mangrove ecosystem.

2	Land acquisition	(a) reduction of arable land (b) deforestation and land degradation (c) floods due to blockage of natural water courses as a result of construction of access roads and other facilities.
3	Gas flaring	(a) air pollution (b) loss of safe habitat (c) land becoming arid in near areas of gas flares (d) reduction of animal population in areas near gas flares (e) extinction of natural herbs and other non-timber forest products in areas near gas flares.
4.	Environmental Impact Assessment Studies	Communities are generally ignorant about EIA activities. They seem not to be aware of public forum for the discussion of the EIA report.
5	Other environment-related problems and general problems of the oil industry (such as poor compensation, rural-urban migration, increasing city population, etc.)	(a) pressure on existing resources (b) high cost of living (c) reduction of rural population (d) social vices and crime (e) corroding roofing sheets of buildings (f) population pressure on the "oil city¶

Source: Interviews conducted in selected Oil-Producing Communities. The Communities include Ogbotobo (Bayelsa), egbema-Angalabiri (Bayelsa), Obotobo (Delta), Ellu (Delta), Sokebolou (Delta), Bensidede-Ojobo (Delta), Otu-Jeremi (Delta), Odumukenro-Kogun (Ondo).

During the early 1990s, the military administration had created the Oil Mineral Producing Areas Development Commission (OMPADEC) to take care of some of the resultant issues. In retrospect, it could be said that OMPADEC did not achieve much in the area of sustainable development projects in the oil producing areas. Indeed, it never paid any serious attention to its environmental mandate before it was wound up, and implicitly replaced by the Niger Delta Development Commission (NDDC). One of the reasons why OMPADEC failed as an interventionist body was that its policies and programmes were not based on appropriate development models in relation to the problems of the Niger Delta (Ekuerhare 2002). This again underscores the need to always get the people involved in development matters that concern them.

15.6. Identifying areas of cooperation and sustainable partnerships

One critical flaw observed in the environmental management strategies adopted by oil companies in Nigeria is the tendency to view such strategies mechanistically. The existing strategies do not put the environmental impacts of oil development activities squarely within the domain of man as the ultimate victim, and a being who must desire the maintenance of a qualitative and sustainable environment. Since the 1980s, there has been a paradigm shift, originating at the international level in favour of sustainable development, which accords man and his condition of living a central focus in development policy debates, and to ensure that today's development does not jeopardise future development needs.

A key feature of the sustainable development approach is that it recognises the principle of consultation in policy formulation and implementation. Given the extent of knowledge about the nature of the environmental consequences of oil industry activities in Nigeria's oil communities, and the role which the local population can play in safe-guarding oil facilities, it is our very strong view that it could be useful, if communities are recognised as partners in development by petroleum operators. If this happens, then there might be fewer cases of environmental incidents attributed to sabotage. This should also help to build confidence in the relationship between oil companies and host-communities. Specifically, sustainable partnerships between oil companies and host communities directed at reducing environmental hazards and related crisis in the Niger Delta could be fostered to take care of the following areas:

- Compensation programmes for pollution and the use of natural resources
- Environmental impact assessment of petroleum development projects, including impact evaluations
- Management of spills
- Surveillance of pipelines and other oil facilities
- Resolution of environment – related social conflict
- Information flows and management, and environmental education
- Identification, implementation and monitoring of oil-related community development projects, and
- Recognising the *subsidiary* principle in activities directed at pollution abatement.

The Sustainable Partnership Framework (SPF) should be regarded as a veritable instrument for fostering better working relationships between intrinsically–opposing parties. Three critical stakeholders are recognised in the petroleum industry – the Government, oil companies and the host communities. Among these three, the communities are the most disadvantaged in terms of the power to control the petroleum resource. In one vein therefore, we can indirectly assess the most social outbursts in the Niger Delta as a response meant to increase the bargaining power of a minority and less muscled party at the negotiation table. In practice, the SPF should take the form of implementable Memoranda of Understanding (MOUs) between oil companies and host communities, in which opposing views are equitably balanced towards the resolution of the environmental and other related crises. We therefore highlight the significance of the factors mentioned above in the SPF, so as to bring about better relationships between communities and oil companies operating in the Niger Delta, and help to reduce the negative impact of environment on the economic and social life of the people.

15.7. The issue of compensation programmes

It has been shown that disagreements arising from compensation rates paid by oil companies for the use of natural resources such as land acquired during the process of oil exploration, and for the environment polluted by oil spills and other wastes disposed have been one of the most virulent sources of crisis in the Niger Delta (Onosode, 1997). In the past, particularly during the 1960s and 1970s, oil companies consulted directly with communities before compensation is paid. This function has been taken over by compensation consultants, who as attorneys, negotiate on behalf of affected communities. Several crises have resulted from the current arrangement, with the allegation that individuals whose rights to property have been trampled upon do not get full worth for such property after the consultants have taken their fees. It is therefore necessary to build into the new SPF a mechanism that enables victims of environmental pollution or any other permanent or injurious use of resources by oil companies be part of the assessment process

15.8. Environmental impact assessment of petroleum development projects

At present, the citizens of oil-producing communities are hardly aware of environmental impact assessment (EIA) studies being carried out in their communities for petroleum development projects – apart from the occasional engagement of local youths as data collectors by environmental consultants.

This is apart from the fact that the EIA methodology adopted in the petroleum industry generally pays less attention to human and social factors (Ikporukpo, 1998). It has been shown that community participation in the environmental assessment of projects that affect them has several advantages (Odusola, 1996).

Consequently, an approach which gives definite recognition to the local people in the process of conducting the EIA study would therefore go a long way in allaying their fears and help to reduce social tension. The same should apply to environmental evaluation (post-impact) studies, which are usually commissioned after major environmental incidents have occurred.

15.9. The management of spills

Communities directly affected by spills, particularly those of catastrophic dimension should be allowed to have a say in remediation plans as well as efforts to reduce attendant social tensions. The SPF must therefore address the issue of confidence building, so that all stakeholders can be actively involved in pollution damage and compensation assessment.

15.10. Surveillance of pipelines and other oil facilities

The SPF should work out an arrangement whereby communities can be directly involved in the surveillance of pipelines and other oil facilities, by making them have a sense of "co-ownership". It is interesting to note that a number of oil companies and communities have already gone into such arrangements. In all probability, the cost of maintaining such relationships with communities in which oil facilities are located will be less than the cost of pollution damage, and attendant social tension, including the effect of adverse publicity on corporate image of oil companies.

15.11. Resolution of environment–related social conflict

Alternative conflict resolution (ACR) measures for resolving environment-related social conflicts are preferable within the SPF – as opposed to unilateral actions such as going to the court and kidnapping of oil company staff. Again, this requires confidence building, in order to create the proper environment for negotiation between oil companies and the local people.

15.12. Information flows and management, and environmental education

The flow of information and how it is managed are crucial to the social effects that environmental hazards generate. The SPF will therefore recognise the need for all parties to be cautious in making public statements. Specifically, the SPF will prohibit the making of defensive press statements. If there is the need for a press statement, it should be jointly made. A framework for environmental education of the people will also be an important aspect of the SPF.

15.13. Identifying, implementing and monitoring of community development projects

As already noted, projects of community development embarked upon by oil companies in their areas of operation implicitly compensate the people for depriving them of the use of their resources, and the damage imposed on the environment. In doing this, what is being provided for them must be based on what they need. This therefore underscores the need for the SPF to recognise the need to adopt a "bottom-up" approach in the identification of projects for the communities, if such projects are to be sustainable. The right of communities to monitor such projects should also be recognised by the SPF, in order to ensure that implementation proceeds are as predetermined. It is important to note that a number of oil companies, notably Shell Petroleum Development Company (SPDC) and Chevron are already doing this in certain areas

15.14. Recognising the subsidiary principle

The *subsidiary* principle in environmental policy analysis is based on the need to decentralise responsibility for some kinds of environmental action for impacts that are unquestionably local (Zylicz, 2000). The SPF should explicitly recognise this principle, and apply it with some care to the solution of local environmental problems. For example, the local people can be made to be responsible for minor clean-ups without sending distress and tension calls to the headquarter-offices of oil companies. This would be one way of creating responsibility for local groups, and allowing them to take part at some level of decision-making and action.

15.15. Conclusion

In this paper, we have made an attempt to identify the nature of the environmental problems associated with oil industry activities particularly in the Niger Delta where the bulk of Nigeria's oil and gas is produced. We observed that the role, which the oil-producing communities can play in the resolution of the environmental problems has not been explicitly recognised by the existing statutory regulations relating to environmental policy and management in the industry.

The authors therefore suggest putting in place a sustainable partnership framework (SPF) which explicitly recognises the role which communities can play in the resolution of environment-related compensation problems, EIA studies, surveillance of oil pipelines and other oil facilities, the management of spills (both of catastrophic and local dimensions) and information management.

It is the hope of the authors that the adoption of the SPF will help significantly in reducing oil-induced environmental diseconomies in the Niger Delta, including second-round consequences that undermine peace in the region.

References

Awobanjo, S.A. (1981) "Oil spillage in Nigeria: 1976-1980", Proceedings of the 198 1International Seminar on the Oil Industry, Lagos, NNPC.

Chukwu, L. O.; Brown, C. O., and Nwankwo, D. I. (1998) "The impact of oil pollution on the hydrochemistry and biota of the tidal creeks and canals in Ondo State", 9th International Conference on the Petroleum Industry and the Nigerian Environment, Abuja, November, 538-576

Department of Petroleum Resources (DPR), (1991, 1999) *Environmental Guidelines and Standards for the Petroleum Industry in Nigeria,* Lagos: Ministry of Petroleum Resources.

Ekuerhare, B. U. (2002) "Sustainable Development Models for the Niger Delta Region", Proceedings of Fourth Memorial Programme in Honour of Professor Claude Ake, CASS; Port Harcourt.

Ekuerhare, B.U. and Orubu, C. O. (1996) *Economics of the Niger Delta,* Technical Report, Niger Delta Environmental Survey, June.

Eromosele, Victor E. (1998) "Costing Niger Delta's oil spills: A joint stakeholder's approach", 9th International Conference on the Petroleum Industry and the Nigerian Environment, Abuja, November, 358-368.

Famuyiwa, B. A. (1998) "Seabed survey of the impact of oil based drilling fluid system on offshore environment", 9th International Conference on the Petroleum Industry and the Nigerian Environment, Abuja, November, 461 – 489.

Federal Republic of Nigeria (1989) *Constitution of the Federal Republic of Nigeria,* Lagos: Government Printers

Federal Republic of Nigeria (FRN), (1990) *Laws of the Federation of Nigeria (L.F.N.),* Lagos.

Federal Government of Nigeria (FGN), (2000) Report of the Special Committee on the

Review of Petroleum Products Supply and Distribution, Abuja, The Presidency. Grevy, P. (1995) *The Niger Delta, Nigeria,* Draft pollution assessment study, submitted to the World Bank

Ikporukpo, Chris (1998) "Environmental impact assessment and human concern in the petroleum industry: Nigeria's experience", paper presented at 9th International Conference on the Petroleum Industry and the Nigerian Environment", Abuja, DPR.

Moffat, Daniel and Olof Linden (1995) "Perception and reality: Assessing priorities for sustainable development in the Niger Delta", AMBIO: *Journal of Human Environment,* Vol. 24, Nos. 7-8, pp 527-538.

Niger Delta Environmental Survey (NDES, 1997) Phase I Report, Vol. 1, *Environment and Socio-economic Characteristics,* Port Harcourt.

Nigerian National Petroleum Corporation [(NNPC), (1997) *Annual Statistical Bulletin.*

Nwankwo, J. N.; Ifeadi, C. N., and I. I. Orubima (1998) "The Environmental Guidelines and Standards for the Petroleum Industry in Nigeria: The 1998 Revision", Paper presented at 9th International Conference on the Petroleum Industry and the Nigerian Environment, Abuja, DPR

Odusola, A.F. (1996) "Community Participation in Environmental Assessment", in P.C. Egbon and B. Morvaridi (eds.), *Environmental Policy*

Planning, National Centre for Economic Management and Administration (NCEMA), Ibadan, Nigeria, p. 152–174.

Olomo, R.O. and Omene, L. (1995) "Environmental impact of oil spillage: A case of the operations of Shell Petroleum Development Company in Bonny, Rivers State", paper presented at the 38th Annual Conference of Nigerian Geographical Association, University of Benin, May.

Onosode, Gamaliel (1997) "Petroleum, development and the environment: NDES Perspective" in Victor E. Eromosele (ed), *Nigerian Petroleum Business: A Handbook,* Lagos, Advent Communications, 287-291.

Onwiodukit, E.A. (1998) "An alternative approach to efficient pollution control in Nigeria", in Akinjide Osuntokun (ed.), *Current Issues in Nigerian Environment,* Davidson Press and Friedrich Ebert Foundation, Lagos, Nigeria.

Orubu, C.O. (1999a) "The exploitation of non-timber forest resources in the Niger Delta: Problems and prospects", Technical paper, Niger Delta Environmental Survey, Port-Harcourt.

Orubu, C.O. (1999b) "Oil wealth and the derivation principle: The need for a new fiscal imperative towards oil-producing states", *Calabar Journal of Politics and Administration,* Vol. 1, 189-211.

Orubu, C. O. (2001) "Compensating for trampled surface rights in the Nigerian petroleum industry: suggestion for a sustainable livelihood approach", The *Nigerian Economic and Financial Review,* vol. 6 (1), 57-8 1

Orubu, C.O.; A. Fajingbesi; Odusola, A. and Magbagbeola, N. (2002) *Environmental Regulations in the Nigerian Petroleum Industry: Compliance Status and Implications for Sustainable Development,* Research Report, ACBF/NCEMA.

World Bank (1995) *Defining an Environmental Development Strategy for the Niger Delta* Vol.II Washington D.C. Industry and Energy Operations Division (West Central Africa Department)

World Bank (2000/200 1) *World Development Report,* Washington D.C.

Zylicz, Tomase (2000) "Goals, principles and constraints in environmental policies", in Henk Former and H. Landis Gabel (eds), *Principles of Environmental and Resource Economics,* Cheltenham, Edward Elgar, 130-156.

16. The roles of non-farm employment and rural development in the resolution of the niger delta crisis

Okojie, E. I. & Ailemen, M. I.

16.1. Introduction

Economic development encourages aggressive exploitation of natural resources. In the rural areas where the occupation is mainly farming, there is dependence on the natural environment for economic gains. In Nigeria, the need to maximize revenue to obtain the desired level of growth has intensified the exploitation of crude oil, which is the mainstay of Nigeria's economy. This has however orchestrated an environmental crisis, which has turned the oil rich Niger Delta into a scene of major confrontations between the people and oil prospecting companies. Conflicts over communal land are also on the increase in the area.

The rural dwellers in the Niger Delta are predominantly farmers. Available evidence shows that the people are generally poor and their economic conditions have continued to deteriorate (Ekuerhare & Orubu, 1996) In spite of increased revenue to the region from oil through the 13% derivation base as a criterion for revenue allocation in the Nigerian federation, the situation of the rural dwellers is yet to experience a considerable change. It is clear that if the antecedents of the oil companies, the governments and the non governmental and non state organizations in the area are anything to go by, it would take a considerable length of time before the rural farmers would be affected. Poor incomes, unemployment and poverty have been pervasive in the region. The only cheap means of income to the unemployed youths, in the context of massive oil wealth that does not touch their lives has been to carry out sporadic assaults on the oil companies.

The nexus of the environmental damage, which seriously threatens the agricultural practices of the people and rural life, has become a matter of great concern. If the issue is not immediately addressed, the danger of environmental problems arising principally from oil spillage and gas flaring such as biodiversity depletion, loss of soil fertility, water pollution, flooding and land degradation would reach crisis proportions. Already the development of the area, which is more or less rural in nature, is in disarray. Despite the fact that about 90 per cent of Nigeria's national income is sourced from Niger Delta, the area has suffered from mass poverty and unemployment. Oil exploitation based degradation has led to fisheries decline and agricultural loss and socio-economic disruptions of primary economies all of which have accentuated poverty.

In addition, increasingly scarce land and water resources and the problem of compensation by oil companies have accentuated conflicts over ownership and fuelled communal and inter-community conflicts over land and compensation (Egborn and Okoh 1999). Arising from these problems situated in the oil economy, there has been an increasingly high level of social problems in the region (Okoh & Egbon 1999; Onwuka and Ailemen 2002). Particularly, there has been a huge problem of youth militancy and violence. The youths have resorted frequently to violence to press home their demand for improved benefits from the oil economy and improved living conditions. Egborge (2000) asserts that it is the damage to the ecosystem in the Niger Delta on which the livelihood of the people depend and the uncommitted attitude of the authorities that have ignited temper, violence and youth restiveness against the government and oil companies.

The paper addresses the place of non-farm employment as a panacea to the youth crisis in the Niger-Delta. The remaining part of the work is divided into four sections. The next section discusses oil exploitation and the developmental crisis in Niger Delta. Section two deals with the result of the survey. Section three contains the policy options for solving the Niger Delta Crisis while the last section presents the conclusion and recommendations.

16.2. Oil exploitation and developmental crisis in the niger-delta

Most environmentalists and economists are showing concern over the effects of economic activities on the environment especially in rural areas, where economic activities are causing environmental damage and constraints to future development. Since the United Nations Conference on the Human Environment in 1972, the right of all peoples to a decent environment has increasingly been recognized as inalienable and inviolate (United Nations, 1997).

Since the discovery of crude oil in Nigeria in 1958 at Oloibiri in the present Bayelsa State, oil has remained almost exclusively exploited in the Niger Delta region of the country. The Niger Delta comprises Delta, Bayelsa, Cross Rivers, Rivers, Akwa Ibom, Edo and Ondo states with a landmass of 70,000 square kilometers. it is an area of high ecological value. It has about 200 oil fields with over 400 oil production storage facilities. There are about 500 rural communities with few urban centers (Eromosele, 1998; Okoh & Egbon 1999; Orubu 2000; Aboribo 2002).

What is unique about the area is the fact that the environment is fast deteriorating due to oil exploitation and exploration, thereby endangering land, water, air and aquatic life. The inhabitants of the area live in mostly rural settlements and the economy is largely primary. Since a large portion of the land mass is occupied by oil exploitation production, transportation, storage, refining and utilization facilities, the remainder land available to the rural communities have become insufficient. Intensive use of existing land and water for their own economic activities has manifested in forest loss, decline in the fertility of the soil and income loss, thus aggravating the unemployment situation and poverty. In many ways, the people are hindered from pursuing their traditional occupations of farming, hunting and fishing (Orukele, 2000).

The extent of under-development of the Niger Delta is indeed a thought provoking challenge to development planners, policy makers and administrators both within and outside the region. Although, the area is rich in land, people, oil and natural gas resource, the people can hardly eat, drink or clothe themselves. They are largely unhealthy and uneducated (Okunmadewa 1998). The unemployment rate is uncommonly high because of the social dislocations of primary economies and social disruptions of life because of the pervading conflicts and violence. Furthermore, employment opportunities available in the region are in favour of non-indigenes (Aboribo 2002). The ethnic majorities dominate the employment of most of the oil companies such as Shell, Chevron and Elf.

The people of the region are plagued with poor infrastructures, facilities and services and lack of access to productive inputs. Above all, the wealth from their land is spent on developing the major cities as Lagos, Abuja, Kano and Kaduna. The outright neglect of the area for a long time has compounded the poverty and dislocation of the people (Orubu 2000). The situation has generated considerable anger, frustration, disillusionment and alienation in the region. Of course, these conditions have led to the restiveness and the mobilization of the people in civil organizations and youth groups such as the Movement for the Survival of the Ogoni People, Urhobo Study Group and Ikwerre Youth Convention. There have also arisen militant youth groups and militias. Aboribo (2002) has noted that the major cause of conflict in the Niger Delta is the impoverishment of the inhabitants.

It is obvious that fighting environmental dislocation occasioned by oil exploitation and exploration activities in the Niger Delta is in part a way of resolving the crisis in the area. But there has to be other programmes that can impact on incomes, employment, production and rural development. Most crisis managers have always suggested dialogue, which actually is needed to resolve environmental and resource problems. But it is also necessary to formulate programmes of restoration and economic empowerment that are acceptable and patronized.

The acquisition of vocational skills as well as the promotion of co-operatives are avenues of tackling rural unemployment problems (Okojie et al 2000). Agricultural growth that can enhance rural development and touch a large proportion of rural producers will be difficult within degradated environments. As such, growth must go beyond agricultural productivity especially in areas like the Niger Delta

(Okojie 2001). Recently, the policy emphasis in the World Bank rural development strategy; "from vision to action" is on the totality of the rural economy and not just the agricultural sector (World Bank, 2000).

16.3. Results of survey

Data presentation

The paper focuses on Non-farm Employment as a panacea to the Niger Delta Crisis. The records of impact of oil exploitation on the environment and development of the region and the frequent incidences of communal conflicts most of which are traceable to land disputes, with the attendant disruption or destruction of economic life of the rural people, have informed the need to explore non farm based options and programmes of economic empowerment, employment and income generation.

The study seeks to identify the relevance of non farm employment as a strategy for resolving the region's crisis through the investigation of opinion and perceptions. Two research instruments; the questionnaire and personal discussions were utilized. Four communities were randomly selected in Edo (Ologbo) and Delta states (Sapele, Oghara and Koko) as the study areas. The questionnaire was designed to capture details of such issues as personal data of respondents, effects of oil exploration and exploitation in the rural areas of the Niger Delta, the level of awareness of the people about the effort of government and oil companies towards the development of the region and the relevance of Non-farm Employment in resolving the crisis. A simple random sampling technique was adopted in the administration of the questionnaire. Though the focus was on both adults and youths in the communities, the majority of the respondents were youths. Part of the reason was that it was difficult to administer the questionnaires to the elderly, most of whom could not read and write. Out of the 200 questionnaires distributed, 165 questionnaires were returned. Data on the causes of conflict and supplementary information were collected through personal discussions with some indigenes of Koko and Ologbo.

The respondents in the field survey were largely males (71.5%) and single (64.8%). About 60% of them were educated at the post secondary level (Table 1). Table 1 below also shows that 5.5% of the respondents are farmers. This explains the extent of the effects of oil exploitation on the rural occupation of the communities. Though most of the people were into petty farming, farming was no longer the dominant occupation of people. About 14 % were teachers, 43.6% were students while 6.9 % were craftsmen. The remainder 30% of the respondents fell into the unspecified occupations such as trading and public service employment. The age range also shows that the majority of the respondents are youths. Only 25.4 % of them were above 36 years old. The bias of the questionnaire toward the youths is not disturbing because they are at the centre of crisis in the region. What looked surprising is that 60 per cent of the respondents have tertiary education, which is not typical of the average rural areas in Nigeria.

Table 1: Personal data of respondents

Data	*Options*	*Respondents*	*Percentages*
Sex	Male	118	71.5
	Female	47	24.5
Marital status	Married	58	35.2
	Single	107	64.8
Educational Background	No formal education	8	4.8
	Primary education	7	4.2
	Secondary education	51	31
	Tertiary education	99	60
Age	15-25 years	62	37.6
	26-35 years	61	37
	36-45 years	31	18.8
	46 years and above	11	6.6
Occupation	Farming	9	5.5
	Teaching	23	14
	Student	72	43.6
	Craftsmanship	10	6.9
	Others	48	30

Source: Field Survey 2002 and *Authors' Calculations*

The oil companies located in the area compete with the rural dwellers for the limited and available land. The operations and activities of the oil companies involve seismic investigations, drilling and dredging as well as pipelines, flow stations, ramps and camps, borrow-pits and construction of access roads. Table 2 shows that the majority of the respondents agree that oil is sourced from their environment and it is associated with environmental damage such as land degradation, air and water pollution as well as oil spillages. About 75 per cent of the respondents assert that oil exploitation and exploration has had adverse consequences on their environment.

As indicated in the responses, oil exploitation and exploration has threatened

Table 2: The effects of oil exploitation and exploration on rural areas in the niger delta

Effects	*Responses*	*Percentage*
Land degradation, Air & Water Pollution, oil Spillage	124	75
Destruction of fishermen's business	152	92
Unemployment in the communities	158	69

Source: Field Survey 2002

the agricultural life of the rural people. Okoh and Egbon (1999) assert that oil exploitation activities have exposed the Niger Delta waters to pollution thereby endangering aquatic life. The field survey reveals that 92 % of the respondents agreed that fishermen in the region are out of business due to environmental damage. In a region where the average dweller relates to fishing either as a full time or part time occupation or past time, this is a big problem. This seems to explain the unemployment and general impoverishment in the region.

The responses also indicate that there is considerable youth unemployment (96%) in the region, part of which has been caused by the oil economy. What level of development can a community achieve given the high proportion of unemployment in the area? The incidence of poverty and unemployment are twin sisters, which have been on a visible upward trajectory in the region. Available statistics show that over 72 per cent of households in the area are within the poverty brackets. This means that majority of the region's inhabitants live below the World Bank benchmark definition of poverty (Aboribo 2002).

Table 3 shows the level of awareness of the efforts and contributions of government and oil companies, towards alleviating the adverse effects of oil exploitation and exploration on the people. About 33 % of the respondents are not aware of any contributions that government and the oil companies have made or are making toward alleviating the suffering of the people. However, about 30% are aware of some provision of infrastructural facilities. About 21 % acknowledged that oil exploration firms pay compensations to the inhabitants for environmental damages. Only 3% of the respondents acknowledged that government provides farm input to farmers in the region. The response indicates that the agricultural sector support or assistance from the government and oil companies is very low even when the region comprises mainly farming communities.

The results indicate that the rural people are not fully informed of the activities of government and the oil companies in the oil producing areas. Supplementary information collected show that even the little benefits in the form of revenue allocation and special development efforts by government are not sufficient given the extent of damage in the region. Then such benefits are not properly channeled to the grass roots. Furthermore, benefits such as provision of infrastructural facilities will not suppress the restiveness in the region because they do not reduce the level of unemployment, poverty and impoverishment.

Table 3: Awareness of the people about government and oil companies, contributions to the development of the Niger Delta Region.

Contributions	*Respondent*	*Percentage*
Payment of compensations	34	21
Provision of infrastructural facilities	50	30.3
Granting of scholarship to students	21	12.7
Provision of farm inputs	5	3
None of the above	55	33

Source: Field Survey 2002

The investigation also reveals that the causes of the conflict in the region are varied (Table 4). About 94.6% of the respondents claimed that youth unemployment is the cause of youth restiveness and aggression as reflected in such activities seizure, vandalization and sabotage of oil rigs, wells and flow stations. We noted earlier the severity of the youth unemployment problem in the area. A good proportion of them have basic education, but the vacancies available in oil firms is nothing in relation to the swam of job seekers. It is therefore necessary that government and concerned agencies should begin to think in the directions of encouraging cottage industries and skill acquisition.

A negative consequence of the youth involvement in the conflict is that he youths in the area are not ready to go to school, but rather prefer to make fast money from the oil operators. But though the adult youths may not condescend to the tutorials and long time requirements of formal education, a short term skill acquisition programme could make a difference. About 38% of the respondents strongly agreed that youth militancy and restiveness could be suppressed if opportunities are made available while 45.5 % agreed. (See TABLE 4).

The rural communities are usually the losers in most conflicts in the region. Confrontations between communities or against oil operating firms create hardships to the rural dwellers, in terms of loss of lives, displacement and the destruction of houses or available infrastructural facilities, natural resources as well as disruptions of productive activities. The consequences of these are severe for their survival particularly in terms of drastic falls in their income. Table 4 indicates that

Table 4: Causes of Conflict in the Niger Delta Region

	Options	*Respondent*	*Percentage*
Youth unemployment leads to the seizure of oil rigs, vandalization, sabotage and assault of oil companies	No	9	5.6
	yes	156	94.6
Insufficient land causes communal clashes	No	112	67.9
	Yes	53	32.1
Do you agree that militant groups can be suppressed with employment opportunity	SA	63	38.2
	A	75	45.5
	D	14	8.5
	UN	13	7.8

Source: Field Survey 2002. AS – Strongly Agreed, A – Agreed, D – Disagreed, UN – Undecided.

about 32.1% of the respondents claim that the people of Niger Delta actually suffer from insufficient land, while 67.6 per cent of the respondents assert that the issue is not insufficient land but lack of recognition of value of the available land to them.

The situation is complex, if we understand that the oil industry is highly capital intensive and therefore requires skilled manpower. Most job seekers in the area especially the displaced farmers and fishermen are not able to get jobs in the oil firms due to illiteracy. Skill acquisition will definitely be a good option in the situation. Table 5 shows the relevance of Non-farm Employment in solving the problems of the region.

Table 5: The Relevance of Non-farm Employment to Solving Niger Delta Crisis

	Options	*Respondent*	*Percentage*
Do you prefer Non-farm employment to farming in the region?	Yes	129	78.2
	No	36	21.8
Farming is not profitable due to oil exploitation and exploration activities	True	106	64.2
	False	15	9.1
	Fairly True	44	26.7
Will you want to acquire skills to do other work apart from farming	Yes	157	95.2
	No	8	4.8

Source: Field Survey 2002.

As Table 5 indicates, about 78.2% of the respondents prefer non-farm employment to farming. Only 21.8% still prefer farming to other occupation despite the damage done to the environment by oil exploitation. They assert that they still hold their land as the only valuable asset for the coming generations. Therefore, natural resources in their environment need to be efficiently managed. Furthermore, although farming is no longer profitable in the region due to oil exploitation, the absence of any other occupation necessitates that they still derive their daily income from it.

About 64.2 per cent of the respondents claim that the farming is not profitable in the area. This gives credence to the fact that farming is becoming uninteresting in the region. In fact, about 95.2% of the respondents claim that they will want to acquire skills for self-employment if given the opportunity. As the majority of the people particularly in the rural areas rural Niger Delta are farmers, it is obvious that the development of the people will be hindered, if the development strategies do not actually touch on their traditional occupation. Since their traditional occupation is constantly threatened, alternative development strategies will serve best. For the incessant conflicts in the region to be suppressed, the people should look beyond their traditional occupation as a means of livelihood because their natural resources are continuously depleted. The higher the exploitation and exploration by oil companies, the higher the environmental degradation and consequently the lower the agricultural productivity. This scenario is calamitous as it accentuates misery and poverty.

The search for alternative development strategies impels on the government to encourage cottage industries in the area. Skill acquisition programs like sewing, catering, knitting, soap making, electronic repairing, mini-processing of available resources in the area, plumbing, carpentry and painting would substantially improve the lots of the people. These will possibly redirect the strength of the youth away from crime and harassment of oil companies and its workers, since they have skills that can provide then the means of livelihood. Some available efforts by government in this direction can be seen in table 6.

The women development and education unit of Warri North L. G. A. focuses on community development with the aim of skill acquisition such as fashion, catering, hair dressing and soap making as ways of upgrading and equipping school dropouts, the poor and the rural women in the area. The units lack equipment and support that is necessary and adequate to resolve problem of youth unemployment. The only direct contribution came from United Nations Development Programme (UNDP), which assisted the centre with sewing machines. In view of the urge to solve the alarming problems in the region, this centre and many more should be encouraged by multinational agencies, individuals, government and non-governmental organizations.

Table 6: Skill Acquisition Centre, Warri North L. G. A. Women Development Education Unit.

VOCATIONAL PROGRAM	*ATTENDANCE*	*INSTRUCTOR*
Fashion	40	1
Knitting	9	2
Typewriting	9	2
Catering	25	1
Adult education	29	2
Soap making/Hair dressing	-	–, – –

Source: Women Development Unit of Warri -North LGA.

16.4. IV policy options for solving the niger delta crisis

Table 7 summarizes the policy options needed for resolving the crisis, which has caused so much pains and misery among the people of the region. The major group involved in the conflict is the youths. About 78.2% of the respondents are of the view that government and oil companies as a matter of urgency and deliberate policy should embark on youth empowerment and development to ensure employment opportunities. This will based the youth based conflicts in the region. In particular it could also reduce the vulnerability of the youths to manipulation and violence.

The people realize the relevance of oil companies in the area, as none of the respondents wants the oil companies to leave the region. About 15.8 % of the respondents emphasize that to enable youth development, government would need to increase the percentage of resource allocation based on derivation to the region from the federation account. The youth can be oriented away from militancy. About 97.6 % of the respondents are of the opinion that the youths can embark on enlightenment campaign among themselves and around the communities on the need for conflict free environment for oil companies to operate successfully and for development and empowerment programmes to be undertaken.

16.4. IV policy options for solving the niger delta crisis

Table 7: Policy Options for Solving the Niger Delta Crisis.

	OPTIONS	*RESPONDENT*	*PERCENTAGE*
What government needs to do to solve the problem of frequent conflict	I & II	129	78.2
	I, II &IV	26	15.8
	III	*0*	*0*
	NONE	*10*	*6*
What the youth in the region can do to ensure peace	*A*	*161*	*97.6*
	B	4	2.4
In view of rural development in Niger Delta, the following options(s), should be adopted.	C	24	14.6
	D	3	1.8
	E	15	9.1
	F	17	10.3
	G	105	64.2

Source: Field Survey 2002.

I Ensuring youth educational development

II Make the youth to be fully employed

III Oil companies should leave the region

IV Government should increase the derivation based resource allocation

A Enlightenment Campaign

B Provide vigilante and discourage Militant Youth Groups

C Provide access roads in the area

D Provide water for the communities

E Encourage skill acquisition

F Dialogue with the various communities

G All of the above

Our field investigation revealed that the level of development reached in the rural areas is under threat of destruction by warring youths seeking their share of the national cake. A visit to Koko town in Niger Delta shows the wreckage of burnt houses, companies and infrastructural facilities destroyed in 1998 during the Ijaw and Itsekiri Crisis.

The concept of human development stresses educational attainment, life expectancy and standard of living as indices of development (UNDP, 1997). Going by these, the amount of concern given to youths and women development by both the governments and corporate organizations is still very low in the region. If development is expected, the government needs to ensure the education and skills development of the people. Most of those who are involved in these harassment and aggression against the state and oil companies are school dropouts and the unemployed.

Rural development entails socio-economic transformation of the rural dwellers. About 64.2 per cent of the respondents claim that the efforts of government and oil companies located in the area are still insufficient to bring succour to the rural people. What government and all concerned need to do is to complement youth empowerment is the provision of infrastructural facilities and rural development.

16.5. Conclusion and recommendations

The extent of conflict in the Niger Delta is of great concern to people within and outside the region. The importance of the region to national development cannot be under-stated. The bulk of the nation's revenue is sourced from the area. Most oil companies in Nigeria are located there and engage in oil exploration activities there. There are about 250 oil fields. The region is the richest and largest mangrove and fresh-water swamp. It is one of the largest African Wetland.

The operations of these oil firms have led to environmental crisis, such as land degradation and air and water pollution in the rural areas, where the occupation is predominantly farming and fishing, there is dependence on the natural resources for economic livelihoods. But the discovery and exploitation of oil in the region marked the beginning of their misery and poverty.

As a result, the incidence of unemployment, poverty and conflicts over land and water resources between communities and with the oil companies are on the increase. There are youth restiveness, militancy and considerable conflicts and violence in the region. The consequences of the activities of youths as it relates to the harassment of corporate bodies, hijacking and detaining of corporate staff of oil companies, seizure of oil rigs, wells and flow stations as well as armed confrontations with law enforcement agencies on the region's socio-economic fortune have been quite negative. There has been the relocation of oil companies, flight of capital and investments, economic downturn and increasing poverty and misery in the western axis of Niger Delta region.

As stated earlier, the main focus of this paper is to evaluate the place of non-farm employment as a solution to the Niger-Delta crisis indicated by youth restiveness, unemployment and rural underdevelopment. It is evident from the survey that the youth militancy and conflicts can be traced in part to unemployed youths. There are also insufficient land, inadequate compensations by oil operators, and the misappropriation and embezzlement of the little compensations by some privileged indigenes in the area. The most viable option for mass employment and poverty alleviation for now is non-farm employment through cottage industries in such areas as food processing, soap making, shoe making and textile industries and service centres such as hair dressing, fashion designing and knitting. This would involve skills development and economic empowerment schemes.

There should be comprehensive enlightenment campaigns within and outside the region to facilitate the achievement of the development of skills, economic empowerment programmes, cottage industries and informal sector activites that are are non farm based. There may be considerable need for re-orientations of both youths and the people if the aspirations of government and the people for accelerated development are to be realized. Peace would continue to elude the region until the people shift towards new platforms to generate employment and wealth that is not affected by the decline and dislocations attributable to oil based environmental degradation.

REFERENCES

Aboribo .R. (2002) Mythical Oil Politics in Niger-Delta: A Search for Valid Explanations, A Paper Presented at the National Conference on Taxation, Natural Resources, the Environment and Economic Growth. Ambrose Alli University, Ekpoma.

Aigbokhan, B.E. (1995) *Macroeconomics: Theory, Policy and Evidence* Benin City: Idelojie Publishers.

Ailemen, M.I. & Onwuka, E.C. (2002) Oil Exploitation and Its Impact on Rural Development in Niger-Deltaa, A Paper Presented at he National Conference on Taxation, Natural Resource, the Environment Growth, Ambrose Alli University, Ekpoma.

Egbon, P.C. & Okoh, R.N. (1999) Fiscal Federalism and Revenue Allocation: The Poverty of the Niger-Delta in Aigbokhan, B.E. (Ed): *Fiscal Federalism and Nigeria's Economic Development*, Selected Papers of the 1999 Annual Conference of the Nigeria Economic Society, NES, Ibadan.

Ekuerhare, B.U. & Orubu, C.O. (1996) *Economics of the Niger-Delta*, Technical Report NDES, June.

Eromosele, V. (1998) Costing *Niger-Delta Oils Spills*, Proceedings of the International Seminar on the Petroleum Industry and the Nigerian Environment.

Okojie, C.E.E. (2001) *Determinants of Poverty and effective Planning and Implementation of Poverty Alleviation Programmes*, Paper Presented at the NISER/WB Development Symposium, May 8.

Okunmadewa, F. (1998) "Essence of Micro-Credit on Scope, Dimension and Reduction of Poverty in Nigeria", *Business Concord*, August 12.

Orubu, C. O. (2002) *Oil Industry Activities, Environmental Quality and the Paradox Of Poverty in the Niger-Delta,* A Paper Presented at National Conference on Management of Petroleum and Energy Resources For Sustainable Development in the Niger-Delta, Delta State University, Abraka.

Orukele, A.J. (2000) *effective Poverty Alleviation Programme as a Panacea to the Oil Related Niger-Delta Crisis,* A Paper Presented at National Conference on Management of Petroleum For Sustainable Development in the Niger-Delta, Delta State University, Abraka.

Todaro, M. (1998) *Economic Development,* New York: Addison Wesley Publishers.

United Nations (1997) *Government Views on the Relationship Between Population and Environment,* Department of Economics and Social Affairs, Population Division, United Nations, New York.

UNDP (1997) Human Development Report, UNDP, New York Oxford University Press.

World Bank (2000) World Development Report 2000/2001: Attacking Poverty, Washington D.C.

17. The case for preventive mechanism as a strategy for resolving communal and ethnic conflicts in the niger delta

A. S. AKPOTOR & J. U. AZELAMA

17.1. Introduction

Conflict does ensue when there is exploitation, (between labour and capital) discrimination/marginalization and authoritarian leadership. There is the consensus that these are root causes of the Niger-Delta conflict. This has been corroborated by a lot of sources. Mr. Worfershn, then World Bank President in his 2000 visit to Nigeria made a landmark observation, "The Niger – Delta is a dramatically poor area despite years of projects exploration and production activities of oil companies" (Africa Today, February 2000; 34). That the prevalence of gross neglect can be an elixir to conflict intensification, is well reflected in late Chief Bola Ige (SAN) and then Attorney General of the Federation's comment, thus;

"... if he was from the Niger – Delta he would have long become a guerrilla fighter. That what else would anybody be doing out except becoming a guerrilla leader or fighting a guerrilla war. (New York Times, November 1998)

The 1995 World Bank report gave factual evidence in these words;

"...despite its vast oil reserves, the region remains poor. GNP per capital is below the national average of $280. Education levels are below national average and are particularly low for women. In the urban areas of Rivers state, the cost of living index is the highest in Nigeria. Only one-fifth of rural housing is considered physically sound; water borne diseases are common, electrification (despite the enormous energy resources) is poor, so are water supply and sanitation, with only 12 percent adequate sanitation in Rivers state."

The situation is so harrowing that Dr. Peter Ogbang, the leader of the Poverty Alleviation Team from the Presidency on a visit to the Niger Delta, described the zone as "the zone of death" (The Vanguard 1995).

17.2. Conceptual clarifications and theoretical framework

Conflict refers to a condition in which one identifiable group of humans (whether primordial, ethnic–linguistic, cultural, religious, socio- economic or political) is engaged in conscious opposition to one or more other identifiable human groups because these groups are pursuing what appears to be incompatible goals. Lewis A. Coser (1913), defines conflict as a "struggle over values and claims to scarce status, power and resources in which the aims of the opponents are to neutralize, injure or eliminate their rivals. Judith Asuni (AAPW) argues that in a situation of limited resources, (competition for scarce resources and unfair system of distribution of natural resources), people are more likely to fight for the little that there is. Professor Amoda in his spoliation theory states that power flows from the barrel of the gun (Ruth First, 1970). Life, property, and other material considerations and pursuit in society are usually perceived as booty. The elite (self-seeking politicians) treat the nation as a spoil, because they want eminence. Thus human rights are abused and expansionist/imperialist actions persist, and so move to the chosen faction stage. In this stage, the elite (victors) treat the conquered like subhumans thus breeding racialism as it existed in South Africa and exists presently in the Sudan.

Mazrui (2000) argues within the same frame when he said, most conflicts are a function of "a clash of civilizations between the lovers of land and the lovers of animals". The lovers of animals have a knack for seizing power. He named the Fulanis (Nigeria) and the Tutsis (Rwanda/Burundi) as examples; while the lovers of land are people of Southern Nigeria and the Hutus who have the knack for creating wealth. He argued further that the conflicts get intensified when the lovers of animals with the power they acquired try to acquire out of the wealth of the lovers of land (Nigerian Tribune, June 2000:3). These were the situations in Uganda, the Ibo Pogrom that led to the Nigerian Civil War (1967-1970), Somalia, Angola, Mozambique, Rwanda/Burundi, Liberia and Sierra Leone. The same scenario characterizes the Niger – Delta conflict.

Conflict is a part of life, and often a part of growth and change. We can live with it, which implies learning about it, making choices, and seeking to transform it in a creative way; or we can live in it, and allow it to become destructive.

Nigeria with its more than 250 ethnic groups grafted together from varying backgrounds, through colonialism presents a good case of communities in a flux. With regards to the Niger Delta, the conflict is a complex mix of historical development and contemporary political and sociological factors enrooted in the existential needs of various ethnic communities (Ogonis, Andonis, Ijaws, Itsekiris and Urhobos). As Eric Fromm (1973:26) explains;

> "Whether man,s dominant passion is love or whether it is destructiveness depends largely on social circumstances, these circumstances, however, operate in reference to man,s biologically given existential situation and the needs springing from it and not to an infinitely malleable, undiferentiated psyche, as environmentalist theory assumes".

From the above context, the dynamics of conflagration in the Niger-Delta (as in most communities expresses both defensive or "benign" aggression in the service of the survival of the individual or his communities and "malignant" (cruelty and destructiveness) aggression often associated with character-rooted reaction formations such as sadism – the "passion for unrestricted power over another sentient being" and of necrophilia – the "passion to destroy life and the attraction to all that is dead, decaying and purely". The Niger-Delta conflict is a dramatic extreme manifestation of what Dennis Sandole (1980) calls "manifest conflict processes", that is, a "situation in which at least two actors and their representatives try to pursue their perceptions of mutually incompatible goals by undermining, directly or otherwise, the goal seeking capability of one another.

The growth, expansion, and intersection of the ethnic groups in the past four hundred (400) years and the differential administrative policy of the British colonial power engendered "catastrophic balance" between contending ethnic forces in the Niger- Delta and shape the structures of politics in the area. The reciprocal trajectories of communal violence between the Ogonis/Adonis, Ijaws and Itsekiris and Urhobos since the early 1990s have resulted in the phenomenon of violence trap where identity and violence reinforce each other.

Ethnic conflict has a primordial affiliation, in the sense that it is connected to the things people cannot live without, among them traditionality, the persistence of the past into the present, and a sense of collective self-consciousness (Shils; 1957; 1995 Geertz, 1963; Connor, 1993; 1994; Isaacs, 1975). A sense of community of this sort – Gemeinschaft rather than Gesellschaft – necessarily generates awareness of other communities, which spills over into conflict and violence. Ethnic conflict also ensues from ancient hatreds – the Urhobo/Itsekiri/Ijaw; Ife/Modakeke; Tiv/Hausa; or from the clash of cultures – religious crisis between Christians and Muslims.

Ethnic conflict in Nigeria has been more of economic cleavages than cultural cleavages (Nnoli, 1978). This supports Karl Marx who argues in the manifesto of the Communist Party that economic differences are at the heart of political disputes. This is reflected in economic competition between ethnically differentiated segments of the working class (Itsekiri /Urhobo/Ijaw) or between ethnically differentiated trades (the Hausa/Yoruba traders in Lagos and Shagamu in the 2002 crisis). It has also brought conflict between whole ethnic groups against government as the cases indicated by ethnic resolutions and statements in the Ogoni Bill of Rights, the Kaiama Declaration, the Ikwerre Charter of Demands, the Asaba Declaration, Urhobo Economic Summit Resolutions, the Oron Bill of Rights and the Alaka Declaration. These calls have been extended to the unanimous South-South or Niger – Delta demands for Resource control.

17.3. Theoretical statement

Our basic analytical tools in this discourse are David Easton" s system theory and the frustration-aggression/ failed expectation theory. Easton (1965) described the political system as that system of interactions which is concerned with the authoritative allocation of scarce values. Every political system exists within and is affected by the social, economic and cultural ingredients of its environments. What is paramount is the ability of the political system to manage stress that may emanate from within. Stress of course evolves from demands within and the survival of the system depends on the capability of the authorities to confront and process the sources. Thus the authorities must evolve policies of resource allocation that should be satisfactory and binding. Ethnic tensions therefore are stresses, which the political authorities must process and convert into political decision if the system must survive.

The Nigerian political system in the latter part of the last century was weighed down by tensions within and between groups and between the Niger Delta groupings and the Nigerian state. These tensions have created serious problems of resource distribution, insecurity and survival of the state. The contention is that Nigeria could be described as a failed state, for it has been unable to authoritatively create systemic balance and maintenance in the Niger-Delta, hence the conflict persists. And this is unfortunate despite the fact that the region produces over 90% of the nation's resources.

It is within this context that we also want to situate this discourse within the framework of frustration/aggression and failed expectation theory. John Dollard (1993), a leading exponent argues that aggression is always a consequence of frustration. He further argues that the strength of the instigation to aggression may be expected to vary with; (i) the strength of instigation to the frustrated response; (ii) the degree of interference with the frustrated response; (iii) the number of frustrated response sequences.

In other words, aggression occurs when one's goal is obstructed or one's emotion is thwarted and the tenacity of the aggression would depend upon the importance or weight attached to the goal denied. Eze Robinson, Ekpeye, the Logbono of Ekpeye land confirms this when he said, the deprivation in the Niger Delta is the remote cause of chieftaincy tussles in the area. When a man is denied of every good thing of life, he gets angry with his neighbour, himself and every object around him [Human Rights Defender 2000].

In Africa and some other less developed parts of the world, the post independence expectations have created a socio–psychological climate capable of producing mass discontent and conflict. The Niger-Delta specifically bears this. Despite the revolutionary pressures *(Adaka Boro)* and others, with the consequent creation of states, and setting up of Development Commissions e.g. Oil Mineral Producing Areas Development Commission *(OMPADEC)* and Niger Delta Development Commission *(NDDC)*, the people are yet to see the "*el dorado*'. That notwithstanding, the Niger-Delta people's awareness has increased and found that their poverty is as a result of the opulence of the major ethnic groups who incidentally holds the political leadership of the country. That the observed opulence of Abuja the nation's capital, during the Two Million Man March in 1997 for the self-succession bid of General Sani Abacha when compared to the poverty of Niger-Delta spurned the youths towards restiveness and militant actions is a testimony to this. Consequently, the conflict rather than get minimized has been exacerbated. We can therefore say that the Niger Delta Conflict can be situated within the context of neglect, deprivation and environmental abuse *(See Table 2 below)*.

Given the above preface, our contention is that if government had a systemic approach to its governance style, that is equitable development projects with serious emphasis in the Niger-Delta region, from which over 90% of the nation's resources are derived, (preventable steps) the crisis we witness today in the region would not be at its present level. The consequent frustration from failed expectation vis-à-vis the opulence in other communities ignited the conflict and determines its dynamism.

17.4. Conflict management

Conflict management connotes the way we respond to and deal with conflicts. Because many conflicts cannot be readily resolved, we use the term conflict management rather than conflict resolution. Efficient conflict management enables parties to co-exist amicably and deal with their disputes without necessarily resolving the underlying conflict (Salem, 1973:14). The conflict management circle, comprises of three main stages of responses: – (a) Conflict prevention or peace promotion and consolidation, (b) conflict control or abatement; and (c) conflict resolution. These three stages have different techniques. While (a) above is the preventive, (b) and (c) constitute the curative measures.

The technique adopted in this study is to critically analyze the curative measures with the aim of justifying the preference of preventive mechanism.

17.5. The curative measures (cm)

Curative measures in conflict management evolved from the belief that fundamental causes of conflict should be identified, isolated, investigated and cured. It

Table 1: Ranking of Major Environmental Problems, Social Issues and Priorities in the Niger Delta

ISSUES	*PROBLEMS*	*RANKING*
Natural Environ- ment	Coastal/river bank erosion	Moderately high
	flooding	Moderate
	Sedimentary/situation	Low
	Substance	
	Exotic (water hyacinth)	
Develop- ment	Land degradation/soil fertility loss	High
	Agricultural decline/shortened	High
	fallow	High
	Shortened fallow	High
	Delta forest loss	High
	(Mangrove loss)	High
	Bio-diversity depletion	High
	Fisheries decline	Moderate
	Oil spillage	Moderate
	Gas flaring	High
	Sewage and waste water	Moderate
	Other chemical	
Socio- economic problems	Poverty	High
	Unemployment	High
	Community-oil company conflict	High
	Inter-community conflict	High
	Intra-community conflict	Moderate
	Conflict over land	High
	Inadequate compensation	High
	Displacement	Moderate
	Decay in societal values	High
	Poor transportation, high cost of	High
	fuel, housing pressure and	High
	infrastructure decay	
	Crime	

Source: NDES. 1997 (See Okoh and Egbon, 1999:413).

postulates that understanding a Conflict System involves analyzing conflict structures and process(es). This is so because the problematique of conflict negotiation and settlement is predicated on a number of considerations, which include; (i) Whether a "prescription will be unacceptable if it were not logically related to its analysis and (ii) whether a prescription based on faulty analysis will be unlikely to produce the desired consequences" (Watt, 1959)

Thus the fundamental premise for curative measures is that conflict and destruction are not just "reaction formations" rooted in a philogenically programmed innate instinct which seeks avenues for discourse and waits for proper occasion to be expressed; but more importantly is a product of a complex historical, sociological and existential needs of people (AAPW Report on Warri Crisis 1999). Once cured therefore, the problem generated from the cause would no longer rear its head. It is unlike the preventive measure that seems to treat the symptoms rather than the cause.

Those who believe in the curative method argue that the preventive measure tends to keep alive the causes of conflict, and that no matter how long the prevention; the conflict will come out in a most destructive way. It is like saying that those who make peaceful change difficult, creates room for violent change. There is of course the argument that the general conflict level in Africa and in particular the ethnic conflicts within the Niger-Delta result from government's unwillingness to look at the fundamental causes of the problems and treat them. Governments have tended to utilize the fire-brigade approach of sending forces consequent upon bloody clashes with a follow-up of Commission of Inquiries, whose reports are seldom made public and implemented; or resorted to avoidance. Avoidance, which is a way of pretending that everything is okay, is predicated on the assumption that the problem would soon fade away by itself with time (Sharia in Nigeria has refused to fizzle out as thought by President Obasanjo). Its main danger is that it can allow a problem to grow unchecked. People often use this approach to make others change their minds, but it has its cost.

In a fundamental sense, the curative method is germane. This is anchored on the belief that since conflict is inevitable in human relations, it could have some functions. Thus, though some believe that conflict is destructive and abnormal and should be avoided, others believe that conflict is not necessarily bad, abnormal or dysfunctional. It can in fact be productive. Fraser (1984:3) and John Burton, (1987) for instance, regards conflict as an essential creative element in human relations, and that it is the means to change, and by which our social values of welfare, security, justice and opportunities for personal development can be achieved. Hence they claimed that if conflict is suppressed, society would become static. James (1912) argued, "Awar-less world with no scorn, no hardness, no valour any more could be "a cattle yard of a planet". However, whether or not conflict would play a functional or dysfunctional role in human relations would depend on the manner such a conflict is managed and resolved. If a conflict is badly managed, it leads to escalation. It could become harmful as it leads to unnecessary dissipation of scarce resources (Imobighe2003: 38)

It needs to be stated here that ethnic and inter-tribal strife has historical root with added sociological importance. Hence when government thinks that peace has come after some quarrel, without really getting to the root of the crisis, they later re-occur in greater dimensions. For instance, why has the Warri crisis, which started as far back as the 1950s, resurrected in the 1990s and continued into 2004? Why is the Ogoni crisis still lingering despite government high-handedness? Why did the Ife/Modakeke crisis after a peace dividend had been achieved break out a few months later? Why the re-occurrences of quarrels between the Ikwerres and the Ibos in Port-Harcourt, why the conflict between Southern Zaria and the Muslims in Kaduna State; why the persistent call for a true Delta State with Warri as Capital; or why the re-occurrence of the BIAFRAN spirit through MASSOB; or why the call for confederation and national restructuring in Nigeria?

17.6. The sanctity of preventive mechanism

Prevention means stopping or forestalling conflict from happening. It is an established axiom in Medicine that "prevention is better than cure" As preventive medicine creates the conditions that support health, making disease less likely and surgery unnecessary, so preventive mechanism creates the conditions that support peace, making war less likely and deterrence unnecessary. And if conflict is a social disease with all the characteristics of an illness, by implication it is better to prevent it from occurring at all, than curing it. If we consider the multifarious complex issues involved in conflict or communal clashes, from the causes to the mode and dynamics, as well as its consequences in terms of time, energy, mental and psychological strain and money, then we should better prevent it.

The United Nations also believes in conflict prevention. Hence in Article I of its Charter, it calls "for effective collective measures for the prevention and removal of threats to the peace". And according to its Current Secretary-general – Kofi Annan, "There is no higher goal, no deeper commitment and greater ambition than preventing armed conflict. Conflict prevention, therefore should be one of the organization's deepest commitment. Therefore it is germane for the UN member states to consider the appointment of special mediators or special commissions to look into the sources of disputes, build confidence and recommend practical solutions" in situation of potential or actual conflict (This Day 04:12:2000; 14). But rather, vast resources are spent in efforts to "cure" conflicts, when for many of the victims it is already too late (UN Annual Report, 1998:9).

Preventing potential conflicts from crossing the threshold of violence require early warning of situations with the potential for crisis, proper analysis, an integrated preventive strategy, and the political will and resources to implement such a strategy. It is important for example that states must initiate intimate relationship between social justice and material well being; reduce poverty and promote economic development and democratization – including electoral assistance and civil education. These efforts are preventive peace building because they attack the root causes of many conflicts. Kofi Annan's journey to Baghdad in February 1998 to put pressure on Saddam Hussein to allow UN inspectors visit nuclear sites, which led to an agreement, forestalled US invasion.

Robert McNamara, a leading American scholar on Defense Studies also shares this belief. According to him;

> [3]In this age of high technology weapons, crisis management is dangerous and uncertain. It is not possible to predict with confidence the consequences of military action between nations equipped with such weapons. Therefore, we must direct our attention to crisis avoidance (World Bank 1991-97)

For the United States, the best security policy is one that prevents conflict, hence her programme of preventive defense rest on its premises that fewer weapons of mass destruction in fewer hands makes America and the world safer. In the post cold World war security environment, U.S. strategy for managing conflicts rests on three basic lines of defense. The first line of defense is to prevent threats from emerging; the second is to deter threats that do emerge; and the third, if prevention and deterrence fail, is to defeat the threat using military force (Perry, 1996:65).

From his experience of the conflict he managed between the then Governor of Ogun State Chief Bisi Onabanjo and the Awujale of Ijebuland, HRH Oba Sikiru Adetona, General Obasanjo (1993) enthused: "we must develop the culture of conflict prevention and resolution in order to live at peace and thereby ensure a progressive society". Although it has its own problem in that it does not allow real grievances to come out for detailed study and analysis with appropriate solution(s), our contention is that most of the violent conflicts could be avoidable if government and the communities resort to available preventive measures.

17.7. Application of preventive measures to the management of communal and ethnic conflicts in the niger delta

The conflicts in the Niger Delta are of two types. The generalized region wide conflict between the communities, ethnic groups and their civil groupings on the one hand and the Nigerian state and her joint venture partners, the multi-national oil companies on the other. This comprise a generalized restiveness, agitation and tension and specific local or community /ethnic based conflicts with the oil companies and the Nigerian state as mainfested by Ogoni, Ikebiri, Odi and Olugbobiri violent conflicts (Table 2). The second is a growing regime of intra and inter-communal and ethnic conflicts as indicated by such conflicts as Ijaw/ Itsekiri, Itsekiri/ Urhobo, Ijaw /Urhobo, Ogono/Ndoni and Nembe/Nembe (Table 2).

17. The case for preventive mechanism as a strategy for resolving conflicts

Table 2 below is a checklist of ethnic conflicts in the Niger Delta.

TABLE 2: A CATALOGUE OF SOME MAJOR ETHNIC CONFLICTS IN THE NIGER DELTA AND THEIR CAUSES

S/N	*Ethnic Groups/Communities Involved in Conflict*	*Nature*	*Cause*
1.	Ineme-Ethnic group and others in Etsako, LGA,Edo	Non-violent	Cultural Segregation
2.	Ijaw-Bini (Edo State)	Non-violent	Land
3.	*Itsekiri – Urhobo*	*Violent (1952-* 2003)	*Chieftaincy/Land*
4.	Ijaw-Itsekiri (Warri, Delta State)	Violent (1997-200	Land & Local Govt. reorganization
5.	Ijaw-Urhobo (Aladja)	Violent (1997-2001)	Land
6.	Auchi –Uzairue	Violent (1981)	Land
7.	Okrika-Eleme (Ogoni)	Violent (1999)	Land
8.	Nembe – Nembe	Non-violent	Identity related
9.	Ogoni – Ndoni	Violent (1994)	Land
10.	Efik- Oron	Non- violent	Land
11.	Usaka Anang – Ubaka Umuofo	Non- violent	Boundary
12.	Nto Ekpu Ikot – Oboni Upa	Violent (1993)	Land
13.	Iwukem-Azumini	Violent	Land
14.	Ini LGA – Arochukwu & Bende LGA	Non-violent	Boundary
15.	Ekpemiong –Asanting	Violent	Land
16.	Asit Urua-Upenekang	Violent	Land (Oil)
17.	Eastern Obolo L.G.A- Ikot Abasi L.G.A	Violent (1991)	Boundary
18.	Ikot Ekpe – Ikpot Effiong	Violent (1980)	Boundary
19.	Okpoama –Ewoame	Violent (1995)	Land
20.	Oluasire –Elesang	Violent (1993)	Land/Boundary
21.	Ogbolamabiri – Bassambiri Bayelsa State	Violent	L.G. H/qtrs
22.	Otuokpoti – Agbura Bayelsa State	Violent (1990-1998)	Land
23.	Epebu – Okigbene	Non-violent (1990-date)	Land/oil
24.	Ikianibiri – Peremabiri	Non-violent (1957-date)	Land/land
25.	Biseni – Agbere	(Violent 1972-96)	Oil/land

26.	Odukpani- Arochukwu	Violent (1996)	Boundary
27.	Okun – Aku- Ikun	Violent	Land/boundary
28.	Ekori, Yankuuru L.G.A C/Rivers	Violent (1996)	Chieftaincy stool
29.	Ogbe – Ijoh- Aladja	Violent (1996)	Land
30.	Enekorogha – Ekogbene	Violent (1996)	Land
31.	Gbaregolor – Esanma	Violent	(Landlord-Tenant)
32.	Olota, Alagbabiri – Oboro	Violent (1995)	Land
33.	Kpakiama – Olodiama	Violent (1994)	Land
34.	Odi Invasion	Violent (1994)	Political/oil
35.	Olugbobiri (Tebidaba station)	Violent (2000)	Political/oil
36.	Ikebiri Invasion	Violent (1999)	Political/oil
37.	Ogoni crisis	Violent (1990-96)	Political/oil

Source: Radio Delta Report Update on Niger Delta Crisis, 2002. Warri

We should note that the first type is traceable to the neglect, under-development, marginalization, poor compensation and poor benefits in terms of employment that have been meted out by the Nigerian state and oil companies. The second is traceable to land and water boundary crises. We should note that land and water have become increasingly scarce and valuable because of oil exploitation in the region. As a result, the land and water based disputes such as boundary disputes have become common. The compensation from MNCs for polluted lands and water and those appropriated for exploration have caused violent conflicts among and within communities. Within communities, there are also chieftaincy-based conflicts. The other major cause of inter-ethnic conflicts is the alleged disproportionate allocation of state resources and political representation such as the creation of state and local governments and local government headquarters. The Nigerian state, being hegemonic and prebendalist, is partisan and partial in its dealings and distributions and this has generated conflicts in the region.

The preventive measures applicable to the two categories of conflicts are different. With respect to the region wide conflicts linked to the political economy of oil, the conflict is rooted in the distribution of resources and development efforts.

That the gap between expected rapid development and actual slow development which are preventable with appropriate polices/programmes could ignite conflict, is well reflected in the consequence of the two million-man march organized by Daniel Kanu led Youths Earnestly Ask for Abacha (YEAA) in 1997, to garner support for the late H ead of state Gen Sani Abacha's self-succession bid. Those who represented the Niger-Delta states; Akwa–Ibom, Bayelsa, Delta, Edo, Rivers, Cross River, Ondo and Abia, were perplexed at the glittering and impressive level of development of Abuja. Some were momentarily thrown into a state of bewilderment and realization at what they consider as a creation from the resources gotten from the bowels of their land while they wallowed in poverty. This was brought out succinctly by This Day (27:07:2003; 14), a daily newspaper; *Ah, na so Abuja be? And na this Niger Delta oil money dem take shine the town fine, fine so! Our own creeks and villages too must fine like this Abuja!*

Corroborating this, Chief Edwin Kiagbodo Clark, a national leader of the Ijaw said the root cause of the crisis lies in the peoples resolve to make the Niger-Delta look like Abuja in terms of development. For Clark, this instant and spontaneous realization marked the very genesis of a fresh and intense battle for the control of their resources.

In an Advertorial in the Guardian (October 1, 2003:9,) the Urhobo Stakeholders Association enthused that while Ibos, Hausa/ Fulani and the Yoruba had manned Federal parastatals in the Niger Delta region, the Niger-Delta indigenes had never manned such, in their places. It also accused the Federal Government over its shallow handling of the unending Warri Crisis, by not taking to what the people needed by appointing the Minister of Petroleum from that region; a position that the President has continuously held to it unconstitutionally.

In other words, the deliberate policy of marginalizing the Niger-Delta people and the environment in regards to functional appointments into strategic positions and general lack of sustainable developmental project which are the root causes of the crisis could have been avoidable, but for lack of visionary leader(s). It is even more paradoxical to note that these opportunities are made available to the major ethnic groups outside the region. We contend here that if the youths that indulge in kidnapping/holding workers hostage and shutting of oil installations leading to loss of revenue to oil companies and the government, were employed in large numbers, they would have found such activities irrelevant.

The US Marshal Plan was a preventive economic system, which combined a vision of American world leadership with a deep understanding that what happens abroad, affects America's security. With regards to the Niger-Delta and to douse the rising tension, a fair share of oil revenue should be granted to oil producing communities. The Niger Delta region has been advocating for increased derivation funds and resource control. To prevent continuous Niger-Delta agitation for benefits, government should go back to 1953 Hicks 100percent Rent/Royalties law. At the state level, there are agitations for some level of re-distribution of funds to the oil producing local governments from the derivation funds. While some states like Rivers and, Bayelsa gives out some financial handouts to the Local Governments monthly from the 13% derivation allocation, the Delta State government has refused that demand. This has been creating some disaffection between the oil-producing local government areas and the state government. It was in a bid to redress this, that a non-governmental organization tagged, Derivation Front, was formed. The opposition party gubernatorial candidate in the 2003 General Election in Delta state, Chief Great Ovadje Ogboru (AD) pledged to follow the Bayelsa and Rivers State example of giving 10 Million Naira monthly to all the local government councils as part of the 13% derivation.

The Niger-Delta people are calling for extensive development of the region. Government should as a matter of policy embark on substantive development projects such as linking up the oil-producing communities and towns through roads and bridges. Towns such as Warri, Burutu, Escravos, Bomadi Sagbama, Port Harcourt, Brass, and Borny should be linked by roads. If the Gowon Government could bridge dry land in Lagos on the large expanse of the Lagos waters to ease traffic, why can't the Warri, Port – Harcourt rivers not be bridged to reach all communities? An extensive social facility and infrastructural provision should be embarked upon.

Partisan and partial allocation of state policies, resources and representation has caused severe conflicts among the Ijaws, Urhobos and Itsekiris. The early Warri crisis between Itsekiri and Urhobo evolved consequent upon the Action Groups (AG) government in the then western region headed by the late Chief Obafemi Awolowo's change of the title of Olu of Itsekiri to Olu of Warri and the transfer of Land Trusteeship into Itsekiri hands. This is to say that the Itsekiri/Urhobo conflict is traceable to land and chieftaincy matters. This was preventable but because of political partisanship, Chief Awolowo committed a life long mistake, which the British avoided. The British officers who were earlier approached in the 1930s by the Itsekiri to carry out this change in nomenclature refused because of the likely consequences in such a multi-ethnic society. Hence we state that if Chief Awolowo had listened to the voice of reason then, the conflict between the Itsekiri and the Urhobo would not have occurred, and even though it occurred, the magnitude and period would have been less than what it has been.

We also contend that the intensity of the Warri crisis from 1997 to 2003 would have been avoided if the Federal and State Government had followed the recommendations of Al-Hasan Idoko and Nnaemeka-Agu Commissions of Enquiries, that separate local governments be created for the three ethnic groups in Warri. This recommendation evolved from the fact that the only local government council in Warri was being manipulated by the Itsekiris, to the detriment of the Ijaws and Urhobos. These were in terms of development project spread, appointment into strategic positions and chieftaincy meetings, where Itsekiri language became the official language with the non-recognition of the Ijaw and Urhobo Traditional rulers. The latter two withdrew from the council meeting.

Rather than follow the Commission's recommendations, the Abacha Administration created Warri South-West local council with headquarters at Ogbe-Ijo, an Ijaw community. But consequent upon the lobbying influence of the Itsekiris, the Federal Government re-located it to an Itsekiri enclave – Ogidigben. Thus the Ijaws felt slighted and unleashed their anger on the Itsekiris. Since then Warri and its environs have known no peace. The creation of separate Local Governments for the three ethnic groups is a major antidote to the ethnic wars was further confirmed when the State House of Assembly in 2002 tried to visit the Warri town on how to demarcate the wards. Unfortunately, the Itsekiris sponsored a violent demonstration that scared the lawmakers away.

It is quite unfortunate to note that in Africa, most conflicts are traceable to Government. This has been so because of their partisanship. While late Obafemi Awolowo created the seed of discord in Warri in the 1950s for his political survival, the unguarded statement by President Obasanjo during his visit to Warri in September 2003 that the creation of separate local government is not the solution is a confirmation of the Yorubas' affinity with the Itsekiri. Our argument is that if separate Local Government councils had been created for the three ethnic groups – Warri East for the Urhobo, Warri Southwest for the Ijaw and Warri South for the Itsekiri, the crisis since 1997 would have been nipped in the bud.

Land and boundary demarcation has been one of the causes of the conflict in Warri, Ife/Modakeke, Umuleri/Aguleri and else where. Perhaps, the anti-dote here is in the 1998 UN report that urged all Third World nations government to evolve preventive peace-building measures of poverty alleviation, through youth employment, social justice, improved material well-being, economic development and democratisation, since they are the root cause of many conflicts. The multiplier effect of this is that the restiveness and energy of the fighting youths would be re-channeled for positive activities. The idle mind, it is said, is the devil's workshop.

Inequalities in distribution of state resources has caused so much communal violence as the case of the Warri region which is over the creation of local government, local government headquarters and wards indicate. The preventive measure here is for the Nigerian state to halt this degeneration of inappropriateness and inequities in the distribution of state resources. If one of the motivations for violence in the society can also be traced to disproportionate allocation of money and other material things, such as those derived from the oil companies, then the degeneration of values into the material should be halted too.

If the curative measure here is that the state and corporate organizations provide easy opportunity to make money, there can never be equity. The bloody clash between Olomoro and Oleh youths was consequent upon the sharing of oil waste pipes given to them by an oil company. This is why we now call for the stoppage of monetary compensation by the MNCs to communities. Rather, the MNCs should carry out project execution themselves. This is so because the cheaper measure of distribution of largess by the MNCs has given fillip to the emergence of emergency leaders who hijack compensation and cause violent crises. This was what led to the Ogoni Youths killing of four prominent chiefs on May 21st 1994 – Mr. Albert, T. Badey, Chief Edward N. Kobani, Chief Samson N. Orage and Chief Theophilus B. Orage (christened the Vultures). The boomeranging effect led to the state execution of Ken Saro – Wiwa and eight others on November 10, 1995. It was the same scenario that made angry youths of Evwreni – Ughelli to kill their King – Owin Kumane over an alleged N4.2 Million donated to the community by an oil company (Punch 2000).

17.8. Concluding remarks

Our discourse above reveals that the perceived deprivation suffered by the Niger-Delta in resource allocation, persistent or criminal neglect in infrastructural development and social welfare, environmental degradation, large-scale unemployment and under employment with attendant poverty are the major sources of conflict between the oil bearing communities and the state cum oil multi-nationals.

It has also become clearer that the conflict has persisted over the years because of lack of early warning or detection mechanisms (prevention) in identifying potential disputes. The persistence of the conflicts can also be factored on the haphazard, inappropriate, inadequate and the inflexible approaches adopted by the State. These measures are usually reactive, patronizing, ad-hoc, fire brigade and containment oriented, coercive and tend towards faultfinding or legalism. These strategies have also been seen to focus on short-term peace and stability, as against lasting peace and stability. In the process, the disputes are swept under the carpet believing that the various groups or the disputes will either whither away or will die down. Unfortunately, the conflict is aggravated rather than resolved.

Our discourse also reveals the complimentary nature of both curative and preventive measures in inter-ethnic conflict resolution. This not withstanding, the costly nature of wars and the curative measures make the preventive a preferred alternative. Even if war evolves, preventive mechanism can prevent the spread. Early warning and detection signals should be encouraged. The best way to manage conflicts as has been recommended by the UN Article is to prevent them through the early detection/identification mechanisms and nip them in the bud before they escalate into violence.

But our preventive mechanism is both substantive and institutional. The substantive is the developmental implying that Government carry out impressive functional development programmes, which should include infrastructural development and employment generation. These would create opportunities for the indigenes and the restive youths and consequently re-direct them away from violence-prone activities. If government had paid adequate attention to the developmental needs of the people, since the beginning of oil prospecting and production like it has been done in Lagos and Abuja, conflict and violence would have been prevented. The institutional on the other hand, implies the role of the State Security agencies; whose function amongst others include the prevention, detection and investigation of threats of espionage, subversion, sabotage and economic crimes like bunkering should be effective. The activities of the youths and the syndicates in the region, as pertaining to sabotage, and economic crimes no doubt have shaken the security of the nation and stability to its very foundation. It is either that the security agencies are not functioning effectively or those who receive the reports do not implement their recommendations or simply ignore detected signals.

The government and its agencies should therefore pay premium attention to conflict prevention than resolution. Early warning signals should be taken seriously and the appropriate authority directed to tackle such potential conflicts. We recommend also that a conflict mediation, arbitration, prevention and national reconciliation commission be established and empowered to prevent, resolve, and manage conflicts as well as recommend appropriate sanctions on erring groups, organizations and individuals.

References

Adefemi I. & Gaskia, J. (2001) *Ethnic Groups and Conflicts in Nigeria,* The South South Zone of Nigeria, PESS, Ibadan: Programme on Ethnic and Federal Studies, University of Ibadan, Ibadan.

Burton, John W. (1987) *World Society,* Volume 3, Lawhen University Press of America

Coser, Lewis (1956) *The Functions of Social Conflict,* New York: Harcourt Brace, Jovanovich Inc

Crawford, Y. (1993) *The Politics of Cultural Pluralism,* Ibadan: Heinemann Educational Books (Nigeria) Plc.

Daniel Bell (1975) "Ethnicity and Social Change in D. P Moynilian and Glazer (eds), *Ethnicity,* Cambridge, Mass: Harvard University Press.

Easton, David (1965) *A System Analysis of Political Life,* New York: John Wiley & Son Inc.

Fraser, Neil & Hipel Keith W. (1984) *Conflict Analysis Models and Resolution,* New York, Elsewiep Science Publishing Company

Geertz Clifford (1963) *Old Societies and New States,* New York: Free Press

Imobighe, T. A. (2003) *The OA U (A U) and OAS in Regional Conflict Management: A Comparative Assessment.* Ibadan: Spectrum Books Ltd.

Jennet, F.E. (ed) (1973) *Conflict: Productive & Destructive in Conflict Resolution Through Communication,* New York: Harper & Row.

John Dollard, (1939) *Frustration and Aggression,* New Haven: Yale University Press

Kahn, Herman, (1965) *On Escalation: Metaphors and Scenarios,* London: Pall Mall Press
Karl Marx and F. Engels, *Manifesto of the Communist Party*
McNamara, Robert (1968) *The Essence of Security: Reflections in Office,* New York: Harper and Row.
Ochoche Sunday (2001) *Understanding the Dynamics of Peace and Conflict Resolution in Nigeria.* A paper presented on a function organized by the United Nigeria Development Forum (UNDF)
Okwudiba Nnoli (1978) *Ethnic Politics in Nigeria,* Enugu: Fourth Dimension Publishers
Osaghae, E. (1992) " Managing Ethnic Conflicts and Democratic Transition in Africa: The promise, failure and the future" in B. Carom et al (eds) *Democratic Transition in Africa,* Ibadan: CREDU.
Peter, H & Ben, R. (eds.) (1998) *Democracy and deep-rooted conflict: Options for Negotiations:* A publication of Institute for Democracy and Electoral Assistance. Handbook Series, Sweden.
Ruth, First (1970) *The Barrel of a Gun: The Pol itics of Coup d‘ etat in Africa,* London, Allen Lane: The Penguin Press
Sandole, Dennis J. D, and Sandole, Starosle Ingrid (eds.) (1987) *Conflict Management and Problem Solving: Interpersonal to International Application,* New York: University Press
Schermerhorn, R. A. (1979) *Ethnic plurality in India.,* Tucson Arizone: University of Arizona Press
Tamuno, S.O, Girigiri, B.I.C; Anele K.A, Onuchuku O; Obafemi A.O; (1999) *Ethnicity, Conflict and Developmen: Prospects in Nigeria; A Book of Readings:* Lagos Corporate Impressions.
Africa Leadership Forum and Academic Associated Summer Report and Papers presented at the Seminar on Conflict Management Mechanism (1997)
Academic Associate Peace Works (AAPW Warri Crisis Project Report 1999) Makofe Konsult Lectures Series
British Broadcasting Corporation, the film, "*Heat of the moment*' 1994 *CODESTRIA Bulletin,* Vol. 1 No. 3 & 4, 1999,
Federal Ministry of Information and Culture, Lagos Bulletin (1994) The Ogoni crisis ; The Judgment.
........................The Ogoni crisis; "The Truth of the matter" *Foreign Affairs,* November/December 1986
Partnership for Global Community, *Annual Report* on the Work of the United Nations Organization, 1998
Peace Research Abstracts Journal, Vol... 3 1, No.1. February 1994.
"The Oil Companies and Community Relations" The Post Express, December 2, 1996
Ijaw Youth Council (IYC); *Our Resources, Our Life. 100 reasons why the Ijaw nation waits to control its resources* Submitted to Senate Ad-Hoc Committee on Crises, Riots and Conflicts (2002); *A Critical Appraisal of Persistent Communal Clashes in States of the Federation,* A Report.

18. Towards understanding and resolving environmental conflicts through research: an assessment of the niger delta environmental survey

B. A. CHOKOR

18.1. Introduction

For over three decades, the Niger Delta has been the primary source of Nigeria's socio-economic development and transformation, contributing since 1975 over 90% of the country's export earnings and over 70% of the federal budget which currently stands at $9 billion. The country earns over US$14.5 billion yearly from crude oil export with over a third committed to external debt servicing. It is projected that gas production and export would in the coming years exceed crude oil. In spite of the immense resources, the region remains in a parlous state of socio-economic development with incipient and advanced stages of environmental degradation reported for several communities, including Ogoniland. Ecological devastation and impoverishment of communities have bred community conflicts, with an increasing intensity over the last decade.

It was against the background of the Ogoni uprising and the general unrest that threatened multinational oil and gas production that the Shell Petroleum Development Company of Nigeria Limited (SPDC) announced in February 1995, the setting up of an independent data gathering programme, the Niger Delta Environmental Survey (NDES), to provide an authentic, authoritative and valid up-to-date scientific information on the state of the environment of the Niger Delta, including the scale of pollution associated with oil exploration and production activities, and the impact on communities. This paper explores the nature of the survey carried out largely between 1995 and 2000 and how it has helped to clarify and resolve some of the contentious environmental issues of the region. We argue that the presentations by various interests on the region have suffered considerably from lack of empirical database and as such insights have been limited. Further, solutions to problems arising from the complex linkages between environmental and economic activities in the region that have been proffered have been predicated on personal perceptions and convictions; such speculative insights have become sources of friction between contending parties (oil companies, government, NGOs and communities). Thus, poor understanding of the issues has deepened the crisis in the region. Second, given the scale of developmental neglect, impassioned views have understandably been on the ascendancy, while politicking over the issues has taken the front seat. Against this backdrop, this paper appraises the essentiality of valid data to the Niger Delta question, whether the NDES offers some scope for reconciling the concerns raised over time and what limitations there were in the process of carrying out its work.

18.2. The need for an independent data base on the niger-delta

Over the years there have been irreconcilable differences between the major stakeholders in the region. The Department of Petroleum Resources (DPR) is the government body responsible for monitoring and enforcing operational standards in the oil industry. Although it is the custodian of reports from industry, it has conducted limited studies of its own. Communities because of perceived bias do not take the works by oil companies seriously. Even studies by international interests such as the World Bank (1995) have been controversial in nature. For example, using a prioritization system based on environmental, human health and economic significance, it rated issues more related to communities. In this rating, agricultural land degradation, flooding, fisheries losses, deforestation, *loss, water hyacinth expansion, untreated sewage, vehicular emissions, municipal solid wastes (also toxic and hazardous wastes) were serious and of high priority for management while oil pollution, industrial effluents, industrial air emissions and wastes; fisheries and habitat degradation were rated as of moderate priority and mangrove degradation and gas flaring were rated as of low priority. Earlier reports (World Bank 1990; FEPA 1992) did not also emphasise oil pollution and gas flaring as serious environmental problems. Although it employed consultative process, there is some contradiction in the World Bank findings. While it concludes that the impact of all oil-related activities on the environment of the Niger Delta is widespread and substantial, it argues that oil pollution by itself is only of moderate priority when compared to the full spectrum of environmental problems in the Niger Delta.

Several questions could be raised. What are the reasons for these varied assessments; is the oil and gas issue a political or an environmental question?[1] Whichever way it is seen, the need for balanced information appears paramount. As Brian Anderson (1996), former Managing Director of SPDC stated:

> "In spite of all … effort being made, we recognise the need for a more complete understanding of the environmental pressures on the Niger Delta. There are not enough facts available to decide how best to manage and sustain the eco-system of the delta. There is need to know more about population growth, migration, deforestation, farming, soil degradation, oil activities, road building and other factors, and how these have effected the delta over time".

18.3. Review of the survey process

The mission of NDES is:
"In concert with communities and other stakeholders to undertake a comprehensive environmental survey of the Niger Delta, establish the causes of ecological and socio-economic change over time and induce corrective action by encouraging relevant stakeholders to address specific environmental and related socio-economic problems identified in the course of the Survey to improve the quality of life of the people and achieve sustainable development in the region".

The Survey was initiated by SPDC and later embraced and funded by oil companies in Nigeria operating under the umbrella of Oil Producers' Trade Section of the Lagos Chamber of Commerce and Industry (OPTS) with minor contributions from other interests. The Survey is managed by a Steering Committee of nine members chaired by a prominent private sector person, with representatives from the Federal Ministry of Environment, Department of Petroleum Resources, Delta State Government, (old) Rivers State Government, Nigerian Conservation Foundation, Conservation Foundation, UK (later resigned), the World Bank, OPTS and academia (a scholar with sufficient interest in community issues). The broad spectrum of the Steering Committee was to give it technical/professional competence, guarantee its independence and attain the widest local and international acceptance of its results. The actual survey work was undertaken by local and external consultants, following a competitive bidding process, while a secretariat of technical staff supervised the day-to- day survey process and reported to the Steering Committee.

The Survey followed some five basic stages: First, the preliminary stage encompassing the setting up of secretariat and definition of Terms of Reference (TOR). The TOR covered scientific evaluation of the state of natural resources, use and changes over time and environmental and health impacts with reference to land, soil and surface and groundwater quality. It included productive engagement with stakeholders in data collection, assessing social and economic developments, measures to mitigate environmental problems and rehabilitate degraded land, and the potentials and constraints to sustainable development. Second, Survey Phase I involved a preliminary assessment of existing data and information on the Niger Delta environment and pilot schemes on Participatory Rural Appraisal/Participatory Learning and Action (PRA/PLA) in three communities. Third, Survey Phase II embraced actual field studies to fill the data gaps identified in Phase I. Because of the shortfall in funding, proposed studies in 32 project areas in which data gaps were to be filled were down-sized to five clusters, namely, biological environment and resources, socio-economic studies, human health, pollution, extended PRA/PLA studies in 32 communities while subsidiary collaborative/bridging studies with FEPA were carried on Flooding and Erosion; Hydrology/Hydrodynamics studies relied on available secondary information especially from oil companies (see World Bank/FEPA, 1998). Biological environment and resources received over half of the direct allocation to study sectors. A Geographic Information System (GIS) was also developed to provide a seamless database such that various comparative issues and indicators of change in the region were not only geo-referenced and stored for retrieval but also offered prospects for simulation and modeling of future environmental impacts and management options. Fourth, an Action Plan on the region based on the study reports and wide stakeholder consultation was produced as a basis for government, oil industry, communities and NGOs to address the problems of the region. In order to assure quality, results from the various studies were independently peer-reviewed by local and external technical/professional resource persons as well as community/NGO interests that were not involved in the initial work.

The study used scientific sampling procedures as well as participatory processes based on PRA/PLA methodology in communities. The last point deserves some comments because of its import for the work. PRA/PLA involved the training of resource persons in PRA/PLA techniques for two weeks. A team each made up of four persons – a leader and three facilitators were stationed in 32 selected communities from across 8 states in the Niger Delta. These lived within the communities for four months and through interactive group discussion, semi-structured interviews with leaders and interests, chance encounter/interviews, community transect walk and mapping, use of stories, parables etc., collected vast community-based information, drawn on checklist of issues such as community natural resources, community social structure, socio-economic infrastructures/community development projects, rural economy, stakeholder analyses, community perceptions, community relationships and impacts, income, health and quality of life, community conflict resolution processes, research, extension and training.

By 2000 studies in virtually all the areas had been concluded, reviewed and submitted to NDES except for some works in pollution initiated late because of funding constraints. The reports are available in electronic form but could not be printed and freely made available to stakeholders as assured in the terms of reference. While efforts are still being made to secure funding for the printing, it remains mysterious why a survey of such local and international importance which is unprecedented in scope, and considered by UNIDO as a fast track model for coastal zone management in the Gulf of Guinea, can no longer be supported at the stage of printing; this is in spite of the enormous resources available to oil companies. The only guess is that there has been considerable realignment of priorities in the last half decade in the Niger Delta, especially with democratisation and the coming on stream of the Niger Delta Development Commission – which ought to be the key beneficiary of the NDES reports. On the other hand, the results of the Survey may not be of interest to the oil industry, particularly from the standpoint of the critical socio-economic analyses of the history of oil production and impacts in the region and suggested actions as contained in the NDES reports.

18.4. Major findings on real and perceived factors of environmental conflicts

There are over forty volumes of reports from NDES Phase II work. The findings from participatory assessments, socio-economic surveys and environmental health studies revealed the level of deprivation in the region and the conflict-inducing circumstances under which people of the region live. The review here relies more on PRA/PLA studies, which represented the major medium of interaction with communities most affected by developments in the region (see NDES 1997a; NDES 2000a).

Environment – development indices

Mean annual income in the Niger Delta was some N29, 175/household, that is \$0.8/day that is below absolute poverty line of \$ 1/day by United Nations standard. Only 15.7% of households own television and radio; 23.4% have modern beds. Unemployment was 22-27% in the Warri and Buguma area. Scattered small settlements of 5,000 people represent about 93% of 3,807 settlements in the delta. The settlement pattern is predominantly primate urban formation; this by itself is a major barrier in terms of provision of infrastructure and amenities if resources were actually made available. The Niger Delta accounts for only 2% of Federal roads. This situation is not entirely new. Pessimism on prospects for development has been expressed long ago by the British Colonial Office Commission of Inquiry under Sir Henry Willink (1957-1958) on the Niger Delta: "The country in which they live is divided by creeks and inlets of the sea and of the Niger into many small inland, which nowhere rise far above the highest tides and floods: transport by water and the construction of roads or railways will be prohibitively expensive. Theirs is a country which ... has been neglected and which is unlikely ever to be highly developed". (Willink Commission Report 1958, p. 8).[2]

Some 30% of houses are mud walls, 53.8% corrugated iron roofs; pier toilets are dominant in the swamps and coastal areas. Less than 30-40% of the settlements in the region have electricity. Only 19-39% of randomly selected communities have a formal health service while existing primary health care facilities in some parts of the region serve as little as 2% of the population. Most facilities were not functional and 64% of households never got the medicine prescribed at the facilities they attended while 40% of households could not afford the cost of the usual means of transport to the facilities nearest to them.

Socio-economic dislocation in communities

Sexual Perversion amongst young girls is a major issue; many have turned sex workers living on the rich oil workers and mobile police guards leading to social distortions and cultural quaking in communities as well as upsurge of illegitimate children and sexually transmitted diseases (e.g. gonorrhoea, syphilis, AIDS). There is general rise in smoking of cigars, hemp and taking of hard drugs. Sexual and physical harassment by mobile police increases ill feeling, especially amongst the unemployed youth. There is breakdown of family values; societal discipline is weak; marital infidelity and instability have been heightened by oil activities.

On the community economic front, oil business rates in dollar induce inflation and raise expectations; traditional occupations can hardly engage labour because of high cost and communities even lack the capital to do so or engage in more lucrative long distance trading to market their products. The poor road network in the region has given much advantage to middlemen who purchase products from poorly accessed communities at low prices. Although they produce the oil, the cost of refined products and riverine transportation are the highest in the country – the later over five times that of road. Fuel dumps from major oil marketing companies are few and far between, most products are sold sometimes double or triple the government-fixed price for petroleum products; the people are therefore embittered.

Attractive remuneration packages and parade of affluence by oil workers within the sea of poverty represent a significant intrusion, leading to general rise in cost of living, and other socio-economic changes with destabilising influence on community economic and social life. While the people are disaffected and brood over what has befallen them, the emergence of intra- and inter- community conflicts and rivalries – between benefiting and disadvantaged interests, in terms of contracts and compensations reflects the kind of distortion and dislocation induced by oil activities over time. Further, alienation from resource areas through federal laws means that communities are merely playing the role of bystanders or onlookers rather than partakers in the multi-billion-dollar economic activities taking place around. Social rules of resource appropriation have been defined without community consent and representation; the direct experience of the benefits accruing to other regions raises community consciousness of the need to alter relation; more so, as the people of the region do not feel the impact of governments from local through state to federal levels. Compensation for losses also breeds conflicts. There is protracted haggling involving the use of agents by different sides; this deprives communities of real gains from payments, as the bulk of the money ends up as fees or commission. Remediation is delayed, since necessary interventions to contain spills are stalled by unresolved issues and community protest actions.

Instability in community leadership has become a critical issue because of lack of transparency and the mutual suspicion between youths and elders/chiefs over financial dealings with, or gains from oil companies. In such a climate, the delivery of projects and development benefits are affected because of protracted arguments over goals and means by contending community interests.

Pollution and degradation of ecological resources

Although detailed results are yet to be issued on pollution survey, results from biological environment and resources aspects of pollution survey indicate that air pollution was significant around operational sites; so also is heavy metal pollution in surrounding waters with industrial activities especially around Warri and Port Harcourt. Loss of aquatic life and fish spawning sites is attributed to pipelines and roads leading to oil installations, which block streams, creeks and alter ecosystems preventing migration to community fishponds. Gas flares were perceived to induce acid rain, alter soil pH and nutrient balance, and kill trees and shrubs. Gas flares light up the communities 24 hours and cause noise pollution.

Biodiversity and forest losses are also significant although new species have also been discovered. Losses are largely due to oil activities, seismic operations and land clearance, noise, settlement development, and community farming, fishing and hunting activities. In addition, the Niger Delta suffers losses in outflow of resources and receives several negative externalities or environmental disbenefits. Some 1.7 billion cubic feet (BCF) is flared daily and from 1958 to 2000 over 5.7 Trillion CF (TCF) of associated gas has been produced and over 5.0 TCF or 88% flared.[3] It shows that in terms of socio-economic and environmental considerations the region is deeply impacted and community development projects of industry and government are far too paltry and token to compensate for the magnitude of economic deprivation and social distortions.

18.5. Towards an action plan

The NDES (2000b) recommended an action plan to government, industry and communities with a 3-year life cycle. It had four focuses or pivots: 1. the dire need to address environmental degradation through sound programmes of management; 2. tackling endemic poverty in communities; 3. tackling infrastructure underdevelopment and 4. addressing social injustice and conciliation of interests. The last point was deeply debated by the Steering Committee because of its perceived radical ideological leaning and the tacit assumption that there had been unjust practices in the past and so identified ills ought to be put at the doorstep of the relevant stakeholders. Although it was evident that there were several unjust federal laws such as the Petroleum Act, the Land Use Decree and Revenue Sharing arrangement, as well as the employment structure and pattern of contract awards which alienated communities, some members of the Steering Committee preferred to see them as political issues outside the ambit of the terms of reference of the Survey.

Environmental remediation

Recommendations on environmental management focused on conservation of biodiversity, remediation of polluted sites and waste management. It embraced the listing of protected areas, public enlightenment to reduce poaching, forest restoration programmes and conservation sites for Eco-Tourism.[4] It also included the need to address the impact of canalisation by oil companies and the problems of salt water intrusion into freshwater swamps with accompanying forest and community natural resource losses as in Awoye Ondo State), Opuekeba and Ogidigben (Delta State) and Bonny (Rivers State).[5]

Poverty reduction

To address poverty and deprivation, a range of farming, fishing and trading schemes under community cooperatives with grants and credit schemes were recommended. It included developing small and medium scale industries based on local resources and assets of the people, for example, small enterprise development schemes in crop production and seedling multiplication for rice, cassava, plantain, banana, pineapple, oranges, grapes, sugar cane, Irish potatoes and developing animal husbandry: snailry, rabbitry, piggery, poultry, aqua-culture (fish and shrimps); api-culture (bee-keeping); improving capacity for fishing through provision of engine, boats, nets and gears; provision of credit; processing and storage. The Survey identified opportunities for agro-based food processing industries: e.g. cassava pellets, fruit juices, fish canning, salt extraction etc; gin distilling support. The establishments of Vocational and Technical Schools and Youth Development programmes, sporting and recreational facilities were seen as vital. In terms of direct employment, the implementation of the local catchment policy as practiced in federal government establishments located in various states was commended to oil companies.

Infrastructure and amenities

To address transportation and infrastructural problems, the action plan recommended a new coastal east – west road, the dualisation of the existing East-West Road from Port Harcourt – Warri and other feeder or link roads. Water transportation routes were to be enhanced. Part of the reason for conflict was spatial: the scattered nature of settlement and remote communities and the overall quality of urban development could not support good quality of life. The Survey recommended site-and-service schemes for Port Harcourt, Warri, Yenagoa and the LGA Headquarters.and the creation of nine new growth centres with concentration of infrastructure, investments and amenities in such centres to provide services to remote riverine hinterlands.

Also proposed were the refurbishment and extension of educational and health facilities that were very limited in the region. Educational facilities were most inadequate in riverine communities. A massive School Rehabilitation Programme was recommended and as well as the establishment of Federal Universities in Bayelsa State and Delta State. NDES recommended a maximum of 5 km access or within 1-2 hours travel time to health facilities. It also recommended a Travelling Doctor/ Ambulance Scheme and the refurbishment of existing gas turbines and procurement of new ones; desalination schemes and sinking of deep boreholes for regional water schemes.

Social injustice

In the area of social justice the Survey recommended the abrogation of all unjust laws that affect resource access by communities, local governments and federating states.

18.6. An assessment of constraints and opportunities

Internal Constraints in Defining Methodological Direction

The NDES is stakeholder-driven. This offers some constraints as well as opportunities. NDES not only provided for hard data gathering on scientific and ecological features of the region but offered real basis for social action research and interaction with communities. At various stages of the work the Steering Committee reviewed and assured quality and consulted with communities, NGOs and industry. This by itself though bridge-building, raised further conflict. The Steering Committee with divergent interests could not agree on focus for the Survey, as they differed on the scientific/ecological and humanistic/socio-economic content of the work. NGOs and communities were more inclined towards socio-economic development and saw the Survey as offering scope to inventory on environmental damages for remediation and to initiate community development programmes to redress past wrongs. On the other hand, industry seemed to have favoured scientific and physical ecological studies that could establish the level of degradation as alleged by communities and NGOs without being dragged into new and uncontrolled community development initiatives.

Although there were significant material gains from claims that accrued to some Steering Committee representatives, the threat of NGOs to resign membership, especially following the Ogoni hangings, the funding constraints of NDES, in spite of the assurances at inception that funding would not be a problem, the inability of the NDES to print its final report and present to stakeholders, all underscored the varied perceptions of what the NDES ought to concern itself with. The 'ideological' differences between operators of the Survey elicited much interest; the technical advise of the professional secretariat could not always address the varied concerns. Survey results were subjected to several local and external reviews and amendments to clarify and affirm Members' own position on issues. In one instance of intense protest by Itsekiri interests over propriety of historical facts contained in Phase I Four-Volume Report, a new set of experts and community representatives had to be set up to review the published work. Although renown professors of history did not disagree with the content of the report, nonetheless, the NDES not willing to be a source of conflict agreed that all references to the history of communities, especially on migration, trade controls and inter-ethnic cultural influences, pattern of residence and ownership of areas in the Niger Delta, be expunged and another edition of the reports be published. NDES acting under this pressure cannot be said to have helped in reconciling differences amongst communities through researched facts or data available to it.

Under the circumstance, the Survey swam in difficult waters; the cumbersome process of defining priorities, aside from the very ambitious nature of the Survey, slowed down the work; the initial target of 18 months for its work crept to five years and without a formal end. Because of the consultative process and changes to work programme, the Survey became more expensive in spite of efforts at prudent management. Although the OPTS-approved budget of $5.1 million was described by the Steering Committee as cut-to-the-bone, given the enormity of the task defined for it, and a far cry from the "not less than $10 million as conservative estimate' mentioned in some circle at the inception of the work, it remains true that much could have been achieved by way of scientific data gathering if the scope of work and the methodology for implementation had been left at the hand of experts. In an attempt to provide credibility through stakeholder participation and achieve a wide acceptance of survey result, opportunities for essentially technical decisions became limited. The ambitious scope of work and the uniqueness of the NDES process even on a global level, prompted one international reviewer of the work output to describe it as an unparalleled path in which all will be on a steep learning curve.

18.7. Reconciling the tension between scientific research and humanistic ideals and deconstructing complex community values

The study could not agree on cartographic and scientific definition of the limits of the Niger Delta because community interests misconstrued the mission of NDES and thought it would bring projects with real benefits to their localities. This was one evidence that a purely scientific inquiry without coming to terms with the human dimensions to the crisis of the Niger Delta will be of limited value. This calls for some form of deconstruction, especially of complex community values and the conciliation of the divergent humanistic and scientific motifs that underlie the NDES research and its findings. Deconstruction shows how language imposes limits on our thinking, in particular how to "read for absences' in the text (Dear 1988). The whole NDES process represents a profound hermeneutical challenge because of the divergent approaches that were adopted to suit various disciplinary and stakeholder interests. Simple expressions were suffused with deeper social meanings and intents that needed to be translated and reinterpreted. As one Chief in Nembe community elucidated in context of conflict with oil companies: we know the Federal Government to be the culprit but they are faraway in Abuja; when you are looking for a father and he is not available you hold the son (the oil company) who is on ground in the hope that someday the father will turn up.

The tension between purely scientific and social science research was evident as scientists could not fully appreciate the social and participatory community-based aspects of the data gathering, and the need for them to reconnect with, authenticate and corroborate their findings with non-scientific groups. This they saw as unscientific imposition and a constraint, although many later could appreciate the insights gained through the more anthropocentric methodologies.

The NDES initiative has also shown that defining the social theory and ideological framework for research and data collection is a vital but difficult task; bridging the interest of pure science, humanistic social science and reconciling with those of NGO activists and communities to attain a comprehensive and acceptable database for public policy was an added challenge that could hardly be met. At best it became evident that objective scientific studies of the environment and development must be sufficiently sensitive and responsive to human feelings and attitudes and the social milieu within which resources are exploited and appropriated.

From the NDES work what emerged was that the social contexts and sources of environmental issues and conflicts, rather than their purely physical, biological and chemical characteristics, were much more fundamental in understanding the problems of the region. It is the unequal structure of relation with communities and the net outflow of surplus from the Niger Delta to support growth elsewhere that represents the root of the crisis. Skewed geographical and social concentration of wealth is invariably the product of unjust relation: we need therefore to probe deeper into the supporting institutions, systems and modes of economic integration and exchanges that facilitates unjust transfers (see Harvey 1973).[6] The overall decline in the proportion of revenue accruing to the Niger Delta from Principle of Derivation over the years is clearly reflective of that fact.[7].

18.8. Community interests and conciliation in the region

In spite of evident constraints, the NDES achieved a lot in terms of understanding the issues of the region, although limited in reconciliation of conflicting interests. Dialogue and peace efforts were somewhat fostered by the Survey to the extent that pubic confidence in its work attracted some level of complaints by aggrieved communities. "Concerned Itsekiris" from Warri, Delta State and the Nembe people of Bayelsa sought and came to dialogue with NDES in order to resolve disagreements on historical origin and community land ownership claims contained in NDES reports. The NDES might be considered by some as a good instrument of public relation for the oil industry that was already alienated from communities but the Survey succeeded in forging a link between the divergent interests within and outside the region, especially given the reputation of its Chairman who was well regarded.

Through informal exchanges, Ken Saro Wiwa showed that he was not opposed to the Survey. Another opportunity offered by the Survey was the Community Savings and Credit Scheme (CS&CS). Three pilot CS & CS in Egbemo-Angalabiri, Bayelsa State; Mbiabet, Akwa Ibom State and Igbuku, Delta State became part of community asset building for small scale investment in agriculture and agro-processing businesses in order to economically empower the people and reduce discontent. The success of the programme encouraged other PRA/PLA communities to start their own schemes. A number of oil companies have also initiated and are supporting revolving credit schemes in oil producing areas, while the European Union is in the process of supporting micro-enterprises as a way of creating sustainable communities. Capacity building for participatory research and development was achieved. Over 100 local and international consultants were involved at various stages of the work; dozens of community youths were trained in PRA/PLA techniques, creating some potential for the dissemination of participatory approaches to development in the region.

Undoubtedly, oil industry has also realized the benefits of participatory engagement of communities as a way of minimising conflicts. In contrast to the past when community development projects were 'handed down' by companies, most, including Shell, AGIP and Elf have evolved PRA/PLA studies in identifying community priorities and for facilitating people-centred approach in implementation efforts. New partnership arrangements with CBOs, NGOs (Living Earth, African Women Empowerment Group etc.), government and international development agencies such as UNDP and UNIDO have also emerged.

18.9. Proactive initiatives

One of the most challenging aspect of its work and which might be considered as being at the heart of community expectation was the proactive initiatives of the NDES, especially in areas where the environmental problems and conflict-inducing situations were sufficiently well defined. Proactive initiatives included those focusing on environmental problems around Utorogu Gas Plant operated by Shell in Delta State which caused permanent flooding and loss of community rubber plantations and fish ponds; the environmental problems associated with Warri Refinery and Petrochemical Company, operated by NNPC, where untreated wastes and sludge were being discharged into the Crawford and Tori creeks around Ubeji Village, causing considerable pollution and affecting the community resource base and livelihood support system. Also the East – West road between Warri and Port Harcourt and the abandoned Benin – Warri dual carriage road were brought to the attention of Government. In all NDES wrote and met with those concerned to address the issues (see NDES 1997b). While most assured that something would be done, the necessary interventions were slow in coming, and most had to follow a course of contract and action that were necessarily slow.[9] These efforts might be considered indicative but NDES was careful in ensuring that all key sectors that operate in the region: oil industry, government and parastatal were focused to provide a sense of fairness and firmness. The real achievement in all this was in creating awareness, more so with the acceptance that NDES represented a collision of interest in the region whose view should be taken seriously.

While much was expected from NDES, it was not a legal entity and it lacked powers of enforcement and could only employ persuasion and the circumstances of its birth to secure some cooperation. NDES also lacked resources to carry out direct developmental activities or to implement its results; and it was not even within its mandate to do so, in spite of the expectations from the people.

18.10. Constraints on implementation

One of the biggest setbacks was the non-acceptance of the Indicative Niger Delta Management Plan (INDMP), which translated later into *Development Priorities and Action Plan for the Niger Delta,* issued in 2000. Using its field studies reports, the GIS database and across the board stakeholder consultation processes in Effurun and Port Harcourt, starting in 1999, the NDES saw the new democratic government as offering an opportunity to commend its work for implementation. It therefore proceeded with intense efforts and much financial commitment to draw up the Action Plan document. Unfortunately, in spite of the wide consultation with government interests and lobbying at the National Assembly and the Presidency, NDES had difficulties selling its programmes to policy makers, not even to the Niger Delta Development Commission (NDDC) whose Managing Director, a former Shell Director, was instrumental in setting up the NDES. Rather than for NDDC to use the reports and build upon them, it initiated another round of surveys and studies on the Niger Delta, which in the last two years have rarely got off the ground. This is gross waste of public fund and reflective of poor coordination in data management and use that seems to characterise the country. Further, many politicians were either in a rush to fulfil other agendas or could not appreciate the value of the NDES process in face of the exigencies of national party politics and material interests through fresh contracts. After five or more years of coordinated efforts, the non-acceptance of the NDES report as a basis for action and the failure to fund its publication by the interests that initiated the Survey represent a major set back. While it calls in to question the sincerity of these interests in initiating the work;[10] it reveals clearly the fact that essentially socio-economic interests define environmental problems and that social forces are critical to the identification and implementation of environmental goals and courses of action.[11]

18.11. Conclusion

Many critical stakeholders had argued at the outset of the NDES that enough information already existed to act on the issues of the region and that no new survey effort was needed. This paper demonstrates the imperative of research/data collection to clearer insights into the environmental problems of the Niger Delta region and its import for public policy to address them. While research and data collection might be seen as vital to the clarification of the interface between people and environment and between economic activities and environmental conflicts, there were two key constraints that the Survey had to contend with, and which it could hardly overcome. These constraints reveal some idealism, and idealistic expectations, in the NDES mission. First, funding of study sectors was skewed in favour of ecology; pollution issues which gave rise to the Survey could not be sufficiently funded and was scaled down because of cost such that evidence of pollution had to be inferred largely indirectly from biological studies while socio-economic studies were limited, including PRA/PLA programme. NDES thus in face of funding constraints over-concerned itself with biological environment and resources in funding allocation for its programmes whereas from Survey results, the ecological environment overly provided the backdrop to the much more fundamental socio-economic issues that were raised by the people. Although there was evidence of pollution and negative ecological changes in the Niger Delta due to industrial and other human activities, in the end, the politics of pollution and ecological devastation were more significant due to the poor socio-economic development of the region. The embedment of ecological thinking and the desire to identify areas for nature conservation in the NDES programme were thus more or less diversionary in face mass deprivation, poverty and the conflict-inducing social conditions under which the people lived, including the stifling societal structures and regulations that overwhelmingly defined and limited the access of communities to oil resources. These serious social questions that constitute the roots of the crisis, ignored at the outset of the work programme, surfaced at the stage of developing the action plan.

Second and related, is the reliance on the good will of stakeholders, particularly government and industry, in accepting and implementing its reports. Since NDES was not set up through a legal instrument of government its report was not binding on any interest. At best it was assumed that such interests might understand that the findings and recommendations arising, if accepted and implemented, would be in the overall interest of the oil industry, communities in the region and the nation, especially in reducing tension and creating for a harmonious environment for oil and gas production. This "common-good' assumption in the action plan suggested by the Survey remains a major failing since to date neither the oil industry nor the government has taken the reports with any interest. This attitude of major stakeholders has limited the capacity of a major research initiative in reconciling the differences between contending interests in the region; a major opportunity may have been missed. However, it shows the need to take seriously the critical interconnection between environment, ideology and policy (see Sandbach 1980). While it has been difficult deconstructing complex community values and expectations and reconciling the divergent humanistic and scientific motifs that underlie research, the paper underscores the need to come to terms with the social processes that drive research work and public policy, and the implementation of any recommended action on the Niger Delta.

End notes

1 There are no doubt evident environmental issues. In recent times, in order to address some of the environmental concerns, for example, mangrove losses and pollution, a number of measures have been adopted, including LNG and gas plants for reduced flares, horizontal wells, reduced seismic paths, replacement and upgrading of terminals, old pipelines, and flow stations, better management of effluents, drilling wastes and produced water and mandatory EIA for all new major projects in the oil and gas industry. In spite of these efforts there has been a groundswell in the level of agitation and protests over environmental issues pointing to the interwoven nature of environment and developmental concerns.

2 The problem of terrain and arguments over quality of development by ethnic minorities prompted the British Colonial Office to set up a Commission of Inquiry under Sir Henry Willink (1957-1958) to investigate and allay the fears of minorities. While its report provided the basis for the establishment of Niger Delta Development Board in 1960 to advise government on physical development of the Niger Delta, one significant outcome was its conclusion on difficult prospects for development.

3 Yet, the poor development of the region predicated on the difficult physical terrain is not supported by the fact that in the remotest swamps, oil companies have developed modern facilities of electricity, water, housing, telecommunication, recreation etc within installations, most to international standard, which sharply contrast the dearth and decay witnessed in host communities that adjoin them.

4 In energy terms: about 2.8 billion barrels of oil equivalent (cumulative oil production to date amounts to about 16 billion barrels). Flare amounts to about 0.3 million barrels daily (loss of over $7.5 million per day (at $25 a barrel). From flaring each square kilometre of an oil producing area is exposed to 1.52 million kilowatts of heat per day. Nigeria accounts for some 25% of gas flared worldwide... Fine of 2 Kobo per thousand cubic feet of gas flared in 1979, 50 Kobo in 1990 and 10 Naira in 1998 has not encourage gas re-injection although government and companies have set 2008 as target for extinguishing flares.

5 Sites included for Rivers State – Upper Orashi, with Monkeys, yellow-backed Duiker and speckled-throated Otter, Parrots and Waders from Europe; the Andoni Reserve for Elephant and Hippos; in Bayelsa State – Ramos-Dodo-Pennington-Digatoro for Turtle; Taylor Creek: White-collared Mangabey; Chimpanzee, Elephants and monkeys and Apoi Creek – Red Colobus monkey; in Delta State – Ikodi for Parrots.

6 Sites recommended for remediation/restoration from spills, dredge spoils by impacting companies include: Ubeji, Bomu, Enwe, Rumuekpe, Eruemukohwoarien, Olomoro, Uzere, Taylor Creek as

well as the establishment of proper waste collection and disposal systems and landfill sites

7 Ben Nanaen (1995) in appraising the development in the region argued that there had been internal colonialism whereby the ethnic-based political domination by major ethnic groups in collaboration with multinational oil companies which they also control was used to expropriate the resources of minority oil communities for the benefit of the dominant groups and areas leading to peripherisation of Niger Delta communities and, living behind trail of environmental degradation.

8 There have been over ten revenue arrangements to date from colonial times all largely post-1970 and targeted at oil from the Niger Delta. Colonial government supported derivation. The Hicks-Phillipson Commission of 1953 and the Binn Commission of 1964 recommended 50% derivation and 35% to the regions; 15% to the central government. Chick Commission of 1953 had recommended 100% derivation. In 1958 the Raisman-Tree Commission recommended derivation (5 0%), regions (3 0%) and the Central Government (20%). At the end of the civil war, Decree 13 of 1970 put derivation at 45% and Federal Government controlled offshore oil revenue. Two bodies: Dr. Pius Okigbo Commission of 1979 and Prof. Ojetunji Aboyade Technical Committee of 1977 recommended that derivation principle be abolished. Derivation dropped to 1.5% in 1980s from 45% of 1970 and increased to 3% in 1990 after community complains of marginalisation and subsequently to 13% in 1996 after Ogoni hangings but implemented from 1999 after the 2002 Supreme court ruling. A policy to take over off-shore oil by the Federal Government had only been recently amended following massive opposition and resentment but yet to be signed into law by the President.

9 The author through a pioneering Rockefeller Leadership for Environment and Development (LEAD) International fellowship programme in the Environment Policy Department of the British Foreign & Commonwealth Office secured the support of the British government to fund the training of 23 Niger Delta environmental managers with a split between local and UK training. The programme of training was completed in 2002 and a network of trainees has been formed to coordinate activities amongst government, industry, communities and NGOs.

10 In one instance one individual from Isoko land whose farm was impacted by seismic activities of a multinational oil service company turned to NDES as an avenue to seek redress which NDES consistently took up with its management, although it was later found out that he had accepted in poverty and ignorance, a settlement that was exceedingly minor compared to the magnitude of loss described.

11 Shell alone was going to fund the Survey; there is no reason why a larger interest – OPTS, made up of Shell, ChevronTexaco, Mobil, AGIP, TotalFinaElf, Addax, Consolidated, Dubri, Pan Ocean etc. could not fully support such an important work and its publication even if NDES exceeded its the cut-to-the-bone budget. The fact that NDES had to be chasing some of the companies to pay up their quotas tells a story.

12 All is not lost, NDES could review its management, talk with OPTS, DPR and NNPC NNDC

and other government interests as well as external organisations with interest in the region to fund and sustain this very important information endeavour on the region. The coming years certainly represent tremendous challenge to oil and gas sector with the expansion of on-shore and off-shore activities. The sector is also very strategic to Nigeria's survival, sound information base provides some base for dialogue between stakeholders which should not be ignored.

References

Anderson, B. (1996) Environmental Issues and Management Strategies *Paper Presented at the International Seminar on Petroleum Indus try and the Nigerian Environment, Port Harcourt*

Dear, M. (1988) "The Postmodern Challenge: Reconstructing Human Geography" Transactions Institute of British Geographers New Series *13, 262-2 74*

FEPA (1992) *Transition to Sustainable Development in Nigeria* Lagos: Federal Environmental Protection Agency.

Harvey, D. (1973) *Social Justice and the City.* London: Arnold

Naanen B. (1995) Oil-Producing Minorities and the Restructuring of Nigerian Federalism: The Case of the Ogoni People *Journal of Commonwealth & Comparative Politics* 33, 46-78

NDES (1997a) Niger Delta Environmental Survey Phase I Report, Volumes I – IV Lagos: Niger Delta Environmental Survey

NDES (1997b) *Briefing Note 2* Lagos: Niger Delta Environmental Survey

NDES (2000a) Niger Delta Environmental Survey *Phase I Reports on: Biological Environment and Resources; Socio-Economic Studies; Human Health; Pollution; PRA/PLA; GIS Database* Lagos, Niger Delta Environmental Survey

NDES (2000b) *Development Priorities and Action Plan for the Niger Delta,* Lagos: Niger Delta Environmental Survey

Sandbach, F. (1980) *Environment Ideology and Policy,* Oxford: Basil Blackwell Willink Commission (1958) *Willink Commission Report* London: Colonial Office

World Bank (1990) *Towards the Development of an Environmental Action Plan for Nigeria* Washington D.C., The World Bank

World Bank (1995) *Defining an Environmental Strategy for the Niger Delta* Washington D.C., World Bank, Volumes 1 & 2

World Bank/FEPA (1998) *Flooding Study of the Niger Delta* Federal Environmental Protection Agency, Abuja.

Bibliography

Aboribo, R.I. (2001) "The Political Economy of Environmental Pollution in the Niger Delta: From Compensation to Participation," *African Journal of Environmental Studies,* Vol. 2, No. 1

Abu, A.(2002) Environmental and Behavioural Risk Assessment for Childhood Diarrhea Intervention. Unpublished Ph.D. Thesis, Department of Geography, University of Ibadan, Nigeria.

Academic Associate Peace Works (AAPW) Warri Crisis Project Report 1999) Makofe Konsult Lecture Series.

Achebe, C. (1983) *The Trouble With Nigeria,* Enugu: Fourth Dimension Publishers.

Adamolekun, L. (1991) "The Federal Solution: Assessment and Prognosis for Nigeria and Africa." The Journal of Federalism (Vol. 21) Fall.

Adedapo, A. (2002) *Nigeria: The Case for a Peaceful and Friendly Dissolution,* Lagos: Unitype Enterprises.

Adefemi I. & Gaskia, J. (2001) *Ethnic Groups and Conflicts in Nigeria,* The South South Zone of Nigeria, Ibadan: Programme on Ethnic and Federal Studies, (PEFS) University of Ibadan, Ibadan.

Adekanye, J. B (1999), "Conflict Prevention and Early Warning systems." S. Klason, E. Rothschild et al (eds.) *Common Security and Civil Society in Africa.* Nordiska Afrika Institutet.

Adesina, O. (1998) "Revenue Allocation Commission and the Contradictions in Nigeria's Federalism" in K. Amuwo et al (ed.s) *Federalism and Political Restriction in Nigeria.* Ibadan: Spectrum Books.

Adewale, O. (1989) Oil compensation claims in Nigeria: Principle, Guidelines and Criteria, *Journal of African Law* 33 (1).

Agin, C. (1997) "Understanding Community Relations in Nigeria's Oil Industry". In V.E Eromosele (ed) (1997) *Nigerian Petroleum Business: A Handbook. Lagos: Advert Communication Limited.*

Aina, E.O.A (1996) Towards Environmentally Sustainable Development in the Petroleum Industry in Aina E.O.A. and Adedipe N.O. (eds) *The Petroleum Industry and the Environmental Impact in Nigeria.* (Ibadan) F.E.P.A Monograph pp. 1-6.

Ake, C. (1981) *A Political Economy of Africa,* London: Longmans Ltd.

Ake, C. (1993) What is the problem of ethnicity in Africa? *Transformation* 22

Ake, C. (1994) *Democratization of Disempowerment in Africa,* CASS Occasional Monograph No. 1, Lagos: Malthouse Press Limited.

Ake, C. (1996) Is Africa Democratizing? Port Harcourt: Centre for Advanced Social Science. Monograph 5

Ake, Claude (1982) *Social Science as Imperialism. The Theory of Political Development,* Ibadan: Ibadan University Press.

Akhakpe, I. (2002) "Federal and Resource Control Agitations in Nigeria: Towards an Alternative Framework for Peace" in The *Constitution, A Journal of Constitutional Development,* Vol. 2, No. 3.

Akinyele, R.T.(1998)"Institutional Approach to the Environmental Problem of the Niger Delta" in Osuntokun, (ed) *Current Issues in Nigeria Environment.* Ibadan: Davidson Press.

Aluko, M.A.O. (2000) "Social D imensions and Consequences of Environmental D egradation in the Niger Delta of Nigeria: Suggestions for the next Millennium" in A. Oshuntokun (ed) *Environmental Problems of the Niger Delta,* Lagos: Friedrick Ebert Foundation.

Amadi, S.N and Tamuo, S.O (2001) Oil Exploration in Nigeria: Its Socio-Economic Impact on Host Communities, *Journal of Environment Studies* 2 (1)

Armstrong, R. W., (1983) Medical Geography and Risk Assessment. In R. Akhtar and Y. Verhasselt. (eds) *Disease Ecology and Health* Readings in Medical Geography Rawat Publications, Jaipur India.

Anyaegbunan, E.O. (2000) "Niger Delta: A Case for Regional Contigency Plan" in Osuntokun, A. (ed) *Environmental Problems of the Niger – Delta.* Lagos: Friedrich Ebert Foundation

Ate, B. (1992) "Towards An Integrative Approach to Nigeria's Security Relations with is immediate neighbours" in Ate B and Akinterinwa (eds) *Nigeria and its Immediate Neighbours: Constraints and Prospects of Sub-Regional Security in the 1990s,* Lagos: NIIA

Ayoade, J.A.A (1997) *Nigeria and the Squandering of Hope.* An Inaugural Lecture delivered at the University of Ibadan, 4th October…

Ayoob, M. (1995) *The Third World Security Predicament:* State making, Regional *Conflict and the International System.* Colorado: Lynne Rienner Inc, 1995.

Bayart, J. F. (1986) Civil Society in Africa in P. Chabal (ed) *Political Domination in Africa: Reflections on the limits of power.* Cambridge: Cambridge University Press.

Bentham, G. (1992) "Global Climate Change and Human Health" *Geojournal* Vol. 26, No. 1

Boele, R. (1995) Ogoni: Report of the UNPO Mission to investigate the situation of Nigeria, The Hague: Un-represented Nations and People Organization

Burton, John W. (1987) *World Society,* Volume 3, Lawhen University Press of America

Buzan, B. (1992) *People, State And Fear,* New York: Harvest Wheatsheaf Press.

Cable, S. and Cable .C. (1995) *Environmental Problem, Grassroots Solutions* The Politics of Grassroots Environmental Conflict. New York: St Martins.

Carr, S.O.D and Onyeaguch. U (2001) The Ogoni People's Campaign over oil Exploitation in the Niger Delta in A. Thomas, S. Carr and D. Humphrey (eds.) *Environmental Policies and NGO Influence* London: Routledge

Chazan, N. (1994) "Engaging the State: Associational life in Sub- Saharan Africa" in J. Migdal, A Kohic & V. Shue (eds) *State Power and Social Forces: Domination and Transformation in the Third world, Cambridge:* Cambridge University Press.

Chokor, B. A. (1996) Assessment of the Environmental and Socio-Economic impact of the Petroleum Industry: A conceptual and methodological framework in Aina E.O.A.

and Adedipe, N.O. (eds) *The Petroleum Industry and the Environmental Impact in Nigeria*. (Ibadan) F.E.P.A Monograph.

Clerk, R. T. (1984) *Mining and Environment:* Ibadan: ONAKS Press

Colson, Elizabeth (1971) "The Impact of the Colonial Period on the Definition of Land Rights' in Turner, Victor (ed) (1971) *Colonialism in Africa 1870-1960 Vol. 3 Profiles of Change: African Society and Colonial Rule*, London: Cambridge University Press.

Constitutional Rights Project (1999) *Land, Oil and Human Rights in Nigeria's Delta Region*, Lagos: Constitutional Rights Projects.

Cormack, D. (1983) *Response to Oil and Chemical Marine Pollution*, London: Applied Science Publication.

Coser, Lewis (1956) *The Functions of Social Conflict*, New York: Harcourt Brace, Jovanovich Inc.

Cowling, B. E. (1991) Acid Rain and other Airborne Pollutants: Their Human Causes and Consequences. In Davis K. and M.S. Bernstam (eds) *Resources, Environment and Population:* A Supplement to Vol. 16, 1990.

Crawford, Y. (1993) *The Politics of Cultural Pluralism*, Ibadan: Heinemann Educational Books (Nigeria) Plc.

Daniel Bell (1975) "Ethnicity and Social Change in D. P Moynilian and Glazer (eds), *Ethnicity*, Cambridge, Mass: Harvard University Press.

Department of Petroleum Resources (DPR), (1991, 1999) *Environmental Guidelines and Standards for the Petroleum Industry in Nigeria*, Lagos: Ministry of Petroleum Resources.

Diamond, C. (1994) Rethinking Civil Society: Towards Democratic consolidation.*Journal of Democracy* 5 (3)

Dollard, J., Doob, L. Miller, Mowrer O., and Seers, R. (1939) *Frustration and Aggression*. New Haven: University Press

Dudley, B. (1982) *An Introduction to Nigerian Government and Politics*. London: Heinemann.

Easton, David (1965) *A System Analysis of Political Life*, New York: John Wiley & Son Inc.

Ebene, R. T. (1983) "Health Risks to the Population of Gezira/Sudan Following Irrigation of Cotton Fields" *Geojournal* Vol. 7, No. 3,

Edmond, K. I. (1983) "The State, Public Policy and The Mediation of Ethnic Conflict" In Donald and Victor (eds) *State versus Ethnic Claims: African Policy Dilemmas*, Boulder; West View Press.

Egbon, P.C. & Okoh, R.N. (1999) Fiscal Federalism and Revenue Allocation: The Poverty of the Niger-Delta in Aigbokhan, B.E. (Ed): *Fiscal Federalism and Nigeria's Economic Development*, Selected Papers of the 1999 Annual Conference of the Nigeria Economic Society, NES, Ibadan.

Ekeh, P.P (1992) The Constitution of Civil Society in African History and Politics in B. Caron, A. Gboyega and E. Osaghae (eds) *Democratic Transition in Africa* Ibadan: CREDU Documents in Social Sciences and Humanities No. 1.

Ekeh, P.P. (1989) "Emergent Political Culture." In P. P. Ekeh et al (eds.) *Nigeria Since Independence: The First 25 Years*. Ibadan: Heinmann Ltd. Pg. 1.

Elaigwu, J. I. (2002) "Federalism in Nigeria's New Democracy Polity" in *Publius; The Journal of Federalism,* 32:2 (Spring, 2002).

Eskeland, Gummar S. and Emmanuel Jimenez (1992) "Policy Instruments for Pollution Control in D eveloping countries." *World Bank Research Observer.* Vol. 7, No. 2 (July)

Eson, P.O. (2000) "The Political Economy of Oil Extraction in Nigeria" in Raji, W. A.C, and Akinsola E. (eds) *Boiling Point* Lagos CDHR Publication.

Falana F. (2002) "Crises in the Oil Producing Communities in Nigeria: A Welcome Address" in Raji, W. Ale, A and Akinsola, E (eds) *Boiling Point* Lagos: Committee for the Defence of Human Rights [CDHR].

Fanon, F. (1968) Wretched of the Earth New York: Grove Press. 1968.

Fatton, R. (1999) " Civil society revisited: Africa in the Millennium' *West Africa Review* 1,1 Federal Republic of Nigeria (1987) *Report of the Political Bureau,* Abuja: MAMSER.

Federal Republic of Nigeria (1989) *Constitution of the Federal Republic of Nigeria,* Lagos: Government Printers

Federal Republic of Nigeria (FRN), (1990) *Laws of the Federation of Nigeria (L.F.N.),* Lagos.

FEPA (1992) *Transition to Sustainable Development in Nigeria* Lagos: Federal Environmental Protection Agency.

Florini, Ann M. (1999) "Does the Invisible Hand Need a Transparency?" In Boris Pleskovic and Joseph Stiglitz (eds.) *Annual World Bank Conference on Development Economics.* Washington D.C. Work Bank, pp. 163-184.

Foster, H. D. (1992) Health, Disease and the Environment. CRC Press Inc. Boca Raton Florida.

Fraser, Neil & Hipel Keith W. (1984) *Conflict Analysis Models and Resolution,* New York, Else view Science Publishing Company

Friedman, Milton (1981) *The Invisible Hand in Economics and Politics,* New York: Institute of South East Asian Studies.

Frynas, J (2000) Oil in Nigeria: Conflict and Village Communities, Hamburg and New Jersey: LIT VERLAG and Transaction Publishers

Frynas, J. (2000) Oil in Nigeria: Conflict and Litigation between Oil Companies and Village Communities, Hamburg: Lit.

Galbraith, John Kenneth (1973) *Economics and the Public Purpose.* Boston: Houghton Mifflin Company.

Gamble, A. and P. Walton 1976, *Capitalism in Crisis:* London: Macmillan Press.

Gbafade, U. and P. Gowon (2002) Gas Flaring: Why burning the Resources? *Atmospheric Review,* Vol. 14 (2): 82-103

Geertz Clifford (1963) *Old Societies and New States,* New York: Free Press

Gellener, E. (1995) The Importance of being modular in John Hall (ed) *Civil Society: Theory, History, Comparison.* Cambridge: Polity Press.

Greenpeace (199 4) "Shell Shocked: The Environmental and Social Costs of Living with Shell in Nigeria, Amsterdam: Greenpeace International.

Gwexe, S.G. (1999) "Prospects for African Conflict Resolution in the Next Millennium: South Africa's View" *African Journal on Conflict Resolution,* 1(1)

Hall, J. A (1995), In search of Civil Society in John Hall (ed) *Civil Society, Theory, History, Comparison,* Cambridge: Polity Press.

Harris, Richard (1975) *The Political Economy of Africa.* Cambridge: Schenkman Publishing Company.

Hearn, J. (1999) Foreign Aid, Democratization and Civil Society in Africa: A Study of South Africa, Ghana and Uganda Institute of Development Studies *Discussion Paper 368*

Human Rights Watch (1999) *The Price of Oil: Corporate Responsibility and Human Rights Violations in Nigeria¶s Oil Producing Communities,* printed in USA.

Ibeanu, O (1999) "Ogoni-Oil Resource Flow and Conflict in Rural Nigeria", in T.Granfelt (ed.) *Managing the Globalized Environment,* London: Intermediate Technology.

Ibeanu, O. (1997) *"Oil Conflict and Security in Nigeria: Issues in the Ogoni Crisis,*AAPS Occasional Paper, Vol. No. 2.

Ibeanu, O. (2001) Our Mothers¶ Courage silenced all Guns: Women and Conflicts in the Niger Delta, Paper presented at "Listen to Africa Conference", Chicago, 12-14 September 2001.

Ibeanu, O. (Not dated) "Janus Unbound: Antimonies of Petrol business and Petropolitics in the Niger Delta" unpublished commentary.

Ibeanu, O.O. (1999) Oil Resource Flow and Rural Nigeria in T. Granfelt (ed.) *Managing the Globalised Environment Local Strategies to Secure Livelihood:* IT Publications.

Idowu ,A.S. (1993) clash as govt. deploys soldiers to Ogoni The Guardian May 4.

Ihonvbere, J and Shaw, T. (1998) The Illusions of Power: Nigeria in Transition. New Jersey: Africa World Press.

Ihonvbere, J.(2000)"A Recipe for Perpetual Crises: The Nigerian State and the. Niger-Delta Question" in Raji W; Ale, A and Akinsola, E (eds) *Boiling Point* Lagos: CDHR Publication

Ihonvbere, J.O. (2001) Constitutionalism in Africa: The Issues, Challenges and Opportunities A keynote Address in Segun Jegede, Ayodele Ake & E. Akinsola (eds.) *State Reconstruction in West Africa.* Lagos: Committee for the Defence of Human Rights (CDHR).

Ikeaguani, F.D. (1984) Petroleum Exploration and Exploitation: Impact on the Socio – Economic life of Nigerians living in oil producing areas. *A Publication of the Nigerian National Petroleum Development Company,* Inspectorate Division, Lagos.

Ikelegbe, A. O. (2001A) Civil Society, Oil and Conflict in the Niger Delta Region of Nigeria: Ramifications of Civil Society for A Regional Resource Struggle *Journal of Modern African Studies* 39, 3.

Ikelegbe, A.O. (2001B) The perverse manifestation of civil society: Evidence from Nigeria *Journal of Modern African Studies* 39, 1.

Ikelegbe, A. O. (2003) Civil Society and Alternative Approaches to Conflict Management in Nigeria in T. A. Imobighe (ed) *Civil Society and Ethnic Conflict Management in Nigeria,* Ibadan: Spectrum Books Limited.

Ikporukpo, C. O. (1980) "Managing Oil Pollution in Nigeria: Towards an Alternative Approach" in P. O. Sada and F. O Odemerho (eds). *Environmental Issues and Management in Nigerian Development.* Ibadan: Evans.5

Ikporukpo, C. O. (1983) Environmental Deterioration and Public Policy in Nigeria. *Applied Geography* Vol. 3.

Ikporukpo, C. O. (2002) Spatial Engineering and Accessibility. An Inaugural Lecture,University of Ibadan

Imobighe, T. (1998) *The Management Of National Security,* Ekpoma: Edo State University Publishing House.

Imobighe, T. A. (2003) *The OA U (A U) and OAS in Regional Conflict Management: A Comparative Assessment.* Ibadan: Spectrum Books Ltd.

Innocent, A. O. (2000) "The Nigerian State and the Environmental Question" in Osuntokun (ed) *Environmental Problems of the Niger Delta,* Lagos: Fredrich Ebert Foundation.

International Institute for Democracy and Electoral Assistance (IDEA) (2000) Democracy in Nigeria-*Continuing Dialogue(s) for Nation-building.* Stockholm

Iyayi, F. (2000) "Oil Corporations and the Political of Community Relations in Oil Producing Communities" in Raji, W; Ale, A and Akinsola, E (eds) *Boiling Point.* Lagos: CDHR Publication.

Iyayi, F. (2000) "Oil Companies, Community Relations and International Interests: Oil Corporations and the Politics of Community Relations in Oil Producing Communities" in (eds) *Boiling Point: The Crises in the Oil Producing Communities in Nigeria,* Lagos: CAHR.

Jackson, R. (2002) "Violent Internal Conflict and the African state: Towards A framework of Analysis", *Journal of Contemporary African Studies* 20(2)

John Dollard, (1939) *Frustration and Aggression,* New Haven: Yale University Press

Joseph, Richard (1996) *Democracy and Prebendal Politics in Nigeria: The Rise and Fall of the Second Republic,* Ibadan: Spectrum Books.

Kenneth, W. (1979) *The Theory of International Politics,* Addison: Wesley Reading Mass Khan, S. (1994) *Nigeria: Economy of Oil* Oxford: Oxford University Press.

Kinako, F. D. S. (1981) "Short-term Effects of Oil Pollution on Species Numbers and Productivity of Single Terrestrial Ecosystem" *Environmental Pollution,* 26

King, M.C. (1988) *Localism and Nation-building,* Ibadan: Spectrum Books. 1988.

Kirk-Greene, A.H.M. (1971) *Crisis and Conflict in Nigeria: A Documentary Source Book 1966-1970,* Vol. 1, London: Oxford University press.

Lemmy, O. (2000) "Political Economy of Resistance in the Niger Delta" in Ola (ed) *The Emperor has no clothes: Report of the conference on the Niger Delta and The 1999 Constitution,* Benin City: Environmental Rights Action/Friends of the Earth Nigeria.

Luard, E. (1988) *The Blunted Sword: The Erosion of Military Power in Modern World Politics,* London: Turis and Co. Ltd.

Magbagbeola, N.O. (1999) "Economic Growth and Environmental Development in Nigeria: Impact of Climate Change," *Nigerian Journal of Economic and Social Studies,* Vol. 41, No. 2,.

Maier, K. (2000) *This House has Fallen: Nigeria in Crisis,* London: Penguin Books.

Makumbe J. (1998)" Is there a Civil Society in Africa" *International Affairs* 74 (2)

Mama, A. (1999) Dissenting Daughters? Gender politics and Civil Society in a militarized State Dakar: CODESRIA Bulletin 3 & 4.

Mamdani, M (1995) A critique of the State and Civil Society Paradigm in Africanist

Studies. In Mamdani M. & Wamba-dia-Wamba, E (eds) *African Studies in Social Movements and Democracy.* Darker: CODESRIA Book Series. Pp 602-616

Manby, Bronwen (1997) *The Price of Oil: Corporate Responsibility and Human Rights Violations in Nigeria's Oil Producing Communities,* New York, Washington, London, Brussels: Human Rights Watch.

McNamara, Robert (1968) *The Essence of Security: Reflections in Office,* New York: Harper and Row.

Michael, T. K. (1998) "Redefining Security: The New Global Schisms" in Bernard Martin (eds) *Violence and Terrorism, 1998/99 Connecticut:* Dushin Publishers.

Miller, D. (1981) "Configuration of Strategy and Structure," in R.E. Miles and C.C. Snow (eds), *Organisational Strategy, Structure and Process,* New York: McGraw-Hill.

Mkandawire, M. and N. Bourenane (1999) "The State and Agriculture in Africa: Introductory Remarks", in M. Mkandawire and N. Bourenane (eds.), *The State and Agriculture in Africa,* London: CODESRIA Books.

Mkandawire, T. (2000) "The Terrible Toll of Post Colonial Rebel Movement in Africa: Toward an Explanation of the Violence against the Peasantry" , *Journal of Modern African Studies* 40(2)

Moffat, Daniel and Olof Linden (1995) "Perception and real ity: Assessing priorities for sustainable development in th e Niger Delta", AMBIO: *Journal of Human Environment,* Vol. 24, Nos. 7-8.

Naanen, B. (1994) "Oil Producing Minorities and the Reconstruction of Nigerian Federalism: The case of the Ogoni People" *The Journal of Commonwealth and Comparative Politics* 33 (1).

Naanen, B. (2001) "The Niger Delta and the National Question" in E.E Osaghae and E. Onwudiwe (eds.) *The Management of the National Question in Nigeria.* Ibadan: PEFS.

Nabudere, D.(2000) *Globalisation and the Post-colonial African State* – Harare: AAPS Books.

NDES (1997a) Niger Delta Environmental Survey Phase I Report, Volumes I – IV

Niger Delta Environmental Survey (NDES) (1997) Environment and Socio-Economic Characteristics: Vol. 1, Phase 1. Port Harcourt.

Nnoli, O. (1978) *Ethnic Politics in Nigeria,* Enugu: Fourth Dimension.

Nwankwo, J. N.; Ifeadi, C. N., and I. I. Orubima (1998) "The Environmental Guidelines and Standards for the Petroleum Industry in Nigeria: The 1998 Revision", Paper presented at 9th International Conference on the Petroleum Industry and the Nigerian Environment, Abuja, DPR

Obi, C. (1997A) *Oil, Environmental Conflict And National Security In Nigeria: Ramifications of The Ecology-Security Nexus For Sub-Regional Peace.* Arms Control, Disarmament And International Security Occasional Paper, University Illinois At Urbana-campaign.

Obi, C. (1997b) "Resources, Population and Conflicts: Two African Case Studies." Being Paper Presented to Laureates of the Council for the Development of Social Science Research in Africa (CO D ESRIA) on the theme: "The Political Economy of Conflict in African' Dakar, Senegal, August 18-22.

Obi, C. (1999A) Globalization and Environmental Conflict in Africa." *African Journal of Political Science,* vol. 4 (1) June.

Obi, C. (1999B) "Oil, Environmental and Conflict in the Niger-Delta." *Quarterly Journal of Administration* vol. 30, No. 3 &4 September – January.

Obi, C. (2001) Oil Friendships Are Greasy: A Political Economy of Corruption in Oil Company – Community Relations in the Niger Delta. Paper Present at International Political Science Association (IPSA) RC4mid-term Conference, Abuja, 15-17 October2001.

Obi, I C. (2000) Globalized Images of Environmental Security in Africa *Review of African Political Economy* No. 83

Obi, I.C. (1993) Oil and the Nigeria State: An Overview *National Seminar on Conflict Containment Strategies for Oil Companies and Oil Communities* Port Harcourt, April 6-7

Obi, I.C. (1998) Environmental Scarcity in Africa: Some Theoretical Concerns and Emerging Issues, *Africa Insight,* Vol. 28, No.1/2

Obi, I.C. (1999C) The Crisis of Environmental Governance in the Niger Delta 1985-1996, *AAPS Occasional Paper Series,* Vol.3 No3

Odjugo P.A. O. (2000) Global Warming and Human Health: Current And Projected Effects. Environmental Analar No.5.

Odu, C. T. I. (1977) Oil Pollution and the Environment. *Bulletin of the Science Association of Nigeria.* Vol. 3.

Odusola, A.F. (1996) "Community Participation in Environmental Assessment", in P.C. Egbon and B. Morvaridi (eds.), *Environmental Policy Planning,* National Centre for Economic Management and Administration (NCEMA), Ibadan, Nigeria.

Ofonagoro, W. Ibekwe (1979) *Trade and Imperialism in Southern Nigeria 1881-1929,*Lagos: Nok Publishers International.

Ogbemi-Ifediora, A. I. (2002) "Ethnic Intolerance, Human Rights Violation and Terror in a Nigerian City: Warri 1990-2001' *Paper present at 1st Annual National Conference of the South-East Zone of Historical Society of Nigeria* at Imo State University, Owerri, December

Ogisi, O. D., Omotor, D. G. and Dada (1999) "Determinants of Rural to Urban Migration: A Case Study of Warri, Nigeria" in Egbon, Peter C. & Orubu, Christopher O. (eds.) *Critical Issues in Nigeria's Development* Abraka: Faculty of Social Sciences, Delta State University, Abraka.

Ogunsanya, Adegboyega (1991) *Strategy and Public Relations,* Richmond Books Ltd.

Oil Mineral Producing Area Development Commission (OMPADEC) (1994)Quarterly Report Port-Harcourt Vol.1, No.2 March.

Ojefia, I.A (2000) "The Nigerian State and Environmental Question" in Osuntokun, A (ed.) *Environmental Problem of the Niger-Delta.* Lagos: Friendrich Ebert Foundation

Ojiako, J. (1981) *Nigeria: Yesterday, Today and....* Onitsha: Africana. Feb Publishers.

Ojo, O.J.B. (2002) *The Niger Delta: Managing Resources and Conflicts,* Ibadan: Development Policy Centre

Okoh, N. V. (2000) Environmental Impacts of Oil Exploration, *Atmospheric Review,* Vol. 12 (1).

Okoko, K. & Nna, J. (1997) "Emerging Trends and Community Perception in the Nigerian Oil Industry" in *Nigerian Journal of Oil and Politics,* Vol. 1, No. 1.

Okonta, I. & Douglas, O. (2001) *Where Vultures Feast: 40 Years of Shell in the Niger Delta,* Benin City: Environmental Rights Action/Friends of the Earth.

Okumagba, Paul O. (1999) "Strategies for Combating Youth Unemployment: The Nigerian Experience¶ in Egbon, Peter C. & Orubu, Christopher O. (eds) *Critical Issues in Nigeria¶s Development* Abraka: Faculty of the Social Sciences Delta State University.

Olagoke, C. (2001) Nigeria Economy: The Oil Sector, *OPC: Bulletin No. 12.*

Olajide, I. Fajonyomi, B.Akhape I. And Mudasiru S.U. (2000) *Nigeria Oil Pollution, Community Dissatisfaction and Threat to National Peace and Security,* Zimbabwe: African Association of Political Science (AAPS) Vol. 4 No. 3

Olayiwola, P. (1987) Petroleum and structural change in a developing country: The case of Nigeria New York: Praeger

Olomo, R.O. and Omene, L. (1995) "Environmental impact of oil spillage: A case of the operations of Shell Petroleum D evelopment Company in Bonny, Rivers State", paper presented at the 38th Annual Conference of Nigerian Geographical Association, University of Benin, May.

Olukoshi, A. and Agbu, O. (1996) "The Deepening Crisis of Nigerian Federation and the Future of Nigerian State" in Olukoshi and Laakso (eds) *Challenges to the Nation-State in Africa,* Uppsala: Nordiska African Institute.

Omari, A.H. (1995). *Regional Security: One view from the front line states.* Arusha Working Paper Series in Southern African Security,

Omotor, D.G. (2000) " Poverty, Environment and Sustainable Development: A Review " *African Journal of Environmental Studies,* Vol. 1

Omotor, G., Ogisi, D. & Pessu, (2000) "Youth Restiveness and the Development of the Niger Delta: An Exploratory Analysis of Selected Communities in Delta State Nigeria" in Orubu, Christopher, (ed.) *Nigeria Petroleum and Energy,* Abraka: Delta State University

Omoweh, D. (1995) Environmental Pollution, Culture and Health in Nigeria: The Sad Plight of the Ughelli Oil Communities *Africa Spectrum* 3 (2).

Onosode, G. (2000) "Environmental Management and Sustainable Development in the Niger Delta" in A Osuntokun (ed) *Environmental Problems of the Niger – Delta,* Lagos: Friendrich Ebert Foundation.

Onosode, G.O (1996) Community Expectations Versus Health Safety and Environmental and Economic Stewardship of Joint Venture Partners: Finding the Right Balance. *Paper Presented at the 8th Edition of the Biennial International Seminar on the Petroleum Industry and the Nigerian Environment.* Port Harcourt November 17-21

Onosode, Gamaliel (1997) "Petroleum, Development and the environment: ND ES Perspective" in Victor E. Eromosele (ed), *Nigerian Petroleum Business: A Handbook,* Lagos, Advent Communications, 287-291.

Orubu, C. O. (2001) "Compensating for trampled surface rights in the Nigerian petroleum industry: suggestion for a sustainable livelihood approach", The *Nigerian Economic and Financial Review,* vol. 6 (1).

Orubu, C.O. (1999a) "The exploitation of non-timber forest resourc es in the Niger Delta:

Problems and prospects", Technical paper, Niger Delta Environmental Survey, Port Harcourt.

Orubu, C.O. (1999b) "Oil wealth and the derivation principle: The need for a new fiscal imperative towards oil-producing states", *Calabar Journal of Politics and Administration,* Vol. (1).

Orubu, C.O. (2001) "Compensation Programme for the Use of Natural Resource by Oil Companies in Nigeria: Implications for Sustainable Development", in The Nigerian Economic Society (2001) *Natural Resource Use, The Environment and Sustainable Development: Selected Papers Presented at the 2001 Annual Conference, Ibadan:* The Nigeria Economic Society.

Orubu, C.O.; A. Fajingbesi; Odusola, A. and Magbagbeola, N. (2002) *Environmental Regulations in the Nigerian Petroleum Industry: Compliance Status and Implications for Sustainable Development,* Research Report, ACBF/NCEMA.

Orukele, A.J. (2000) *effective Poverty Alleviation Programme as a Panacea to the Oil Related Niger-Delta Crisis,* A Paper Presented at National Conference on Management of Petroleum For Sustainable Development in the Niger-Delta, Delta State University,Abraka.

Osaghae, E. (1992) " Managing Ethnic Conflicts and Democratic Transition in Africa: The promise, failure and the future" in B. Carom et al (eds) *Democratic Transition in Africa,* Ibadan: CREDU.

Osaghae, E.E. (1995) "The Ogoni Uprising: Oil Politics, Minority Agitation and the Future of the Nigerian State." *African Affairs (94).*

Osaghae, E.E. (1997) "The Federal Solution in Comparative Perspectives." *Politics* Vol. 16 (1).

Osaghae, E.E. (1999) "Exiting from the state in Nigeria." *African Journal of Political Science.* vol. 4 (1) June.

Osuntokun, A. (2000) " The Niger Delta and Nigeria's Environmental Problems" in Osuntokun, A. (ed) *Environmental Problems of the Niger Delta. Lagos:* Friedrich Ebert Foundation.

Pal, L.A. (1997) Beyond Policy Analysis, Ontario: ITP Nelson.

Pearson, S.R. (1970), *Petroleum and the Nigerian Economy,* Standard/California: Standard University Press.

Pedersen, D. (1996) Disease Ecology at Crossroads: Man-Made Environment, Human Rights and Perpetual Development Utopias. *Social Science and Medicine,* Vol. 21, No. 2.

Pegg, S. (1999) "The Cost of Doing Business: Transnational Corporations and Violence in Nigeria: *Security Dialogue,* Vol. 30, Number 4.

Peter, H & Ben, R. (eds.) (1998) *Democracy and deep-rooted conflict: Options for Negotiations:* A publication of Institute for Democracy and Electoral Assistance. Handbook Series, Sweden.

Philips, A. (1971) "Nigeria's Federal Financial Experience" *Journal of Modern African Studies,* 9 (3)

Philips, A. (1975) "Revenue Allocation in Nigeria 1970-80" *The Nigerian Journal of Economics and Social Studies,* 17 (1).

Pietrzyk, D. (2003) Democracy or civil society? *Politics* 23. 38-45.

Redclift, M. (1987) Sustainable Development: Exploring the Contradictions, London: Meth. Roberts, N. (1998) Managing the Minorities and Development Problem in Nigerian Federalism. The OMPADEC Initiative. Ibadan, NISER Monograph Series No. 11

Roberts, N. (1999) The State, Accumulation and Violence: The Politics of Environmental Security in Nigeria's Oil Producing Areas. Ibadan: NISER Monograph Series No. 17

Robinson, D. (1997) Ogoni: The Struggle continues. Geneva and Nairobi, World Council of Churches and All African Council of Churches

Rowell, A. (1996) *Green Backlash*, London: Routledge.

Rowell, A. (1994) Shell shocked: The Environmental and social Costs of Living with Shell in Nigeria. Amsterdam: Green peace

Ruth, First (1970) *The Barrel of a Gun: The Politics of Coup d' etat in Africa*, London, Allen Lane: The Penguin Press

Sandbach, F. (1980) *Environment Ideology and Policy*, Oxford: Basil Blackwell

Sandole, Dennis J. D, and Sandole, Starosle Ingrid (eds.) (1987) *Conflict Management and Problem Solving: Interpersonal to International Application*, New York: University Press

Saro Wiwa K. (1992) Genocide in Nigeria: The Ogoni Tragedy, Port Harcourt: Saros Publication

Saro Wiwa K. (1993) Community Reaction to Oil Exploration. *Paper Presented at the National Seminar on Conflict Containent Strategies for Oil Companies and Oil Communities* Port Harcourt, August 6-7

Saurin, J (1996) International Relation, Social Ecology and Globalization of Environmental change in J. Vogler and M. Imber (eds.) *The Environmental and International Relations*, London: Routledge

Southhall, Aidan (1971) "The Impact of Imperialism upon Urban D evelopment in Africa' in Turner, Victor (ed) (1971) *Colonialism in Africa 1870-1960 Vol. 3*

Strudsholm, J. (1993) The Oil Field in Nigeria *Africa Insight 29 (1).*

Tamuno, S.O, Girigiri, B.I.C; Anele K.A, Onuchuku O; Obafemi A.O; (1999) *Ethnicity, Conflict and development: Prospects in Nigeria; A Book of Readings:* Lagos Corporate Impressions.

Tandon, Y. (1998) Globalization and Africa's Option AAPS Newsletter,

Teriba, O. (1996) "Nigeria Revenue Allocation Experience" *The Nigerian Journal of Economic and Social Studies* 8 (3)

The Africa Fund (1999) Nigeria: Blood and Oil: A Special Report by Michael D Fleshman, Human Rights Co-ordinator. 7 September.

Todaro, M. (1998) *Economic Development*, New York: Addison Wesley Publishers.

Turner T (1981) "Commercial Capitalism in Nigeria: The Pattern of Competition in D. Cohen and J. Daniel (eds.) *Political Economy of Africa*, London: Longman

Turner, Terisa (1980) "Nigeria: Imperialism, Oil Technology and the Comprador State" in Peter Nore and Terisa Turner (eds.) *Oil and Class Struggle*, London: Zed Press.

Turshen, M (2002) Introduction ACAS Bulletin: Its About Oil, *Review of Africa Political Economy* 91, 151-162

Uche, J. I. (1997) Air Pollution. In Odjugo, P.A.O. and J.E. Evwoh (Eds). *Physical Components of the Environment.* Agbor: Pon Pubs.

Ukaogo, V. (2000) "Transnational Business Ethics, Government Policies and the Crisis of Pollution and Underdevelopment in the Niger – Delta" in Osuntokun. A (ed) *Environmental Problems of the Niger – Delta.* Lagos; Friedrich Ebert Foundation,

Umoren, Rose (2001) *Economic Reforms and Nigeria's Political Crisis.* Ibadan: Spectrum Books.

UNDP (1997) Human Development Report, UNDP, New York Oxford University Press.

United Nations (1997) *Government Views on the Relationship Between Population and Environment,* Department of Economics and Social Affairs, Population Division, United Nations, New York.

Van Horn, C. and Van Meter, D. (1977) "The Implementation of Intergovernmental Policy" in Nagel, Stuart (ed) *Policy Studies Review Annual,* Beverly Hill and London, Sage Publications, Vol. 1

Verhasselt, Y. (1987) Environmental Changes and Health in Tropical Africa *Revenue Belge Geographie 111e annee 1987, Fascicules 3-4.*

Yogt, M. A. (1990) "Nigeria's Defence Policy: An Overview" in Ekoko A and Vogt M.A (Eds) *Nigerian Defence Policy: Issues and problems,* Lagos, Mathouse Press Limited.

Wardley – Smith, J. (1979) "Sources of Pollution" in Wardley – Smith (Ed) *The Prevention of Oil Pollution* London: Graham and Trotman Ltd.

Warpner, P. (1996) Environmental Activism and World Civil Politics. Albany (NY): State University of New York Press

Watts, M. (1999) Petro Violence: Some Thoughts on Community, Extraction, and Political Ecology, Berkeley: *Berkeley Workshops on Environmental Politics Working* Paper No. WP 99-1

Willink Commission (1958) *Willink Commission Report* London: Colonial Office

Woodhouse, P. (1992) Environmental Degradation and Sustainability in T. Allen and A. Thomas (eds.) *Poverty and Development in the 1990s.* Oxford: Oxford University Press

Woodman J. N. and Cowling, E. B. (1987) Airborne Chemicals and Forest Health. *Environmental Science and Technology* 21: 120-126.

World Bank (1990) *Towards the Development of an Environmental Action Plan for Nigeria* Washington D.C.,The World Bank.

World Bank (1995) *Defining an Environmental Development Strategy for the Niger Delta* Vol.II Washington D.C. Industry and Energy Operations Division (West Central Africa Department).

World Bank (2000/200 1) *World Development Report,* Washington D.C.

World Bank/FEPA (1998) *Flooding Study of the Niger Delta* Federal Environmental Protection Agency, Abuja

Yates, A.D. (1996) *The Rentier State in Africa: Oil Rent Dependency and Neo-Colonialism in the Republic of Gabon.* Trenton & Asmara: Africa World Press Inc.

Zylicz, Tomase (2000) "Goals, principles and constraints in environmental policies", in Henk Former and H. Landis Gabel (eds), *Principles of Environmental and Resource Economics,* Cheltenham, Edward Elgar, 130-156.

Contributors

Amos Abu is a Lecturer, Department of Geography, University of Lagos, Lagos. Nigeria

A.A.Agagu is Professor, Department of Political Science, Universi of Ado-Ekiti, Ado -Ekiti, Nigeria

A.S. Akpotor is Professor, Department of Political Science, Ambrose Alli University, Ekpoma, Nigeria

Julius Azelama is Professor, Department of Public Administration Ambrose Alli University, Ekpoma, Nigeria

Akhakpe Ighodalo is a Lecturer, Department of Public Administration, Lago State University, Ojo, Lagos, Nigeria

M.I. Ailemen is a Lecturer, Department of Economics, Ambrose All University, Ekpoma, Nigeria

William Ehwarieme is Senior Lecturer,Department of Political Science, Delta State University, Abraka, Nigeria

B. A. Chokor: is Professor, Department of Geography and Regional Planning, University of Benin, Benin City, Nigeria

Banji Fajonyomi is Associate Professor, Department of Public Administration Lagos State University, Ojo, Lagos, Nigeria

Audu, Gambo is Senior Lecturer,Department of Political Science, University o Jos, Jos, Nigeria.

Dauda S. Garuba is a Programme Officer, Centre for Democracy an Development, Abuja, Nigeria.

Augustine Ikelegbe is Professor, Department of Political Science and Public Administration, University of Benin, Benin City, Nigeria.

Onovughe Ikelegbe is a Senior Lecturer, Department of Geography and Regional Planning, University of Benin, Benin City, Nigeria.

Christopher Isike is a Lecturer, Department of Political Science and Public Administration. University of Benin, Benin City, Nigeria.

Remi Medupin is Senior Lecturer, Department of Economics, University of Lagos, Lagos, Nigeria

Orji Nkwachukwu is a Research Fellow, Institute of Development Studies, University of Nigeria, Enugu Campus, Nigeria

P. A. O. Odjugo is Professor, Department of Geography and Regional Planning, University of Benin, Benin City, Nigeria.

Ayodele Odusola is with the National Centre for Economic Management and Administration (NCEMA), Ibadan, Nigeria.

Anthony I. Ogbemi-Ifediora, is Senior Lecturer, Department of Economics, College of Education, Warri, Nigeria

Christopher O. Orubu is a Professor, Department of Economics, Delta State University, Abraka, Nigeria

E.I. Okojie is a Lecturer, Department of Economics, University of Benin, Benin City, Nigeria

Olojede Iyabo is a Professor, Department of Public Administration, Lagos State University, Ojo, Lagos, Nigeria.

Index